Tropical Forestry

Series Editor
Michael Köhl, Hamburg, Germany

More information about this series at
http://www.springer.com/series/5439

Ratan Lal Banik

Silviculture of South Asian Priority Bamboos

Springer

Ratan Lal Banik
NMBA (National Mission
on Bamboo Applications)
New Delhi
India

Series Editor
Michael Köhl
Department of Wood Science
University of Hamburg
Hamburg, Germany

ISSN 1614-9785
Tropical Forestry
ISBN 978-981-10-0568-8 ISBN 978-981-10-0569-5 (eBook)
DOI 10.1007/978-981-10-0569-5

Library of Congress Control Number: 2016941929

Printed on acid-free paper

This Springer imprint is published by Springer Nature
The registered company is Springer Nature Singapore Pte Ltd.
The registered company address is: 152 Beach Road, #22-06/08 Gateway East, Singapore 189721, Singapore

Foreword

Recent years have shown that bamboo silviculture and management is entering a new era. Considering the vast variety of bamboo species, the basic requirement for the purposeful cultivation of bamboo is the knowledge and comprehensive understanding of the growth conditions and characteristics of individual species.

As editor of Springer's Tropical Forestry Series, I am especially pleased to have been able to engage Dr. Ratan Lal Banik as author for a book on bamboo species. He is a proven expert on the subject and has a record of numerous projects and activities in the field of bamboo silviculture in South Asia. With this book, Dr. Banik offers his overwhelming knowledge and expertise to scientists and practitioners.

I am confident that the current book will largely meet existing needs and wish it well deserved success.

Hamburg, Germany　　　　Michael Köhl

Preface

Bamboo is a versatile group of plants, capable of providing ecological, economic, and livelihood security to the people: "It is to shelter, to fashion tools, to weave baskets, to help water obey, to provide beauty and sounds." In the tropics, especially the rural areas in different countries of South Asia, most of the houses are made of bamboos. In the hilly areas of Bangladesh, Bhutan, Nepal, and India, the tribal people take bamboo shoots as one of their major food items since prehistoric days. Certainly, it saved many lives of our forefathers. Thus, bamboo has been identified as a symbol of life and became "the poor man's timber" to the Indians, "the friend of the people" to the Chinese, and it is "the brother" of the Vietnamese. With high productivity and grass-like leaves, bamboo plants have been liked by most of the herbivore animals, such as elephants, wild cattle (*Bos gaurus* and *B. javanicus*), and various species of deer. The red panda in the Himalayas, primates, pigs, rats and mice, porcupines, and squirrels are also important incidental feeders on Southeast Asian bamboos. However, it is not that all bamboo species are liked by these animals; rather, they have some selection about the species. Reforesting and managing forest of these selected species is also important for sustainability of ecosystem and fauna of the region.

There has been a growing awareness in recent years about the values of bamboo, being an important means of economic growth and for improving the socioeconomic conditions of the rural poor. Bamboo as an industrial material can substitute wood and that too at low cost. Due to increasing demand and squeezing of bamboo area, the plants have been overexploited and the quality and quantity of resource alarmingly getting depleted. Besides, many new bamboo-based industries have come up which also urgently require uninterrupted supply of species-wise bamboo resource. The South Asia region has been bestowed with more than 300 bamboo species with enormous diversities at species, ecological, and genetic levels. People from their age-old experiences have selected only some of these bamboo species for their socioeconomic, specific ecologic, and modern industrial needs, and they started cultivating them with priority. A number of such priority bamboo species are found to be common among countries of the region, indicating their wide range of ability to adjust to the environmental conditions of these countries and various utilization potentials. Both government and private planters in the countries of South Asia have started allocating funds, land, and other logistics to raise large-scale plantation of desired bamboo species. Often, they have queries to know the specific local and modern industrial uses of each bamboo species, how to recognize them at the field, traditional vernacular and correct scientific name of the bamboo species for making local and international trade contacts, what are the flowering (seeding) intervals and seed availability, how to have sufficient number of quality planting materials (QPM), and details of planting and management techniques. In many occasions, it has not been able to answer these queries to the satisfaction of the clients due to the lack of information. This book has been drafted to find out answers of these queries mostly based on my field observations on each of the bamboo species and knowledge learned from the indigenous people living with bamboos in different parts of Southeast and South Asian countries. During the last 45 years of my association with bamboo plant, I had the opportunity to observe flowering, seeding, and seedling of more than 30 bamboo species, and the relevant information are reflected in this book. The incidences of flowering in *Bambusa balcooa* and *B. vulgaris* has been reported to be very rare and without any seed production. Such rare flowering event of these two most commonly grown bamboos in the rural areas of the region was also luckily observed and included in this treatise. The production and nursery management of different types of planting material is a major bottle neck in bamboo cultivation. This has forced greater attention to bamboo propagation practices and techniques. Over the years, many new propagation techniques have been developed, tested and gradually being made suitable for field application. Detail step-wise practical notes along with pictorial guides have been drafted as Appendix-I in the book for obtaining better success in production of bamboo planting materials. Additionally, I tried to collate the available documented information, especially the culm wood properties of each of the bamboo species and added in this book for their proper engineering utilization. The overall purpose of this book is to make available all possible information on the above queries of each important bamboo species of the region and serve these in one tray to the consumers.

I believe this monograph would be interesting and useful to bamboo professionals, foresters, horticulturists, field level extension workers, nurserymen, planters, industrial entrepreneurs, and ecologists, and would be a valuable source of reference to the relevant researchers and students in the region.

New Delhi, India
January, 2016

Ratan Lal Banik

Acknowledgments

With great honor and gratefulness, I like to remember Mr. Syed Mortuza Hasan, the then silviculturist of BFRI, Chittagong, where I was working as a scientist as he had motivated me in 1969 to learn all about the amazing bamboo plants. Many innovative ideas and local ethnic knowledge I learned was gathered from the gracious support and sharing of wisdom and know-how about bamboos by local people living with this plant resource. I am indebted to my research and academic colleagues, foresters, and local indigenous people from the different parts of South Asia (especially BFRI, Chittagong; Tripura Department of Forests; and GBP University of Agriculture and Technology, Pantnagar, Uttarakhand) for their support and assistance while I was studying bamboos in the field.

I gratefully acknowledge the support of Prof. Dr. Michael Köhl, Center for Wood Sciences, World Forestry, University of Hamburg, Germany, and for his generous consent of editing the manuscript and inclusion of the draft for publication as a book in the Tropical Forestry Series, of which he is the editor. Further, I am indebted to Prof, Köhl for his untiring efforts in critiquing, reviewing, and editing the manuscript, and I am also expressing my deep gratitude to him for making it possible to publish this book as a monograph on priority bamboos under the aegis of Springer.

My special thanks go to Prof. Dr. Walter Liese, University of Hamburg, Germany, a living legend on bamboo anatomy and physical and mechanical properties of bamboos, for inspiring and overall criticizing the draft manuscript; and to Dr. Cherla B Sastry, former DG of INBAR and currently Adjunct Professor at University of Toronto, for his inspiration in writing the book. Also thankful to Mr. Sudhir Kumar Pande, Advisor of NMBA (NECTAR), New Delhi, and Ex-DG of Forest of India; Mr. S T S Lepcha PCCF and Ex-CEO, Uttarakhand Bamboo and Fibre Development Board (UBFDB); and Dr. Selim Reza of TRIBAC, Tripura, India, for their encouragement and support.

I also acknowledge the sincere efforts of Ms. Aakanksha Tyagi and others of Springer IN, Delhi, for helping me in all official matters for fulfilling the requirement in publishing the book. Also acknowledging the untiring support of

Mr. M. Pradeepkumar of SPi Global, Chennai, for assisting me in proof reading and correction works.

With heartfelt respect, I am expressing gratefulness to the departed soul of my beloved parent for their fountain of blessings to complete the book.

New Delhi, 2016 Ratan Lal Banik

Contents

Part I
Introduction

Chapter 1
Introduction to South Asian Bamboos

In the globe, South Asia or Southern Asia represents the southern region of the Asian continent, which comprises the sub-Himalayan countries, adjoining countries to the west and east. South Asia is bounded on the south by the Indian Ocean and on land by West Asia, Central Asia, East Asia and Southeast Asia. The current territories of Bangladesh, India and Pakistan form the central region of South Asia, while the mountain countries of Nepal and Bhutan in the north and island countries of Sri Lanka and Maldives in the south are generally included in the region. Often Afghanistan and Myanmar are also added.

South Asia is largely divided into four broad climate zones (Olive 2005).

- The northern Indian edge and northern Pakistani uplands have a dry subtropical continental climate.
- The far south of India and southwest of Sri Lanka have an equatorial climate.
- Most of the peninsulas have a tropical climate with variations:
  - Hot subtropical climate in northwest India
  - Cool winter hot tropical climate in Bangladesh
  - Tropical semi-arid climate in the centre
- The Himalayas have an alpine climate

Maximum relative humidity of over 80 % has been recorded in Khasi and Jaintia Hills of Meghalaya state of Northeast India and Sri Lanka, while the area adjustment to Pakistan and Western India records lower than 20–30 %. Climate of South Asia is largely characterized by monsoons. South Asia depends critically on monsoon rainfall. Two monsoon systems exist in the region (Tyson 2002).

- The summer monsoon: Wind blows from southwest to most of parts of the region. It accounts for 70–90 % of the annual precipitation.
- The winter monsoon: Wind blows from northeast. Dominant in Sri Lanka and Maldives.

R.L. Banik, *Silviculture of South Asian Priority Bamboos*, Tropical Forestry,
DOI 10.1007/978-981-10-0569-5_1

The warmest period of the year precedes the monsoon season (March to mid-June). In the summer, the low pressures are centred over the Indus-Gangetic Plain, and high wind from the Indian Ocean blows towards the centre. The monsoons are second coolest season of the year because of high humidity and cloud covering. Moderately vigorous monsoon depressions form in the Bay of Bengal and make landfall from June to September.

A brief note about bamboo vegetation and utilization in different countries of the South Asia region is presented below.

1.1 Bangladesh

The country is bounded by India on the west and north and Myanmar on the east and the Bay of Bengal on the south. Except the hilly regions in the northeast and southeast, some areas of high lands in the north and northwestern parts, the country consists of low, flat and fertile land. Bangladesh is basically a plain land country. However, the hilly areas account for about 12 % of the total land surface covering about 1,733,503 km^2 in the greater districts of Chittagong Hill Tracts (hereafter CHT), Chittagong, Sylhet and some other areas. The CHT, an area of 13,295 km^2, is the south-eastern part of Bangladesh, bordering the Arakan and Chin States of Burma and Tripura and Mizoram States of India.

The country enjoys generally a tropical monsoon climate; the average annual rainfall in Bangladesh varies from 1500 to 5500 mm which falls mostly between last part of May and August and period of dry months is from October to March. The relative humidity is correspondingly high and ranges from 60 % and upwards. The atmospheric temperature ranges from 7° to 41 °C; it rarely falls below 5 °C, mostly remaining within 20° and 35 °C. Maximum rainfall is recorded in the hilly and southern part of Chittagong and northern hilly part of Sylhet districts, while minimum is observed in the western part of the country.

About nine genera and more than 33 species have been found in Bangladesh, out of which seven are occurring naturally in the forest areas either as understorey in association with tree species or as pure stand naturally in the semi-evergreen and moist deciduous forests of the hills of CHTs, Cox's Bazar, Sylhet and northern Mymensingh (foot hills of Garo Hills). Among them, *Melocanna baccifera* is the most widely occurring common bamboo species of Bangladesh (Alam 1995, 2001; Banik 1980, 2000). *Melocanna baccifera* (commonly known as *muli bansh*) constitutes 70–90 % of the total bamboo forests of the country. The other species *Bambusa tulda*, *Dendrocalamus longispathus*, *D. hamiltonii*, *Gigantochloa andamanica* (*syn. Oxytenanthera nigrociliata*) and *Schizostachyum dullooa* occur sporadically either in association with *Melocanna* or in isolation forming small patches of pure bamboo vegetation. As regards abundance, *B. tulda* is next to *muli* bamboo. The remaining two species *Melocalamus compactiflorus* and *Dendrocalamus hamiltonii* are localized only in limited forest areas of the country. Bangladesh, being a part of the subtropics, has only clump-forming bamboos both in forests and in villages. There are no temperate species, that is, non-clump-forming bamboos are available in the country. Among the cultivated species, *Bambusa vulgaris*, *B. balcooa*, *B. tulda*, *B. nutans* and *B. cacharensis* are most common throughout the country.

There are millions of people in rural Bangladesh who depend on bamboo for part or all of their housing, agricultural activities, construction works and also income from selling the green poles. In the hills, the livelihoods, shelter and foods of indigenous people depend almost entirely on the harvesting, processing and selling of bamboo poles, edible shoots and bamboo products such as baskets, mats and handicrafts. The demand of bamboo poles has been increasing in the housing sector (Anon 1983; Banik 2000), especially in construction of tall high-rise buildings for making scaffoldings to plaster and painting the walls. The demand of poles gets further high during religious (Eid, Puja, Buddha and Christmas gatherings), social (marriage, club meetings, etc.) or any other community gathering specially at villages mainly for constructing the pandals and tents. Due to increasing demand and squeezing of bamboo area, the plants have been overexploited, and yearly, there is about 3.0 % loss of bamboo area in the bamboo forests, and simultaneously, the quality and quantity of resource are alarmingly getting depleted (Banik 2000). Besides existing main two pulp and paper mills at Chittagong and Sylhet, many new bamboo-based industries have come up which also require uninterrupted supply of species-wise bamboo resource. The steady supply of required quantity and quality of bamboo is an urgent need to keep pace with such present demanding situation of the resource.

1.2 Bhutan

Bhutan is located in Southern Asia, between China and India. The total area of the country is 47,000 km^2; Thimphu is the capital of Bhutan. The country's climate varies: tropical in southern plains, cool winters and hot summers in central valleys and severe winters and cool summers in Himalaya.

Bhutanese bamboo is principally of Himalayan and Chinese-Japanese origins, with some Southeast Asian and Indian contributions. Bhutan has 15 genera and 31 species of bamboo. Possibly as many as 50 more species exist, but have yet to be identified. Major species found in subtropical areas include *Bambusa nutans*, *Dendrocalamus hamiltonii*, *D. sikkimensis*, *D. patelleries*, *D. strictus*, *Drepanostachyum hookerianum* and *D. intermedium*. Higher elevation species include *Arundinaria racemosa*, *A. maling* (local name: Hima) and *A. polystachya*. Other species are *Dendrocalamus hookeri* and *Arundinaria griffithii* (Griernd dwarf rhododendron trees). In this Himalayan Kingdom, the distribution of bamboos is mainly influenced by the topography and altitudinal variation. In the lower hills, the main species of bamboo is *Dendrocalamus hamiltonii*. Near about 1200 m elevation, *Bambusa nutans*, *Dendrocalamus sikkimensis* and *Chimonobambusa intermedia* make their appearance, while between 1200 and 2000 m, the species seen are *Cephalostachyum capitatum*, *Pseudostachyum polymorphum*, *Dendrocalamus patellaries*, etc. Still higher up between 2000 and 2800 m, *Arundinaria racemosa* occurs with other two species, namely, *Thamnocalamus aristatus* and *T. falconeri* (Sharma 1982). More accurate information is still needed on the distribution, uses and local names of all the species. Stapleton (1994a) published a book on *Bamboos of Bhutan* where the most common bamboo species of the country have been described. In the southern and eastern parts of Bhutan where large-sized bamboos grow, rural life without them is unimaginable. The small bamboos found in central and western part of Bhutan are used for weaving mats and employed as fencing and roofing material. Bamboo finds diverse uses in Bhutan. One of the most well-known uses of bamboo is for making bows and arrows, archery being a national sport and cultural event (Anon. 1995a). Bows are generally made out of *Dendrocalamus hamiltonii*, but only those culms that grow on particular microsites produce good bows. Arrows are made from high-altitude bamboos such as *Arundinaria* species. The species is used for making durable mats for building construction fencing material and high-quality woven handicraft products, such as food and drink containers, hats, arrows, quivers, etc. The leaves are used as livestock fodder during winter and dry season. The species provides shelter and food for endemic fauna in reserve areas. The bamboos are used for collecting, storing and then transporting most agricultural goods. Many rural houses are entirely made of bamboo. Roof, mats and fencing are the main uses at higher altitudes. *Thamnocalamus* species is mainly used for roofing, mats and fencing. In subtropical areas, bamboo is treated as a multipurpose plant from which almost anything can be made. Weaving of thin strips makes baskets and trays.

The shoots of *Arundinaria*, *Dendrocalamus* and *Drepanostachyum* sp. are edible. The culms of *Drepanostachyum* sp. are used for making finely woven domestic and agricultural equipment, such as baskets, trays, mats, sieves, etc. Further, the species is also used for constructing livestock shelters and temporary dwellings and parts of traditional houses. The *Bhutanese* farming system that supports 80 % of the population is highly dependent on forests for sustenance, where bamboo is one of the important crops. Bamboos are harvested and used as part of Bhutanese daily life. People in bamboo houses may also employ bamboo baskets for storage and as water containers. The demand for bamboo is getting increase over time, particularly for use as fodder and other multipurpose uses. There is ample scope for greater bamboo production, especially in the country's higher areas where communities are widely dispersed and agriculture is less profitable, and bamboos have prime role in maintaining the watersheds in catchment of many rivers and controlling landslide and soil erosion and provide food and shelter to the wild lives and preserve hill ecosystem.

1.3 India

India has rich bamboo resources. According to Varmah and Bahadur (1980), there are about 19 principal genera of bamboos in India—*Arundinaria*, *Bambusa*, *Cephalostachyum*, *Chimonobambusa*, *Dendrocalamus*, *Dinochloa*, *Gigantochloa*, *Indocalamus*, *Melocanna*, *Neohouzeaua*, *Ochlandra*, *Oxytenanthera*, *Phyllostachys*,

Pseudostachyum, *Schizostachyum*, *Semiarundinaria*, *Sinobambusa*, *Teinostachyum* and *Thamnocalamus*. Areas particularly rich in bamboos are the northeast region, Western Ghat and Andamans. However, Sharma (1987) reported about 130 species belonging to 24 genera of bamboos from India. Out of these, 20 are indigenous and four are of exotic origin. Then, Tewari (1992) described 23 genera and 128 species. As per the latest compilation, 96 species are native bamboos and 40 species are cultivated ones (Kumar 2011). Among all the species, *Bambusa bambos*, *Dendrocalamus strictus* and *D. hamiltonii* are very common throughout mainland India (north, central, west and eastern part) and have great impact on socio-economy of people of the region. Additionally in eastern India, *Bambusa balcooa*, *B. tulda*, *B. nutans* and *B. vulgaris* are also extensively cultivated as important bamboo crops for rural economy. A special reed bamboo which prefers to grow in marshy land of southern India is *Ochlandra* sp., primarily used in housing, weaving, matting and pulping.

Continuous hills interspaced by vast plains along river valleys are two distinctive features in the northeast region which include seven states of India. Altitudes of the different states vary between 150 and 4521 m above mean sea level. Only three of these states have relatively large areas of plain lands. These plains are located in Assam, Manipur and Tripura states. The other four, Arunachal Pradesh, Meghalaya, Mizoram and Nagaland, are almost entirely composed of hills. The structure of these hills is quite rugged, with steep gorges created by narrow mountain streams and a great many peaks. The northeast zone experiences distinct five seasons in a year, viz. winter (November to February) with little or no rain, where night and mornings are misty; a short spring (March); summer that stretches from April to May; a prolonged rainy season (mid-May to September); and a short autumn (October to mid-November). The monsoon rains normally start around April to May and continue till September, though it may rain at other times of the year sporadically too. The average annual rainfall ranges from 2000 to 6500 mm. The temperature ranges from moderately warm in the plains in summer to freezing cold in the hills in winter. The atmospheric temperature ranges from 2° to 40 °C; it rarely falls below 0 °C, mostly remaining within 20° and 35 °C. The northeastern region accounts for 28 % of total bamboo-growing area of India and produces about 66 % of bamboo, while the rest of the country has 34 % only. Bamboos are found both naturally and cultivated in homesteads and farms. Forest-grown bamboos are natural, owned and maintained by the government. The northeastern part of India, being a part of the subtropics, has mostly clump-forming bamboos both in forests and in villages. Morphologically, forest bamboos are thin walled and comparatively smaller in size to village bamboo species. The forest bamboos are mainly used for thatching, roofing of housing and agricultural purposes.

Phytogeographically, the region is the convergent point of Indo-China vegetation at northeast and Indo-Malayan vegetation at southeast; thus, it has been recognized as one of the hot spots of biodiversity globally. The richness of forest in northeastern region is such that it is called as the storehouse of diversities of trees, bamboos, canes and medicinal plants. In moist deciduous forests of Northeast India, most common bamboo species, like *Melocanna baccifera*, *Bambusa tulda*, *Dendrocalamus hamiltonii*, *Dendrocalamus longispathus*, *Gigantochloa andamanica* and *Schizostachyum dulloa*, have been growing naturally. These bamboos are lifeblood to the local indigenous people of Northeast India for food, housing and agricultural and fishing activities. Some commercially important species like *Bambusa balcooa*, *B. cacharensis*, *B. nutans*, *B. polymorpha*, *B. tulda*, *B. vulgaris* and *Thyrsostachys oliveri* are commonly cultivated in different states of Northeast India. In Meghalaya from Shillong to Cherrapunjee area, *Phyllostachys mannii* has been cultivated in the farmland as shelterbelt against cold wave. This is the only monopodial bamboo growing by a few villages in those areas. *Ph. bambusoides* is another one monopodial species growing by Apatani tribe in northeast corner of Arunachal only.

Bamboos have been integral part of the socio-economic life of the people of Northeast India. The size of the domestic bamboo economy has been estimated at around Indian rupees (INR) in 2000 crore by the Planning Commission. The market potential, however, is estimated around INR 4500 crore, which could grow to INR 26,000 crore by 2015. India's share in the global market is estimated to be $1 billion (about 4000 crore) and expected to increase to $5.7 billion (Bhattacharjee and Chakravarthy 2008).

1.4 Myanmar

Old Burma now Myanmar is located between Bangladesh and Thailand, with India and China to the north; Myanmar covers an area of about 675,000 km^2. Biodiversity distribution of Myanmar forest resources is influenced by a wide range of locations between latitudes 9° 58′N–28° 29′N and longitudes 92° 10′E–101° 10′E, topography traversing from north to south through three major mountain ranges and four major river systems. A major topographical feature of Myanmar is the Irrawaddy River system. Since its deltaic plains are very fertile, it is considered to be the most important part of the country covering about 47,000 km^2. Myanmar weather climate is principally of the tropical monsoon type with three distinct seasons: summer, rainy season and cold season. From mid-February to mid-May are summer months; the rain falls from mid-May to the end of October and the cold season starts in November and ends in the end of February in Myanmar. Generally, Myanmar enjoys a tropical monsoon weather climate. However, Myanmar weather climate conditions differ widely from place to place due to widely differing topographical situations. For instance, central Myanmar has an annual rainfall of less than 40 in. while the Rakhine coast gets about 200 in. in Myanmar. Besides, the average highest temperature in central Myanmar during the summer months March and April is above 43.3° C, while in northern Myanmar, it is about 36.1° C. Temperature of towns varies according to their location and elevation. Generally, in Myanmar, hot season is from March to May, rainy season is from June to October and cold season is from November to February. The tropical monsoon weather in Myanmar is usually cloudy, rainy and hot and humid summers, and during winter less cloudy, scanty rainfall and mild temperatures with lower humidity.

Bamboos are found all over the country in the forests either as an understorey or as a pure stand from sea level to the mountain ranges to about 4000 m. There are sympodial types as well as monopodial types, with sizes varying from 3.0 to 33.0 m in height. The country has about 100 species with a wide diversity of 17 genera and four varieties of bamboos (Htun 1999). The genera are *Arundinaria*, *Bambusa*, *Cephalostachyum*, *Chimonobambusa*, *Dendrocalamus*, *Dendrochloa*, *Dinochloa*, *Gigantochloa*, *Klemachloa*, *Melocanna*, *Oxytenanthera*, *Phyllostachys*, *Pseudostachyum*, *Schizostachyum*, *Sinobambusa*, *Teinostachyum* and *Thyrsostachys*. In Myanmar, language bamboo is called as 'Wa'. It has been estimated that *M. baccifera* occurs as pure stands over 7800 km^2 in Arakan Yoma. Other species like *Dendrocalamus strictus*, *Bambusa longispiculata* and *Thyrsostachys oliveri* occur in scattered locations. Bamboo breaks of *Bambusa bambos* and *B. tulda* are found along stream banks and lower hill slopes. The commercially important species are *Bambusa bambos* (local name: Kyaket-wa), *B. longispiculata* (local name: Tabindaing-wa), *B. polymorpha* (local name: Kyathaung-wa), *Cephalostachyum pergracile* (local name: Tin-wa), *Dendrocalamus brandisii* (local name: Kyalo-wa), *D. giganteus* (local name: Wabo-wa), *D. hamiltonii* (local name: Wabomyetsangye-wa), *D. longispathus* (local name: Wanet-wa), *D. membranaceus* (local name: Waphyu-wa), *D. strictus* (local name: Hmyin-wa), *Dinochloa macllelandii* (local name: Wanwe-wa), *Gigantochloa rostrata* (local name: Waya-wa), *Melocanna baccifera* (local name: Kayin-wa) and *Thyrsostachys siamensis* (local name: Htiyo-wa, Myin-wa).

Bamboo is an important species in the economy of Myanmar. Bamboos are used from toothpicks and chopstick industries to paper pulp and also as food and fodder. For aesthetic reasons, many bamboos are cultivated; examples are *Bambusa bambos* (hedge plants with thorns), *B. vulgaris* var. *striata* (the yellow bamboo with longitudinal green stripes) and *B. vulgaris* var. 'wamin' (short inflated internodes in the lower part of the culms, Buddha belly bamboo), also used for landscaping, house gardens, parks and zoological gardens, in and around big cities in Myanmar. Some species, like *Bambusa longispiculata*, *B. wamin* and *Thyrsostachys siamensis*, are planted in many Buddhist Monasteries, in the villages and gardens in urban areas for local sales. About 100 species grow in large quantities throughout the country. They are major construction materials, particularly in rural areas, and can be used for almost all parts of houses, including posts, roofs, walls, floors, beams, trusses and fences.

People use bamboo to make mats, baskets, tool handles, hats, traditional toys, musical instruments, umbrellas and furniture. In addition, bamboo shoots are edible and pickled-bamboo shoots are becoming very popular. As an industrial raw material, bamboo is commonly used in Myanmar by pulp and paper mills.

1.5 Nepal

Nepal is lying between China and India and located in 28° 00′N and 84° 00′E. The country is very mountainous and hilly. Its shape is roughly rectangular, about 650 km long and about 200 km wide, and comprises a total of 147,181 km^2 of land.

Nepal has great physical diversity, ranging from the Terai Plain—the northern rim of the Gangetic Plain situated at about 300 m above sea level in the south—to the almost 8800 m high Mount Everest, locally known as Sagarmatha (its Nepali name), in the north. Climate varies from cool summers and severe winters in north to subtropical summers and mild winters in south. Eastern Nepal receives approximately 2500 mm of rain annually, the Kathmandu area about 1420 mm and western Nepal about 1000 mm.

Bamboos are widely distributed throughout Nepal, but they are more common in the eastern half of the country, from Dhaulagiri to the Sikkim border. In higher rainfall areas such as those around Pokhara and Ilam, a wider variety of genera and species can be found. One unique feature of Nepal is that it has both tropical and temperate bamboo species. About 11 genera (*Ampelocalamus*, *Arundinaria*, *Bambusa*, *Borinda*, *Cephalostachyum*, *Dendrocalamus*, *Drepanostachyum*, *Himalayacalamus*, *Melocanna*, *Thamnocalamus* and *Yushania*) and more than 30 species of bamboos have been recorded from Nepal (Stapleton 1994b). The bamboos are abundant between the mid-hills and the Terai with most of the species being found in the mid-hills. It is estimated that there are close to 62,900 ha of bamboo stands in the natural forest of Nepal (Anon 1996a). In Nepal, big-diameter bamboos are known as *bans* where smaller ones as *nigalo* and the smallest diameter group as *malingo*. Among different species, *Arundinaria maling*, *Bambusa bambos*, *B. nutans*, *B. balcooa*, *Dendrocalamus hamiltonii*, *D. hookeri*, *D. strictus*, *D. patellaris*, *D. giganteus*, *Drepanostachyum* sp. and *Thamnocalamus* sp. are the most common and useful bamboos (Das 1988). People believe that regular cutting of bamboos promotes good growth, and if not harvested regularly, the bamboo population decreases.

Bamboo is used in Nepal in 180 different ways, the most visible ones being basketing, housing and scaffolding (Poudyal 1991). Bamboos provide a large proportion of renewable material for building, paper, animal fodder and vegetable and cottage industries in many areas of the Terai and mid-hills. While traditional uses continue to consume large quantities of bamboos, new uses are also being developed. Bamboo is a major agroforestry crop widely planted in farms and vacant lands on the periphery of settlements. Most of the households in the rural areas have their own bamboo stands (Anon 1996a). In the eastern Terai region of Nepal, there are a large number of bamboo farms with an area ranging from 1.25 to 2.5 ha each. In the lowlands, natural stands are mixed with deciduous subtropical forest vegetation. The weaving bamboos are the most popular species with about 70 % of the farmers growing them on their farms and homesteads. About 66 % of the bamboos grown are for commercial purposes. Entire culms are used as rafters, pillars and fence posts, while split culms are used for panels or further split for weaving baskets and other articles. Bamboos also find use in making furniture and a host of domestic items. Bamboo-based enterprises are an important source of employment for both the rural and urban workforce. It is estimated that the sector accounts for employment generation of more than 100,000 workdays per annum (Anon 1996a). Out of the estimated 15 % contribution to the national GDP by the forestry sector, bamboo contributes 1–2 %. Estimates show that the annual bamboo production is a little over

3 million culms, of which 2.64 million culms are consumed locally. Young culms are harvested for string making and weaving of bamboo articles. Mature culms are used mostly for construction and furniture making. Most of the bamboo products—such as mats, basketry, household accessories and implements—are manufactured by the farmers and artisans and sold in local markets. The local market, though strong, does not receive products from all regions owning to lack of transportation network and a well-established marketing system. Bamboo is also being increasingly used in rural construction and for making water tanks, tubs, bus stands, etc.

Bamboo fodder plays a significant role in Nepal as one of the main sources of fodder in late winter (Anon 1995b). Over 30 bamboo species are extensively used as fodder. Cattle graze on *Drepanostachyum intermedia*, *Thamnocalamus* spp. and *Arundinaria racemosa* growing at higher elevations in natural fores*ts*. Vitally important for Nepal is the use of bamboos in rehabilitating riverside degraded lands. In Nepal, bamboo plantation has been highly preferred by local community to reduce the impact of landslide (Paudel and Kafle 2012). Investigations are going on in combining mechanical structures and bamboo planting to divert and control river flow in different parts of the country. The use of bamboo strip-based brush dams to replace iron mesh wires is also being tested by different government and nongovernment organizations.

1.6 Pakistan

Pakistan is bordered by Afghanistan to the northwest and Iran to the west, while China borders the country in the north and India to the east. Pakistan lies in the temperate zone. The climate is mostly semi-arid, but arid in the south, characterized by hot summers and cool or cold winters and wide variations between extremes of temperature at given locations—the average daily low of 2 °C in January to an average daily high of 46 °C in June. Half of the annual rainfall occurs in July and August, averaging about 255 mm in each of those 2 months. The remainder of the year has significantly less rain, amounting to about 50 mm/month. Hailstorms are common in the spring.

The arid climate of Pakistan does not favour the natural occurrence of bamboos. No published reports are available on the bamboos of Pakistan. The natural distribution of bamboos in the Indian subcontinent is very scanty towards the western part. It appears from the description of Gamble (1896) that it is likely that a few clumps of *B. bambos* and *D. strictus* may exist near the Indian boarder of Punjab. Bamboos are of limited diversity in Pakistan. Three species grow naturally in Pakistan. *Arundinaria falcata is* in the northwest Himalayas at 1200–2000 m. It occurs in the undergrowth in forests of oak, firs and mixed trees, usually on northern slopes or in ravines. It is part of a wider gene pool through the Himalayas. It is used for making baskets, mats and pipes. *Bambusa bambos* is rare in the Ravi river eastwards. It is absent in the hills. This species is extensively used for construction. *Dendrocalamus strictus* is found in Punjab and Kashmir. It is found also in mixed vegetation on Margalla hills surrounding Islamabad. It is used for construction and a variety of purposes. This forms part of a gene pool extending across India and usually growing below 1200 m. In Pakistan, the bamboo resources are shrinking.

Efforts were made in the 1980s to introduce different species of bamboos from Bangladesh, China, Sri Lanka and Thailand into Pakistan. Some of them are well adapted to climatic conditions of Punjab. In 1980, about a dozen propagules of *B. vulgaris* and *B. tulda* were taken from Chittagong, Bangladesh, Forest Research Institute and four from Thailand and China and then planted in the Punjab of Pakistan (Banik 2000). Some of the species in this regard are *Dendrocalamus giganteus*, *Dendrocalamus strictus*, *Bambusa bambos*, *Bambusa tulda*, *Dendrocalamus hamiltonii*, *Bambusa vulgaris* and *Dendrocalamus longispathus*, and the ornamental species *Phyllostachys aurea* survived and grew well in some pilot plantation plots. Some bamboo species may be grown ornamentally in the urban homesteads, parks and gardens. Bamboo plantations have been raised on good agricultural lands in Sargodha, Jhang, Khushab and Mandi Bahauddin districts of Punjab province in Pakistan. Most of these plantations are on 0.5 acres of lands. In Chunian subdivision of Kasur district, small farmers have also successfully established small bamboo groves of 4–10 acres area. Bamboo plantation area increases or decreases with the market demand in the domestic as well as in the Middle East markets. *Dendrocalamus strictus*, *Bambusa bambos*, *Bambusa tulda* and *Dendrocalamus hamiltonii* are the major species grown on private farmlands.

1.7 Sri Lanka

It is an island in the Indian Ocean, located between five and ten north latitude in southeast of India. It has a total area of 65,610 km^2, with 64,740 km^2 of land and 870 km^2 of water.

The climate can be described as tropical and quite hot. Its global position endows the country with year-round warm weather, moderated by ocean winds and considerable moisture. The mean temperature ranges from a low of 16 °C in Nuwara Eliya in the Central Highlands (where frost may occur for several days in the winter) to a high of 32 °C in Trincomalee on the northeast coast, where temperature may reach 38 °C. The average yearly temperature for the country as a whole ranges from 28 to 30 °C. January is the coolest month, especially in the highlands, where overnight temperatures may fall to 5 °C. May, the hottest period, precedes the summer monsoon rains. The rainfall pattern is influenced by the monsoon winds of the Indian Ocean and Bay of Bengal. The mountain slopes and the south-western sector of the island receive heavy rains. Some of the windward slopes receive up to 2500 mm of rain per month.

Sri Lanka possesses only ten bamboo species according to a recent revision of the group (Soderstrom and Ellis 1988). Of the ten species, *Bambusa bambos* and *Dendrocalamus cinctus* are confined to the dry zone of the country (annual rainfall of less than 1000 mm). A third species, *Ochlandra stridula*, is found extensively in the wet lowlands of the south-western region (2000–5000 mm annual rainfall). The natural habitat of most of the bamboo species is the forest understorey. The bamboo flora of Sri Lanka may be said to approach very nearly to that of the hills of the Western Ghats of Southern India (Gamble 1896). *Dendrocalamus cinctus* and *Arundinaria scandens* are confined to windswept mountain tops, and *A. densifolia* (*Chimonobambusa densifolia*) prefers wet places and grows in marshes at elevations of 2300 m among montane grasslands. A new endemic genus *Davidsea* has also been reported from Sri Lanka (Zoysa Neela de et al. 1988). About 20 bamboo species are supposed to have been introduced into Sri Lanka of which seven are cultivated. Among them, the yellow variety of *Bambusa vulgaris* is most widely cultivated, particularly in the rural areas of the wet low and mid-country and in the vicinity of waterways in the dry zone. *Dendrocalamus giganteus* is cultivated on a small scale in the wet highlands, whereas *Dendrocalamus asper* and *Dendrocalamus membranaceus* are found in the intermediate highlands. During 1964, *Dendrocalamus strictus* was introduced to Sri Lanka, and a few pilot plantations were established in the dry zones (Vivekanandan 1987). Two species of bamboo cultivated for their ornamental value are *Bambusa glaucescens* (synonyms *B. multiplex*, *B. nana*) and the recently introduced *Thyrsostachys siamensis*. The former is found in most parts of the country, while the latter is currently restricted to urban areas. Most of the above-mentioned bamboo species are found in the three Botanic Gardens of the country.

The usage of bamboo in Sri Lanka is largely in the handicraft industry and in the housing and construction sector. Bamboo handicrafts form a traditional cottage industry in the country, employing about 20 % of the people working in the handicraft sector. *Ochlandra stridula* and the introduced *Bambusa vulgaris* are exclusively used in housing and cottage industries (Zoysa Neela de et al. 1988). The supply of *O. stridula* and *D. giganteus* is fast decreasing, while *B. vulgaris*, which is cultivated, is still available in reasonable quantities. The farmers mainly cultivate *B. vulgaris* because of high demand in the construction and housing sectors. The price of 9 m long bamboo culm was rupees 30 in 1988–1989. In early 1995, the price increased to Indian rupees 80–120, and nowadays, it has gone up to rupees 175–200 (2014). Of the people involved in the bamboo sector, 77 % are engaged in collecting and processing raw materials mainly from forest sources, while 11.5 % produce bamboo-based products, and the remaining 12 % includes 9 % in trade and marketing area and only 3 % involved in resource production (nursery, plantation raising and management). About 79 % of the people, who harvest bamboos, process and utilize these themselves (Anon 1994). It is reported that a great number of people are dependent on the existing bamboo resource, particularly *O. stridula* in the natural forests.

From time immemorial, bamboo has played a vital role in the lifestyle of rural areas of South Asian countries. The extensive cultivation of different bamboo species in almost every rural holding throughout the length and breadth of these countries indicates the importance of this plant in the socio-economic life. The uses of bamboo in rural housing and agricultural implements are so common that a homestead in the rural area cannot be conceived without bamboo. Since unknown past farmers in the region have been growing various bamboos in their homesteads through selection from the wilderness. Through experience, field trial and wisdom, people identified suitable phenotypes or varieties in a number of bamboo species in respect to specific utilization. Considering ecologic and socio-economic aspects, some of these were finally domesticated by the people in the region. In addition, some species have been introduced to enhance the productivity, market opportunity and industrial demand. In the South Asian region, the distribution of wet months ranges from 2 to 7 months yearly, with about 1200–4000 mm of annual rain fall. In some areas, the day temperature may even reach 44 °C during the dry season, during April to May, and may drop to –0 to 10 °C in the cool dry season, from December to January. However, the average temperature of the whole region ranges from 25 to 34 °C. The region has many rivers and fertile flat lands where rural people have been traditionally growing bamboo as one of the MPTs in their homesteads and farms. On the hills, indigenous people are also living with the bamboos for food, housing and most of the livelihood activities. Thus, all countries in the region have enormous potentiality of raising extensive bamboo plantation and opportunity for industrialization of bamboo and can provide maximum employment opportunity to the rural community which would upgrade the status of this plant as cash crop, and all the country in the region would be benefitted.

Accordingly, people in this South Asia region started selecting and prioritizing some bamboo species out of the more than 300 naturally grown or cultivated species. Among those selected majorities are common in all the countries of the region. However, some of the prioritized local bamboo species are not considered in the priority list of global context.

Chapter 2
Some Priority Bamboo Species for South Asian Region

To a botanist bamboo is a grass even though apparently it doesn't look like it. Bamboo is a member of the Bambusoideae, a subfamily of the grasses. The subfamily is markedly diverse both in reproductive and vegetative characters. To talk of the habits of bamboos, and of the management of bamboo plants in general, has little meaning and is of no practical use. 'Each species has its own peculiarities and its own requirements…and without a reliable guide, the study of bamboos…would be hopeless' (Brandis 1899).

For centuries, bamboo has been used in fishing, papermaking, landscape gardening, handicrafts, fine arts, food, fodder, building, weapons and hundreds of other things. Some cultures are based on bamboo; the shoots provide a large portion of their food, and the culms are used for building housing and for making products that are sold as their only form of income. Bamboo has long been used in handicrafts and as the raw material for thousands of objects used in daily life and in the pursuit of a livelihood. The qualities of bamboo have been, and are, celebrated in paintings, drawings and verse. The pulp of bamboo is well suited to making fine papers of many varieties and adaptations. In India, Bangladesh and many other countries, bamboo pulp is blended with shorter weaker pulps for making wrappings and fine papers. High-grade bamboo pulp can be, and is, used in its pure state for making coated and uncoated book and magazine papers. The high length-to-diameter ratio of bamboo pulps gives it a special versatility in the papermaking process. *Bambusa vulgaris* and *Melocanna baccifera* rank very high in performance in South Asian countries both in the field and mill though most papers come from other species because of larger and more accessible stands. Two of such species are *Dendrocalamus strictus* and *Bambusa bambos* (syn. *Bambusa arundinacea*). And the people of South Asia region are bestowed with all the four species of bamboos and receiving benefit.

A total of 1482 species of bamboo under 119 genera have been described from different continents of the globe (BPG 2012). Major species richness is found in Asia Pacific and South America while least in Africa, except in Europe which has no native species. The major genera found in Asian countries include *Arundinaria*,

R.L. Banik, *Silviculture of South Asian Priority Bamboos*, Tropical Forestry,
DOI 10.1007/978-981-10-0569-5_2

Bambusa, *Cephalostachyum*, *Dendrocalamus*, *Dinochloa*, *Gigantochloa*, *Melocanna*, *Ochlandra* and *Schizostachyum*. In temperate Asia, as in China, Korea and Japan, *Phyllostachys* and *Sasa* are common. From the tropical moist, primeval forests to the cool mountain foothills, bamboo is a natural partner to humans in all walks of life that to live without it is scarcely imaginable. Out of such huge number of bamboo species growing in the wilderness, only a few have been domesticated and cultivated in different parts of the world. People of each country identified their species on the basis of climatic suitability, social importance and utilization need.

In order to enhance production, especially against a background of over-exploitation, International Network for Bamboo and Rattan (INBAR)'s research networking requires a much sharper focus on a limited number of high-priority species. The INBAR in cooperation with the International Plant Genetic Resources Institute, formerly the International Board for Plant Genetic Resources (IPGRI), canvassed a number of experts (where author was also a member) from different parts of the world to choose priority bamboo species. Experts met together in expert consultation meeting in 1993 and also in 1994 for sharing information, established criteria for choosing species, and the expert group agreed upon groups of bamboo species—as those meriting focused research and wider use (Banik 1995).

From the beginning, it was recognized that a consensus on the major priorities for regional and international action would, of necessity, not include many of the other species which are used locally, many of which are the subject of research by national programmes. It is stressed that research on these should continue to receive attention from national, regionally or sub-regionally important species.

Accordingly, the following 20 taxa of bamboos are accorded high priority for international action, and a document entitled 'Priority species of bamboo and rattan' was published jointly by IPGRI and INBAR in 1998 (Rao et al 1998).

1. *Bambusa balcooa* Roxb.
2. *B. bambos* (L.) Voss
3. *B. blumeana* J A and J H Schultes
4. *B. polymorpha* Munro
5. *B. textilis* McClure
6. *B. tulda* Roxb.
7. *B. vulgaris* Schrad. ex Wendl.
8. *Cephalostachyum pergracile* Munro
9. *Dendrocalamus asper* (Schultes f.) Backer ex Heyne
10. *D. giganteus* Wallich ex Munro
11. *D. latiflorus* Munro
12. *D. strictus* (Roxb.) Nees
13. *Gigantochloa apus* J A and J.H. Schultes
14. *G. levis* (Blanco) Merrill
15. *G. pseudoarundinacea* (Steud.) Widjaja
16. *Guadua angustifolia* Kunth
17. *Melocanna baccifera* (Roxb.) Kurz
18. *Ochlandra* Thw. (spp.)—a number of species

19. *Phyllostachys pubescens* Mazel ex H. de Leh (including *P. bambusoides* Sieb. and Zucc and *P. edulis* Makino)
20. *Thyrsostachys siamensis* (Kurz) Gamble

The publication of the list aroused much interest among bamboo and rattan researchers and conservationists. Many suggested that rating, evaluation and domestication need to be changed for certain species listed in the report. Some of them felt that too few species had been included in the lists omitting a number of others that are yet traditionally used in many countries. Many suggested that bamboos growing in the colder climates like *Arundinaria* spp. should be included since many of them provide fodder, building and biomass materials for people living at very high altitudes.

Additions and amendments are made to cover many of the above points and to accommodate a wider coverage of greater number of economically important species, and accordingly INBAR/IPGRI recognizes this and has noted an additional 18 bamboo species as a second list.

Among the above listed 20, all these are Asiatic except for Genus *Guadua* which is a New World bamboo includes a number of important species. However, about two decades ago, *Guadua angustifolia* has been introduced to a few locations in southern part of India and needs more field trial to get popularity in the region and thus is not discussed in the present document.

Bambusa blumeana of the above priority list is a native of Indonesian islands, cultivated in PNG, northern part of Peninsula Malaysia, Thailand and the Philippines. *Bambusa textilis* is a native and cultivated bamboo in south China.

Gigantochloa apus, *G. levis* and *G. pseudoarundinacea* are not seen in cultivation or natural in Indian subcontinent. However, *Gigantochloa andamanica* is growing naturally in northeast India, Chittagong Hill tracts and other parts of the region and has socioeconomic and ecologic importance to the local people and biospheres, thus included in the priority list of South Asia.

Phyllostachys pubescens and *P. bambusoides* are temperate bamboo species native to China and extensively cultivated in different provinces of China, Japan, Korea and Vietnam and not grown in the South Asian countries. So the *Phyllostachys* spp. are not discussed.

Under genus *Ochlandra*, the species name was not mentioned in the INBAR/IPGRI's above list. However, two species, namely, *Ochlandra travancorica* Benth. & Hook. f. and *O. stridula* Moon ex Thw. are very important bamboo species of South India and Sri Lanka.

Like other regions of Asia and Africa, the people in South Asian countries have been living ecologically with a number of bamboo species and utilizing them economically for their day-to-day requirements. Many of these bamboo species have been growing naturally in the forests of the region, and some are being also cultivated in the villages near the doors for intimacy and easy access. A good number of these regional bamboos obviously cannot have their place in the global list prepared by INBAR and IPGRI. Considering the natural occurrence, ongoing increasing desire of cultivation, new techniques of utilization and increasing demand, threat of genetic erosion and ecological importance in the countries of South Asia, a list of following 25 priority bamboo species is taken for discussion in the book, of which some are already treated as global priority species in the above list of INBAR and IPGRI.

Bamboo species					
1	*Bambusa balcooa*	10	*Dendrocalamus hamiltonii*	19	*Schizostachyum dullooa*
2	*Bambusa bambos*	11	*Dendrocalamus longispathus*	20	*Thyrsostachys oliveri*
3	*Bambusa cacharensis*	12	*Dendrocalamus membranaceus*	21	*Thyrsostachys siamensis*
4	*Bambusa nutans*	13	*Dendrocalamus strictus*	Plus	
5	*Bambusa polymorpha*	14	*Gigantochloa andamanica*	4	Species of Himalayan shrubby bamboo ringal
6	*Bambusa tulda*	15	*Melocalamus compactiflorus*	25	Total species presented
7	*Bambusa vulgaris*	16	*Melocanna baccifera*		
8	*Dendrocalamus asper*	17	*Ochlandra travancorica*		
9	*Dendrocalamus giganteus*	18	*Ochlandra stridula*		

Among the above 25 prioritized species, some (e.g. *Dendrocalamus asper*, *D. membranaceus*, *Thyrsostachys siamensis*, etc.) have introduced earlier in the region and show some promise for future cultivation, and thus gradually field-based information is developing on these bamboo species. These species are economically well known to the bamboo world and also included earlier in the INBAR's second list of priority bamboos.

Additionally, a group of four bamboo species [*Drepanostachyum falcatum* (Munro) Keng f., *Himalayacalamus falconeri* (Munro) Keng f., *Thamnocalamus spathiflorus* (Trin.) Munro and *Sinarundinaria anceps* (Mitford) C.S. Chao & Renvoize] growing in the Himalayan region, locally known as *Ringal bamboo*, are also included in the above list of 25. The ringal bamboo species are socioeconomically very important to the local indigenous people and to this mountain ecosystem. Additionally, these species are also extremely important for wildlife, providing food and shelter for animals growing in the Himalayan ecosystem.

Attempts have been made to collate both field experience and documented information on these 25 bamboo species together and present in the book as an information base for future research and development activities of bamboo resource of the countries in South Asian region.

Part II
Species Silviculture

Chapter 3
Bambusa

Bambusa is Latinized from *Bambu* or *Bamboe*, a word of supposed Indo-Malayan origin and recorded from Dutch colonial sources in the seventeenth century, but is more likely sourced from Indian word *Mambu* used in earlier publications (Piso, Mantiss. Aromat., 1658: 185; Clusius, Exoticor., 1605: 165, 259). Mambu (also called Tabashir or Tabaxir) was originally applied to the plant's stony substance (collected in the hollow joints) used in Asian (Ayurvedic) medicine, but came to be the name by which the plant itself was known in Europe.

The species in this group are diverse vegetatively and in flowering behaviour, but they share a similar structure in having a many flowered spikelet (Dransfield 1980).

The genus is native in the tropics and subtropics of South Asia (from Terai of the Himalayas, northeast and mainland India including Andaman Islands, Bangladesh, Nepal, Bhutan, Sri Lanka, Myanmar, Pakistan) and Southeast Asia (from Thailand, Laos, Kampuchea, Vietnam, mainly southern-central China including Taiwan and Hainan, south Japan, Malaysia, Singapore, Indonesia, Philippines, Papua New Guinea with Solomon Islands, north Australia, and Madagascar). A number of species of *Bambusa* have been successfully introduced in cultivation in Africa, Europe and north and south America in the farms, gardens and parks.

Thirty-nine species are known under the genus *Bambusa* (Ohrnberger 1999). Out of these, the following seven bamboo species are most popular to the local people of the South Asian countries due to their high demand in utilization and trading. Some of the species are naturally growing in the forests, others in cultivation or in both sources. In the next segments, these species are described in alphabetical order.

3.1 *Bambusa balcooa* Roxb.

(Synonym: *Dendrocalamus balcooa* (Roxb.) Voigt, Hort. Suburb. Calcutt., 1845:718)

R.L. Banik, *Silviculture of South Asian Priority Bamboos*, Tropical Forestry,
DOI 10.1007/978-981-10-0569-5_3

3.1.1 General Information

3.1.1.1 Vernacular and Local Names

Baluka bans (Assam-India); *Wah long* (Khasi, Meghalaya-India); *Wamnah*, *Beru* (Garo, Meghalaya-India); *Oti*, *Out*, *Vuteya*, *Awuti*, *Avuthi* (Nagaland-India); *Leewa* (Manipur); *Barak* (Tripura-India), *Wasul* (Kokborok-Tripura); *Balku bans*, *Boro bans* (West Bengal-India); *Boro bans* (north of the Ganges and Terai); *Udal bans* (South Bihar-India); *Haroti* (Samastipur, Bihar-India); *Dhanu bans*, *Ban bans* (Nepal); *Barua*, *Barak* (Sylhet-Bangladesh); *Barak*, *Balku*, *Banbans* (Mymensingh, Comilla-Bangladesh); *Baro bas* (Dinajpur-Bangladesh); *Balkoa*, *Balku* (Rajshahi-Bangladesh); and *Gitaborua* (Tangail-Bangladesh). It is likely that *Bambusa balcooa* has been named from the local name *Baluka bans* of Assam and *Balkoa*, *Balku* of Rajshahi district in Bangladesh and Malda and Nadia district of West Bengal.

3.1.1.2 Natural Distribution and General Habitat

No natural vegetation of *B. balcooa* has been reported, only known in cultivation. This species is said to be a native of northeastern part of India (Assam, Arunachal Pradesh, Meghalaya, Tripura, Nagaland), eastern India (West Bengal, Bihar, Jharkhand, Orissa) and extending up to north India (eastern Uttar Pradesh) and Bangladesh and found up to 1600 m in Nepal and Bhutan. It has also been introduced and cultivated in many other countries of Southeast and East Asia, e.g. in Australia, and in botanical gardens of Indonesia and Sri Lanka.

Bambusa balcooa grows in any type of soil but prefers heavy textured soil with good drainage and pH of about 5.5; occurs in plains and hills at low and moderate elevations 1000–1250 m. It is extensively cultivated in the rural areas having alluvial clay, sandy loam in the plains covering both sides of the rivers Ganges (Uttar Pradesh, Bihar, some parts of Orissa, Jharkhand, Chhattisgarh, West Bengal), the Padma (Bangladesh—Rajshahi, Faridpur, Pabna, Kushtia, Jessore), the Brahmaputra (Assam and Bangladesh—Rangpur, Saidpur, Bogra, Dinajpur, greater districts of Mymensingh and Sylhet) and the Meghna (Dhaka, Comilla, Brahmanbaria) and nearby regions of South Asia.

Bambusa balcooa is the most common bamboo species in the greater Terai region of the Himalayas. This is most common and major bamboo species cultivated profitably in the homesteads of northeast, north and eastern part of India, and also very commonly cultivated in the northern part and scattered all over Bangladesh. The species has been also introduced to many countries of Southeast Asia, tropical Africa, northern South America, Central America, and in the southern USA.

A number of large-size healthy clumps of *B. balcooa* are found to grow in the villages of Kalinagar, Udaipur, about 10 km away from Govind Ballabh Pantnagar Agricultural University, Uttarakhand (India), which is an exotic place for the species. It was learnt from the old villagers that during the time of partition of India a few Bengali farmers while migrating from southern Bengal brought a few offsets of the species from the healthy and high productive clumps along with their belongings and settled in these villages and went on conserving the bamboo through regular cultivation in more homesteads for multipurpose durable construction uses. Similarly in the unknown past, *B. balcooa* has been distributed by human beings throughout different parts of the Indian sub-continent, and thus the species has become one of the major bamboo species both in homestead plantations and farms due to its wide utilization.

3.1.1.3 Climatic Conditions

The plant grows well in somewhat flat lands of alluvial deposits having pH 5.5–7.5 with annual rainfall of 2500–5000 mm and a distinct drought period of 5-7 months from October to March. The plants are very sensitive to waterlogging and cyclone and rarely cultivated in the southern districts (towards the Bay of Bengal) of Bangladesh and India. The species tolerates drier conditions better than many other bamboos, but occasionally suffers from the bamboo blight disease on poorer and ill-managed sites.

3.1.1.4 Uses

One of the best and strongest species for building purposes and scaffolding. Very durable if well seasoned after immersing in water for about 2–2½ months. It is commonly used for structural works for houses (in northeast and eastern India and Bangladesh). In a recent study (Mahzuz et al. 2015), *Bambusa balcooa* was used as both flexural and shear reinforcement, keeping other physical parameters constant, and the variables were flexural and shear reinforcement. The maximum 2.4 times more load with respect to the non-reinforced beams was taken by the beams where bamboo was used as flexural reinforcement, and maximum 3.9 times more load was taken by the beams where bamboo was used as shear reinforcement.

It is a major species for scaffolding and making bridges, used as fishing floats, and also used as floats for timber rafts in north, northeast and eastern India and Bangladesh. Also commonly used in roofing structures, furniture, fencing, toys, novelty items, etc. including agricultural and fishing implements.

Occasionally used for making 'agarbatti' sticks (incense sticks) in Tripura state of India.

Due to its thick-wall nature, *B. balcooa* has been one of the preferred bamboo species for making frames for rickshaw hood, platform of bullock cart, boats, electric poles, ladder, etc. It was estimated in the year 1992 that annually about 0.9 million bamboo poles were used for making rickshaw hood and 3.1 million bamboos were used annually for constructing platform of boats and bullock carts in Bangladesh (Banik 1992, 2000).

The species provides sufficiently a good amount of raw material for pulp and paper mills in northern India. In a number of villages near Pantnagar (Uttarakhand), farmers have been practising clearfelling of clumps after each 8–10 years of clump growth. A good amount of money (Rs 6500–10,000 from a clump, year 2012) is earned at a time by selling all the culms. The contractors as a the middle man buy all the culms (80–120 culms per clump) and transport to the nearby big cities and pulp mills gate by 20 tonne truck. The species also were used as raw materials for pulp mills in Bangladesh.

In Panchagarh and Saidpur districts of Bangladesh, a number of clumps of *B. balcooa* and *B. nutans* are grown closely in one or two lines along the northwestern sides of the rice field as a shelter belt and wind break against the dry and cold wind blowing from Bihar and Nepal. Occasionally in Bengal and Bihar, one to two clumps are being cultivated isolatedly in the vast open rice field surrounding the small water wells and ditches. Farmers at midday hot summer take rest under the shade of clump and drink water from the well. Clumps crown also minimize the drink water loss from wells against scorching sun.

3.1.1.5 Ethnic Utilization

B. balcooa shoots are eaten by local indigenous people in Assam, Meghalaya and Tripura (Banik 1997a), and cost is about Rs 15–18 per kg (in the year 2012). The leaves are used as emergency fodder, especially during floods in rural Bengal and also in the villages of Samastipur, Pusa (Bihar). The species is used in making indigenous bamboo bows in Manipur.

3.1.2 Plant Data

3.1.2.1 Vegetative

Plant Habit

Bambusa balcooa is a stout, large-size, clump-forming bamboo with strong underground sympodial rhizome system, and in adult clumps, the apical part of the culms hangs slightly centrifugally (Fig. 3.1a). The apical part of an elongating young culm is dome shaped covered with diagnostically bright copper or yellowish brown colour culm sheaths visible especially in the growing season (Fig. 3.1b, i).

Fig. 3.1 (**a**) An adult clump of *B. balcooa* apical part of the culms hang slightly centrifugally. (**b**) A developing young culm tip is dome shaped covered with diagnostically bright copper or yellowish brown colour culm sheaths visible from distinctly apart especially in the growing season. Culm sheaths promptly deciduous start dislodging gradually from basal nodes as the culm elongates. (**c**) A light brown whitish pubescent ring present above the node and hairy below. The silvery brown pubescence arranged in lines distinctly visible on internodes of younger culms. (**d**) The culms branched from the base. The branches are strong, thick, stout and elongated with prominent, branch nodes with prominent distinct straw colour buds. Comparatively small and very thin usually leafless, thornlike recurved wiry 1–2 branches are produced from both the sides of these stout branches mostly present at the lower nodes of the culms. (**e**) Clump shows deciduous behaviour as observed in drier site near the side of the road Bihar and UP, India. (**f, g**) The inflorescence is compound and 0.25–1.0 m long, bearing spicate branches with groups, pseudospikelets, spikelet ovoid, lanceolate or flattened, and the green florets having purple tips. (**h**) Rooted culm cuttings are developed in propagation beds. (**i**) A clump of having culms with comparatively shorter internodes, swollen and elevated nodes locally called this type as *Sil barak*. (**j**) A clump type having comparatively thin-walled culms and branches and with long culm internodes is locally called as *Nol barak/Teli barua*. (**k**) *Bambusa balcooa* clumps are grown along the canal bank for stabilizing and binding the soil and to minimize the landslide in flood time. (**l**) Felling of young culms (3 months to 2 years old) from the periphery of clump decreases the clump productivity and vitality and leads to create congestion in clump

a
b
c
d
e

Fig. 3.1 (continued)

Fig. 3.1 (continued)

Culm Height

Tall 15–25 m, may be 31 m.

Culm Diameter

Diameter varies from 8 to 15 cm (at breast height) and has raised nodes. The girth of the nodes are usually 5–6 cm more than those of internodes at basal part, 3.5–4.0 cm at mid-part, and 2.5–3.0 cm at upper part of the culms.

Culm Thickness

Thick walled, 2.7–3.2 cm at basal part, 1.2–2.0 cm at mid part, and 0.3–0.7 cm at upper part of the culm.

Culm Colour

Dark glossy green to bottle green.

Internode Length

Twenty to forty-six centimeter long. The presence of aerial roots is, maximum up to 4–5 nodes, less up to 7 nodes. *B. balcooa* has a series of mid-culm internodes of more or less equal length (38.0–41.0 cm), and therefore the internode length curve in respect of node numbers is broad and flat at the middle section (Banik 2015a). The internode lengths are 15.5–35.0 cm at the lower part and 30.0–36.0 cm at the upper part of the culms. A light brown whitish pubescent ring present above the node and hairy below. The silvery brown pubescence is distinctly visible on internodes of younger culms (Fig. 3.1c).

Branching

The culms branched from the base. The branches are strong and elongated with prominent nodes (Fig. 3.1d). Within 7–9 months of age, the large branch bud on each of the culm node breaks and produces a thick stout branch basipetally. These branches are 3.5–8 m long with swollen nodes, usually stout and thick at the basal ⅓ of the length, and the remaining upper portion is narrow and whip-like. However, except for four to five basal nodes, almost all the buds on the culm nodes produce stout branches within 2 years of emergence. Comparatively small and very thin, usually leafless wiry one to two branches also produced from both the sides of these stout branches. These small branches are thornlike recurved (Fig. 3.1d) and mostly present at the lower nodes of the culms, smaller than the exact thorns present in *B. bambos*. The branches on the younger culms have branch sheaths, with distinct brown hairs.

Leaves

The leaves are comparatively bigger than in many other *Bambusa* spp., oblong-lanceolate, 15–33 cm long, 3–5 cm broad, ventral side hairy especially when young, main vein prominent and shining below. Leaf sheath straight, truncated and has fine white or brown hairs. Leaves of several flushes occur together mainly during spring to rainy summer (March to August). The clumps of *Bambusa balcooa* in some occasions also exhibit pure deciduous behaviour (Fig. 3.1e) in prolonged draught period, also in drier habitat as observed in Bankura, Cooch Behar (West Bengal), near Pusa area of Bihar, Delhi, UP in India and northwestern part of Bangladesh and upper Myanmar.

During a 12-month cycle (annual calendar), clumps of *B. balcooa*, *Dendrocalamus asper* and *D. hamiltonii* produce big and comparatively small-size leaves in warm humid (March to August) and in the beginning of dry winter season (October to November), respectively. Clumps that grow in moist habitat generally have larger leaves than those growing in drier sites (Banik 2015a).

Culm Sheath

Culm sheaths on the lower nodes (from the base to mid-culms) short and broad, densely appressed hairy on the upper surface. The imperfect blades short somewhat triangular. The basal part of the blade edge wavy, auricle very short or absent. Culm sheaths on the upper nodes (from mid to above) long (25–35 cm), broad (20–25 cm) and almost glabrous. The imperfect blades also long (15–20 cm), broad (8–10 cm) and sharp at the apex. *Ligule* broad, 3–7 mm long, margin wavy and finely serrated.

Vegetative Growth

The newly emerged culm is distinctly conical, pointed, dome shaped, covered tightly with sheaths and at this stage used as edible shoot. Culm emergence period May to November, June to August is peak period of emergence (Banik 1993a). Generally at the beginning annually per clump produce 2–5 culms and then it may be 10–23 culms. It is reported that maximum rate of elongation of a culm is up to 77 cm per day; culm attains its optimum height within 75–90 days of emergence (Banik 2000). All emerging culms do not always develop into full-grown culms. The natural mortality of emerging culms was found to be 31–43 % (Banik 1983a). Culm sheaths promptly deciduous (Fig. 3.1b, d) start dislodging gradually from basal nodes as the culm elongates. A clump usually starts producing merchantable size big culms within 5–6 years of planting. Best growth of the species has been observed at Bangladesh and Tripura, lower Assam, northern West Bengal and nearby areas of the sub-continent. The growth study at Chittagong revealed that a number of full-grown culms emerged in the clumps of *Bambusa balcooa* increased up to the 7th year of age and then gradually decreased, if not felled (Banik 1988a). Field trials at different parts of India show that *B. balcooa* can be grown successfully in south and some parts of western India. After 48 months of planting, clump produced 10.92 m tall culms, culm girth 17.7 cm and the culm biomass (green 15.0 kg, dry 9.9 kg) at Kerala, which is not a homely place for the species. However, a culm from an adult (15 years) well-grown clump usually attains 27–30 m height, 12–15 cm diameter and a weight of 75–90 kg in eastern India. A clump may produce 7–19 culms per year.

The normal life of culms in a full-grown clump is 11–14 years (Banik 2000). In a normal situation, a clump dies after flowering. The survival, culm production per clump, culm height and diameter become more or less half if the species is grown on the hills and slopes than those grown in flat lands.

Physical Properties of Culm

The strength properties of *B. balcooa* culm increase with the age and the maximum values obtained when culms are 3 years old. The younger bamboos are less dense than the older ones. The physical and mechanical properties are also not same for base, middle and top position of culm (Table 3.1).

Table 3.1 Some physical and mechanical properties of culms of *Bambusa balcooa*

Observed parameter	Culm		
	Top	Middle	Bottom
Moisture content (%)	66	86	101
Specific gravity (based on green weight)	0.74	0.64	0.57
Specific gravity (based on oven-dry weight)	0.85	0.84	0.79
Shrinkage in wall thickness (%)	4.8	7.6	11.1
Shrinkage in diameter (%)	2.5	3.4	4.2
Compressive strength, parallel to grain (green (kg/cm^2)	506	459	394
Compressive strength, parallel to grain, air-dry (kg/cm^2)	573	536	510
Modulus of elasticity (MOE); green (1000 kg/cm^2)	103	92	72
Modulus of elasticity (MOE); air-dry (1000 kg/cm^2)	127	108	93
Modulus of rupture (MOR); green (kg/cm^2)	624	712	850
Modulus of rupture (MOR); air-dry (kg/cm^2)	696	787	926

Source: Sattar et al. (1991)

3.1.2.2 Reproductive

Flowering Nature

The clumps of *B. balcooa* rarely flower. A few isolated clumps reported to flower in different parts of India during 1849, 1876, 1881 and 1889 and in Bangladesh during 1983 and 1985. A flowering clump completes flowering within 1 year. As recorded, the estimated flowering cycle might be 40 ± 5, 90 ± 5 and 130 ± 5 years (Banik 2000; Table 3.2). Thus, it appears that cultivated plants of *B. balcooa* in the sub-continent are sourced from number of genetic population.

Inflorescence

The inflorescence is compound and 0.25–1.0 m long, bearing spicate branches with groups, pseudospikelets, spikelet ovoid, lanceolate or flattened, 6–12 mm × 4–6 mm; green florets having purple tips (Fig. 3.1f, g) with 4–6 fertile and 0–2 sterile florets. *Anthers* do not produce viable pollen (Banik 1997b).

Seed

Caryopsis is not known (Gamble 1896), sterility common. As the species flowers rarely and seeds are not produced so seedlings cannot be raised (Banik and Alam 1987).

Table 3.2 Estimation of flowering cycle in *Bambusa balcooa* from the flowering records

Country/locality	Flowering date (calendar year)	References of flowering dates	Estimated flowering cycle (year)
India			
Assam, Purnea district	1849	Gamble (1896)	40 (1849–1889)
Assam (Goalpara)	1876	Gamble (1896)	
Assam (Kamrup)	1889	Gamble (1896)	
Assam (Jorhat)	1996–1997, 1998–1999 (2 clumps flowered)	Pathak (2002)	
West Bengal—BCKV[a] (NMBA Expt Plot)	2008 (Isolated clump flowered)	Author observed	
Uttar pradesh (Gorakhpur)	1881	Wood (1881)	
Bangladesh *Mymensingh* (Baldi, Gafargaon) [Assam is not far away from here]	1983–1984 (Isolated clump flowered)	Banik and Alam (1987)	134 (1849–1983) 94 (1889–1983)
Saidpur (Hazirhat) [Assam is not far away from here]	1984–1985 (Isolated clump flowered)	Banik and Alam (1987)	136 (1849–1985) 96 (1889–1985)

[a]*BCKV* Bidhan Chandra Krishi Viswavidyalaya, an agricultural university, Kalyani, West Bengal, India

Vegetative Propagation Methods

The guidelines for producing culm and branch cutting, collection and bagging of wild bamboo seedling and multiplication through macroproliferation and their nursery management are provided in Appendix I, and for details, the reader may consult the INBAR Technical Report No. 6 (Banik 1995).

SL. No	Multiplication method of *B. balcooa*	Success rate (%) and comment
1	*Offset planting*: Rhizome with basal 3–5 node part of a 12–18-month-old culm should be collected during March to April, maintained in sand propagation beds (transit nursery beds) till the time of plantation	70, heavy (8–20 Kg) availability limited, expensive
2	*Culm cutting* (Fig. 3.1h): 2-node segments are collected from 1- to 2-year-old culms during summer to rainy season (mid-February to September) and placed in propagation beds containing sand rooting medium	70–75, within 7–8 weeks
3	*Branch cutting*: Branches are usually collected after pre-monsoon shower (any time after May to June) and placed in propagation beds containing sand rooting medium	65–70 within 7–8 weeks
4	*Seedling macroproliferation*: Not possible as seeds are not produced	Nil
5	*Macroproliferation of cutting*: When a cutting has 3–4 shoots, then the rhizomes are separated; thus, more than 2 cuttings are created and each is then potted separately	40–55, limited numbers
6	*Ground layering*: During rainy season, 1–2-year-old culms are pulled and banded down to ground level and buried below the surface	40–50, cumbersome, need space near the mother clump

Tissue Culture

Bambusa balcooa (explant)	Medium (mg/l) and environments	Results	References
Not mentioned	(a) MS (½) + BAP(4.0) + Kn(1.0) (b) MS(½) + NAA(5.0) + coumarin (0.1)	(a) Multiple shoots (b) Shoot rooted	Farah et al. (1991)
Nodal segments	(a) MS + BAP (11.25 μM) + Kn (4.5 μM) (b) ½ MS + IBA (1.0 μM)	(a) Shoot multiplication (b) Rooting and in vitro regeneration	Das and Pal (2005)
Nodal buds	(a) MS+ BAP (1.0 mg per litre) (b) BAP (1.0–5.0 mg per litre) (c) ½ MS + NAA (1.0–3.0 mg per litre) + IBA (1.0–5.0 mg per litre)	(a) Shoot bud initiation (b) Shoot multiplication (c) Rooting and mass propagation	Islam and Rahman (2005)
Nodal segment from adult clump	(a) MS+ BAP (1.0 mg/l and 1.5 mg/l) (b) Clump of 3-shoot cultured on MS + NAA (3.5 mg/l and 4.0 mg/l)	(a) Nodal bud proliferated (b) Rooted plantlets acclimatized for 1 month, transferred to polybag. 100 % survival in field	Sharma and Sarma (2011)

Note: *MS* Murashige and Skoog growth medium, *BAP* 6-benzylaminopurine, *NAA* α-naphthalene acetic acid, *Kn* 6-furfurylaminopurine, *IAA* indole-3-acetic acid, *IBA* indole-3-butyric acid

Cytology

$2n = 70$ (aneuploid).

Diversities and Conservation

It has been found that the local people in Sylhet, Tripura and lower Assam distinguish the clump of *B. balcooa* as two types on the basis of culm characters, such as internode length, node nature, wall thickness and others. In upper Assam, most of the clumps of *B. balcooa* have thick-walled culms with swollen and elevated nodes; the local people named this type as *Shil barua*/*Sil barak*/*Hil barua* or *Hil barak* (Banik 1994a; Fig. 3.1i). Shil barua culm has many swollen nodes so mechanical elasticity is reduced at nodal part (Liese and Ding 1994) and rarely used for any weaving, thatching and matting works but very much used for house post to bear the load of the heavy roof. In other types, culms and branches are comparatively thin walled and have long internodes and locally named as *Nol barak*/*Teli barua* (Fig. 3.1j). In adult clumps, the apical part of the culms hangs centrifugally, and as a result, the upper clump part is not so compact but rather lightly open. The clump of *Sil barua* or *Sil barak*/*Hil barua* or *Hil barak* is somewhat compact, straight with more number of leafless thornlike (not exact thorns present in *B. bambos*) branches at the lower nodes of the culms. The branches on the younger culms have branch sheaths, with distinct brown hairs.

The word '*sil*' in local language means stony, hard, heavy; thus, *Sil* or *Hil barua/Barak* clump has comparatively stronger, thick-walled culms with shorter internodes and elevated nodal rings than the *Teli barua/Nol barak* clump. Thus, *Sil barua/Barak* is more durable and usually favoured for making props in housing, scaffolding, bridge and other construction works.

A superior clone selected in India from *Bambusa balcooa*, locally known as '*Beema bamboo*', has high calorific value of 4000 k cal/kg, low ash content between 0.4 and 1.0 %, and biomass yield of 100–125 tons/ha/annum. These bamboos are 4–7 cm in diameter and about 12 m tall only, but grow many in a clump. The bamboo is multiplied in large numbers through micropropagation by 'Grow more' at Hasur, a private tissue culture centre in South India, for commercial plantation in India and abroad (Banik 2015b).

The species do not produce seeds after flowering, so progeny with mixed characters of parent is not created. Since the unknown past, the species has been propagated by vegetative means for cultivation, and thus, it is likely that characters of the planted materials are least changed. Additionally, further chances of new genetic diversities are also very limited unless any new individuals with variations are discovered through extensive exploration in the rural areas of the regions as no natural vegetation of the *B. balcooa* is known.

Conservation Status No natural population is known. The species has been mainly conserved in homestead gardens, small farms and plantations. Recently (2002–2012), in northeast India and Bangladesh, large-scale death of different forest bamboo species (*Melocanna baccifera*, *Dendrocalamus hamiltonii*, *Gigantochloa andamanica*, *Schizostachyum dullooa*, *Bambusa tulda*, etc.) due to gregarious and sporadic flowering has created acute shortage of raw material for bamboo-based industries, incense sticks, construction works and housing. Thus, utilization pressure has been concentrated on *B. balcooa* and *B. vulgaris*, two major cultivated (homestead) bamboos of the region. This resulted to overexploitation of these species. Moreover, farmers at Tripura villages (northeast India) are getting attractive monitory benefit to grow more rubber plants, and bamboo groves are, thus, being replaced by rubber. These alarming situations make them especially vulnerable to overexploitation. Characteristically, the species rarely flowers, usually in one to a few clumps, and almost all crops remain in vegetative state without any death.

Effective measures have to be made so that ranges of genetic diversities are stored at community-based ex situ conservation plots, institutional gene banks and bambusetum.

Cultivation Potentiality The species is very productive and high yielding and has been extensively utilized in construction works, pulp and paper making including many new modern bamboo-based industries. Thus, the species has untapped potentiality for cultivation in most of the countries of South and Southeast Asia, except too dry and cooler *areas*.

3.1.3 Cultivation and Management

3.1.3.1 Site Selection and Preparation

B. balcooa prefers to grow in the flat lands and valleys under moderate to high rainfall zones of the South Asian countries. The species can be grown successfully throughout the sub-continent except some drier areas in the west. It requires gentle terrain sites, with deep, loose and fertile sandy loam. Planting site should be moist, well drained or not in waterlogged conditions. Moist environment and sites preferably rich in organic matter are suitable for raising plantation. In the villages, this bamboo is also grown along the river and canal bank (Fig. 3.1k) for stabilizing and binding the soil and to minimize the landslide in flood time. Generally, when the planting site is flat on levelled ground with at least an area more than 10 ha, the overall land can be ploughed thoroughly and deeply as possible. The overall loosening of soil up to 40 cm depth by clearing rocks, tree roots and stumps has to be ensured. This improves the land environment by loosening the soil for better aeration and complete mixing of the organic and supplied inorganic fertilizer with the soil for supporting the plant growth.

If plantation is raised on the hills, lower slopes to valley are good sites; hill top and upper slopes have to be avoided. The strip land preparation is generally adopted on hill slope planting. To prevent the soil erosion and water loss, strip land preparation can be made parallel with the contours. The strip width and distance between strips is about 3.0 m for a big- or medium-size bamboo (*Bambusa*, *Dendrocalamus*, *Gigantochloa* spp., etc.), and depth of loosening soil is about 40 cm.

In steep sloping fields, at desired spacing, the spot can be cleared of weeds and shrubs with 2.0 m radius around the planting spot and then digging of pit is done without much disturbing the surrounding soil.

In the *homestead model*, bamboo clumps are ideally planted in the northwestern corner of the house to afford protection against northwestern storms. New canal banks are also good sites. Farmers occasionally plant one or a few clumps by the side of the well near the paddy and wheat field for taking rest and drinking water at the midday lunch time.

3.1.3.2 Planting Space and Pit Making

Planting space may be 5×5 m under high rainfall and with more weed conditions; wider spacing like 6 m×6 m may be adopted in the areas having lower rainfall and lesser weed problem. Therefore, in a hectare, either 400 or 278 propagules (cuttings or tissue culture plants, seedlings not available as species do not produce seeds) are necessary. In the hill slopes, planting pits may be dug on the contour lines. Further, wide spacing 7 m×7 m to 9 m×9 m may be used for growing agricultural crops during first 2 years and then trees in between the lines.

Planting pits of 60 cm cube are dug at 5 m×5 m or 6 m×6 m spacing for cultivating big-size rhizomes and 45 cm cube for small-size cuttings (branch and culm) of the bamboo species.

Pits should be dug and prepared 10–15 days before planting. All herbaceous and woody weeds within 1 m of the pit should be removed.

3.1.3.3 Fertilization and Soil Preparation

Then soil preparation is to be completed during December to January. When digging a pit (planting hole), the underneath and surface soil should be separated and put at the two sides of the hole. A more effective fertility strategy is to select a moist site with a rich organic soil. Fertilizer should be applied at least 2–3 week before planting. The pits are to be filled with cow dung/FYM (farmyard manure) (10 kg), urea (20 g), triple superphosphate (20 g) and muriate of potash (10.0 g) and soil. The dug earth must be used to fill 2/3 the height of the pit, and during planting, the surface soil must be added near the root zone of plants. The earth in the hole must be thoroughly mixed with farmyard.

During planting, insecticide is given in pits for preventing white ant attack in the cuttings or offsets. Care should be taken to ensure that planted materials are protected from damage (like shaking and browsing, fire, etc.) by people and animals.

After planting, during heavy rains, earth mounting is to be provided at a radius of 0.50 m and height of 10–20 cm, around each of the propagules to prevent water logging at the base of the plants.

3.1.3.4 Weeding and Vine Cutting

After planting in June to August, three to two times weeding and vine cuttings are done in the first year depending on intensity and occurrence of weeds. Then weeding has to be done two times in the second year. Afterwards, only one weeding has to be done around the bamboo plant in the third and fourth year depending on the species nature and weed occurrence. After that, a 3 m radius ring weeding is preferred than complete weeding of land. However, intercropping by seasonal/annual crops may be practised to suppress the weeds and to have best utilization of the land in between the bamboo plants.

3.1.3.5 Aftercare and Tending Operations

The tending operations in the young stand are not quite similar to those of mature stands. The techniques of intercropping, especially with nitrogen-fixing plants, and application of green manure, usually from the intercrop itself, may be applied from the year of plantation. In addition to these crops, farmers also practise agroforestry techniques for 1–3 years to cultivate mustard, potatoes in between the lines (5 or 6 m) of grown bamboo-planting stocks.

Intercropping The practice of intercropping is possible and found to be beneficial to bamboo plants, especially in the early part of plantation before the canopy formation of bamboo clumps is complete. Practice of intercropping also provides opportunity of early earning of extra cash from the first year of bamboo plantation. Some legume crops, like pea (*Pisum sativum*), mung (*Phaseolus aureus*), lentil (*Lens esculenta*), soyabean (*Glycine max*), arhar (*Cajanus cajan*), etc., are grown in between the planting rows. In comparatively wet land, mustard (*Brassica spp.*), *Sesbania* sp., can be grown. Sometimes, rice (*Oryza sativa*), wheat (*Triticum aestivum*) and maize (*Zea mays*) are also cultivated in widely spaced (9 m×9 m, 10 m×10 m) bamboo plantation at the early stage of plantation establishment (Banik et al. 2008). However, crops should not be planted too close to the bamboo plants; usually 1.0 m space is left from the base of a bamboo plant.

The species can be grown satisfactorily with association of some semi-deciduous to deciduous trees (i.e. trees that shed leaves completely once in every year), like simul (*Bombax ceiba*), karoi (*Albizia* spp.), gamari (*Gmelina arborea*), teak (*Tectona grandis*), bel (*Aegle marmelos*), amla (*Phyllanthus emblica*), amra (*Spondias pinnata*), *Dalbergia sissoo*, and narrow-light crown plants like betel nut (*Areca catechu*), *Eucalyptus* spp., Poplars, etc., and other palm plants at wide spacing (Banik et al. 2008).

In the later part of plantation establishment, i.e. when plants become tall and crown comes closer, some shade-tolerant cops like ginger (*Zingiber officinale*), turmeric (*Curcuma longa*), pine apple (*Ananus comosus*) and medicinal plants can be grown successfully.

As bamboo plants grow, the clumps also expand, and simultaneously canopy starts providing shade in between. *Bambusa balcooa* cannot survive in shade, so it fails to grow under evergreen trees. So it is suggested not to plant any umbrageous wide crown evergreen timber/fruit trees (like mango, jackfruit, lichi, jalpai, raintree, banyan tree, etc.) as intercrop in bamboo plantation (Banik et al. 2008). However, in between wide spaced (25–30 m) existing bamboo clumps these fruit trees of graft- origin (where heights are usually controlled) may be planted.

Loosening of Soil Mounding and Mulching Just before the start of dormant dry winter, soil around each of the plants is to be loosen carefully. During every summer or autumn, loosening the soil to a depth of 6–12 cm and removing weeds around bamboo clumps are to be done carefully without damaging the underground rhizomes or buds on them. Soil loosening and weeding could be combined in established stands. In middle February every year, rake the top soil of bamboo clumps in order to expose bamboo buds to higher temperatures and sunlight, stimulate shoot sprouting, prevent bamboo roots from getting entangled and increase nutrient supply for shoot production.

Farmers usually practise mounding and mulching with rice straw or water hyacinth to the bamboo clumps for improving the fertility status and preserving moisture in the ground. Mounding the bamboo groves with earth and mulching promote culm production by nourishing and protecting the mother rhizome. The exposed freshly worked soil needs to be covered by mulch with the organic maters like leaves, twigs, etc. This practice improves quality and grade of tender shoots by protecting against the excessive heat (above 30–46 ° C) and scorching sunlight.

Application of Fertilizer Fertilization increases the productivity of stands, although the amount of increase varies more or less according to the composition of the fertilizer and the season of application. Fertilizer should be applied twice a year. The first application should be done during April to May in north and northwest, but February to March in northeast and south about 2 weeks after raking the soil. Application is directly at the base of the bamboo clumps. About 10–25 kg of composts organic manure or 150–200 kg of pond silt may be applied per clump depending on clump size.

The second fertilizing, called top dressing, is carried out between June and August, when bamboo shooting is most active. Application is in ditches dug around the clumps. Each clump may receive 0.5 kg of chemical fertilizers (urea or ammonium sulphate) or 10–12.5 kg of fermented barnyard manure. After application, the ditches must be covered with earth. Concentrated dose of fertilizers should not be allowed to come in direct contact with young shoots lest the shoots wither and die. Watering should be done at this stage to ensure proper nutrient distribution.

Watering and Mulching On the homestead and in marginal lands, simple irrigation of planted cuttings/offsets by a bucket of water once a fortnight during the drier months of the year is adequate. If there is no rain, regular watering is preferable just after planting.

Pest and Disease The *blight disease* has been recorded in village groves of *B. balcooa* throughout Bangladesh (Boa and Rahaman 1984) and also in some parts of India like lower Assam, Tripura and northeastern part of West Bengal. The disease results in a sequential die back of culms in their first season of growth. The bamboo blight disease is caused by *Sarocladium oryzae*, killing affected clumps within 3–4 years. The length of time varies according to original size of the clump, but may be as short as 2–3 years. The first proper symptoms of blight show as a premature death of culm sheaths. Culm sheaths that are weakened in this way can be easily removed, unlike healthy ones which remain tightly attached until the internode has stopped growing. It is during June to December that bamboo blight attacks new culms. Previous year's culms are not affected.

The confirmed symptom of blight is the partial collapse of the fragile apical region. The culm sheaths associated with these internodes will be more or less dead, and it is only a short time until the apical region breaks off. During this second stage, wet rotten patches develop on the internodes, often associated with insect damage. These necrotic (dead) areas spread rapidly in the cheesy and juicy internodes and eventually join up. At the same time, symptoms begin to develop in the lower, more fibrous internodes and spread slowly downwards ('dieback'). Insect mining may also help in the spread of symptoms in this area; blight only attacks growing culms. The top part of the culm, containing partly developed internodes, is immediately killed. Further spread of the disease downwards is slow and limited.

The recommended control measures of *blight disease* are improved care and tending of bamboo clumps and light burning of dried debris, leaves and twigs at the clump base after light soil works and hoeing. This practice kills the soil-borne fungus and minimizes or may stop the infection in the emerging and developing culms. Further, the movement or utilization of soil from diseased clumps to healthy ones should be restricted, and production of propagules (offsets, cuttings) is only to be made from healthy clumps (Banik 2000).

3.1.3.6 Harvesting

Harvesting of bamboo can begin from the fourth year of establishment of plantation depending on the purpose for raw material. The yield of bamboo increases every year by 10–20 % and gradually stabilizes after 6–8 years. Harvesting of culms during the monsoon months can damage young and emerging shoots and retard the growth of the clump, so post-monsoon season extending through the summer is the best time of the year to harvest. Most of the mature culms (3 years and above) from the central part of the clump have to be cut just above the basal first node on the ground. Felling of young culms (3 months to 2 years old) from the periphery of clump decreases the clump productivity and vitality and leads to create congestion in clump (Fig. 3.11). As a felling tool, either a knife or a pruning saw may be used depending on the thickness of the bamboo to be harvested. The unharvested mature culms should be left scattered throughout the clump to provide mechanical support to the young immature culms against the strong wind and storm (Banik 2015c).

Some important morphological characters of *Bambusa balcooa* for culm-age determination are described below in a tabular form (Banik 1993b, 2000, 2015a). Knowing the culm age in the clump is important in management and harvesting practices of bamboo grove.

Age up to	Morphological description (*B. balcooa*)
1st year	*Culm sheath*: May be present at 2–3 basal nodes of the culm *Bud break and branches*: The large branch bud on the culm node breaks and produces a thick stout branch except on the basal 3–4 nodes. Comparatively small and thin 1–2 branches are also produced from both the sides of the main stout branch. Branch bases and some basal nodes of the culms are generally covered with straw-coloured papery sheaths *Culm*: Brown to whitish pubescent ring is present on the basal 6–9 nodes, and usually basal 4–5 nodes also possess a ring of adventitious roots. Nodes are prominent and ridged. Dark glossy green. Basal 6–8 internodes are lightly covered with minute brownish hairs arranged in many vertical lines closely parallel to each other
2nd year	*Culm sheath*: Usually not present on the culms, may be present on the basal 1–2 nodes *Bud break and branches*: Except 4–5 basal nodes, almost all buds of the culm nodes develop thick stout branches. The auxiliary branches up to basal 5–6 nodes transform into curved thornlike structure. Base of the branches may be covered with thin persistent sheath *Culm*: Brown to whitish ring along with adventitious roots may be present on the basal 3–4 nodes. Not so deep green. Basal internodes are slightly covered with minute brownish hairs

Age up to	Morphological description (*B. balcooa*)
3rd year	*Culm sheath*: Absent *Bud break and branches*: Usually buds on the basal 3–5 culm nodes are dead and rotten. Death is more on the congested clumps. Thin wiry auxiliary branches up to 9–12 basal nodes shed their leaves and transform into curved thornlike structure *Culm*: Dead adventitious root rings may be present on the basal 2–3 nodes of the culm. Deep bottle green (hookers green) and internodal surface is smooth
4th year	*Culm sheath*: Absent *Bud break and branches*: Branches have less number of leaves. Most of the auxiliary and secondary branches transform into curve thornlike structure. Stout thick branches on the basal nodes (up to 6) of the culm are usually dead, and get sheded keeping dead scar on the culm nodes *Culm*: Smooth surface having deep bottle green colour. Black rotten or dried adventitious root ring may be present on the basal 1–2 nodes

3.1.4 Specific Spot Characters for Bambusa balcooa Field Identification

1. Clump big, stout; culms with pendulous tips, thick walled, nodes swollen with supranodal ridges, aerial root rings on the lower nodes, a light brown whitish pubescent band above the nodes, brown hairs below (Fig. 3.1a).
2. Thick and stout branches from all nodes, those from lower nodes bear no leaves, branch nodes prominent with distinct straw colour buds (Fig. 3.1h).
3. Small, thin leafless branches recurved, pointed thornlike but not thorn, specially at the lower culm nodes (Fig. 3.1c).
4. Developing young culm tip dome shaped covered with diagnostically bright copper or yellowish brown colour culm sheaths visible, distinctly apart (Fig. 3.1b, i).
5. Presence of hairs on leaf sheaths.
6. Clumps do not produce any seeds and die after flowering.

3.2 *Bambusa bambos* (L.) Voss

[Synonyms: *Bambusa arundinacea* (Retzius.) Willd.,Sp. Pl.,2, 1, 1799:245. *B. spinosa* Roxb.; *Arundo bambos* Linnaeus, Sp. Pl. [ed.1], 1, 1753:81, without descript, based on pre-Linnean *Arundo arbor*].

3.2.1 General Information

3.2.1.1 Vernacular and Local Names

Kotoha bah (Assam-India); *Wakynta* (Garo-India); *Kanta bauns* (Orissa-India); *Illi, Mula, Pattill* (Kerala-India); *Bidduru, Gatte* (Karnataka-India); *Toncur* (Gujarat-India); *Bongu veduru*, *Mulla veduru* (Andhra Pradesh-India); *Mungil* (Tamil Nadu-India); *Nal bans* (Punjab-India); *Kanta bans, Behor bans* (West Bengal-India); *Kanta bans* (Bangladesh); *Kante bans* (Nepal); *Katu Una* (Sri Lanka); *Cha-kat-wa, Kyakat-wa* (Burmese); *Phai Paak, Phai Namm* (Thailand); *Rai-sai* (Khmer); *May Phaipa, Phaix pa* (Laos); *Russei khlei, Russei prei* (Cambodia); *Tre Ng, Tre Gai* (Vietnam); *Bambu duriori* (Indonesia); *Pring ori* (*Java-Indonesia*); *Indian bamboo* (Philippines); *Spiny bamboo, Bambú espinoso* (Cuba); *Banbu cafia de indios* (Spain); *Bambus, Dorniger* (Germany); *Bambou épineux* (French); and *Thorny bamboo* (English).

3.2.1.2 Natural Distribution and General Habitat

Bambusa bambos occurs throughout India, Myanmar, Sri Lanka and Thailand except in the Himalaya and Sub-Himalayan region and the valleys of Ganges and Indus. Prefers humid tropical climate and grows best along river banks, in river valleys and in other moist sites.

In India, the species is very common in Orissa, Madhya Pradesh, the Cancan and on the Western Ghat Range. In the Deccan, it occurs in valleys in the hills as it does throughout South India, Nilgiris up to 950 m and higher occasionally. The species is infrequently found in lower Assam. This is commonly found under cultivation in Rongrenggiri areas of East Garo Hills and Angratoli Reserved Forest, South Garo Hills (Naithani 2007). Very rarely, the species can be found in other States of Northeast India. It is frequently seen in the drier part of Jharkhand, Bihar and Chhattisgarh. It is rather scarce in Bangladesh except in the northwestern part (Dinajpur, Rangpur districts), where local indigenous people cultivate the species in their homesteads for housing and fencing works (Banik 2000).

Also cultivated in the villages of Indonesia (Java and Sulawesi islands) and in different parts of Indo-China. Introduced to many other tropical countries and regions of Asia including New Guinea and Pacific Islands, northern Australia and New Zealand, tropical Africa and Madagascar, northern South America, Central America including Caribbean islands, Mexico and the southern USA (Florida) (Ohrnberger 1999).

3.2.1.3 Climatic Conditions

The species grows well in humid tropical climatic condition around river valleys in moist and well-drained condition. The species grows on rich to poor soils but prefers acid soils, abundant and thrives well from mixed moist deciduous to semi-evergreen forests. Prefers humid tropical climate, grows best along river banks, in river valleys and in other moist sites. Occurs most frequently in mixed moist deciduous forest in plains and hills at low and moderate elevations usually less than 1000 m in the Himalayas ascending to 1250 m; tolerates –2 °C, also can establish and survive under conditions of low rainfall of at least 600 mm per annum and therefore a preferred cultivating species in drier parts of India (Rajasthan, Gujarat, Punjab).

3.2.1.4 Uses

A much utilized common bamboo species of mainland India. Culms used for construction and many other household works and also an important raw material for pulp and paper industries.

Pruned-out thorny branches are used for fencing and also placed inside the walls of mud houses as reinforcement.

A manufacturing facility has been set up in Jabalpur (Madhya Pradesh) to convert big-size bamboo poles of this species to bamboo-based engineered bamboo products (e.g. laminated bamboo lumber, bamboo veneer, ply bamboo, particle boards, etc.).

Studies in steep terrain of Taiwan showed that thorny bamboo plantation can intercept more rainfall than many other plant species so as to store more water in the watershed; the canopy of thorny bamboo plantation can intercept all of the raindrops when rainfall depth is smaller than 5 mm (Tang et al. 2013). The rainfall interception has been recognized as a hydrological process of considerable importance in the water resource management. The nodes of the thorny branch and rhizomes can densely sprout to keep the plantation canopy with well-closing pattern and thus intercept more rainfall.

Some whitish amorphous siliceous deposits (*Tabashir*) produced in the culm internode used as medicine. Shoots are edible. Leaves used as fodder and also good medicine for blood purification and for treatment of inflammatory conditions. Ideal thorny fence and wind barrier.

It was estimated that quantity of bamboo poles of *B. bambos* used from the rural sector in Kerala during 1987–1988 was around 3.2 million culms.

3.2.1.5 Ethnic Utilization

The shoot is edible. The young shoot of *B. bambos* is metallic purplish green; growing apex blunt, sheath usually coriaceous, glabrous to pubescent with dark brown hairs, ligule continuous with sheath top, auricle inconspicuous. The percentage of (on fresh weight basis) sheath cover and edible portion of a freshly collected shoot are 35.5 and 64.5, respectively. The shoot meat possesses yellowish colour and taste is slight bitter with crisp to tough texture (Banik 1997a, 2000).

Sometimes during campfire night in Araku valley of Andhra Pradesh and Khandua area of Madhya Pradesh,India local people uses bamboo internode as container to cook chicken. One end open of an internode of mid-culm zone is filled up with marinated chicken mixed with salt and spices. The open side is covered with a piece of bamboo and sealed the edges with wheat flour or sticky rice paste. Then the internode cylinder is chirred over the fire and inner mixture of chicken thus cooked then the seal is removed to open the cylinder, and the cooked food is served as delicious dish; such *bamboo-chicken* dish has been considered as a delicacy in the locality (Fig. 3.2a(1), a(2)).

3.2.2 Plant Data

3.2.2.1 Vegetative

Plant Habit

Dense, stout clump. Culms erect, hollow, dark green coloured up to 30 m tall, 15–18 cm diameter, the wall very thick with a lumen; branching at all node, and thorny throughout the clumps (Fig. 3.2b), in some groves, clumps are dense bushy. The compact rhizome system and fibrous roots are distributed within 60 cm below the surface of the ground (Fig. 3.2c).

Culm Height

Strong, cylindrical, erect, tall 18–28 m.

Culm Diameter

Five to twelve centimeter, in fertile land, it reaches a large size up to 18 cm.

Culm Thickness

Very thick, usually 2–3 cm, sometimes in bushy/dwarf type almost solid in upper part of culm and distal part of branches.

Fig. 3.2 (**a**(**1**), **a**(**2**)) A mid-culm internode is filled up with marinated chicken mixed with spices and sealed, then chirred over the fire and inner mixture of chicken thus cooked. The cooked food is *bamboo chicken* served as delicious dish. (**b**) The dense, stout clump, branching at all node, and thorny throughout the clumps. (**c**) The compact rhizome system and fibrous roots are distributed within 60 cm below the surface of the ground. (**d**(**1**), **d**(**2**)) Thorns (branchlets) present at each node of the primary and secondary branches. (**e**) Precocious flowering is also known at the seedling stage of this species. (**f**) Seeds like small wheat grains, covered with glumes. (**g**) Instead of nursery shed, seedlings are placed below the crown of amla trees (*Phyllanthus emblica*) having small leaflets, thus allowing partial light below to the bamboo seedlings, a bamboo nursery of Artisan Agrotech, Harda, Madhya Pradesh, India. (**h**) The common representative of *Bambusa bambos* has neither much bushy or too tall clumps, slightly curved thorns at each nodes. (**i**(**a**), **i** ($\mathbf{b_1}$), **i** ($\mathbf{b_2}$)) Both tall with large culms and dwarf small-culm type with very thorny thickly interlaced branched. (**j**) Pruning of branches from the mid- to lower culm part is useful in management of clumps in the plantation, pruned branches temporarily kept at the clump base to shed off leaves and also to protect the emerging shoots against grazing

a1
a2
b
c
d1
d2

Fig. 3.2 (continued)

Fig. 3.2 (continued)

Culm Colour

Bright green, shining. The intermodal region is covered with a powdery mass or wax.

Internode Length

Thirty-five to forty-five centimeter, basal internodes short; nodes prominent, lower ones sometimes with aerial root ring. Nodes slightly swollen and few lower nodes produce short aerial roots.

Branching

Throughout the culms except 1–3 basal nodes, lower branches are spreading and are 1–3 m long. Those from the lower nodes recurved and bend downwards towards the ground. The primary branch emerges and remains strongly dominant. Other branches develop from its basal buds but do not attain the size of the primary branch. Two to three curved thorns (branchlets) are present at each node and the base of the branches (Fig. 3.2d(1), d(2)). The upper branches are arching and produce a fan-like plume; the upper leafy branches bear small spines.

Leaves

Lanceolate to linear, highly variable in size. The big leaves 13.5–22.5 cm × 1.7–2.4 cm, the smaller leaves 6.5–11.0 cm × 0.7–1.3 cm. Stalks very short.

Culm Sheath

Usually 14–36 × 16–31 cm. coriaceous, glabrous to pubescent with dark brown hairs, deciduous at the time the branches develop; ligule continuous with sheath top; auricle inconspicuous. The young emerging shoot is metallic purplish green with blunt growing apex.

Vegetative Growth

Culm emergence takes place during the months of May to August. Annually, a full-grown clump may produce 5–20 culms. In the forest condition, a seedling may take 20 years to reach a full-grown clump size. Its clump development is also less vigorous but steady. However, the period may be reduced by proper care and management.

Physical and Mechanical Properties of Culm

The average fibre length of *B. bambos* (syn *B. arundinacea*) is 2.7 mm (Liese 1980). A freshly cut green culm contains moisture about 100 % at the bottom, 86 % at the middle and 66 % at the top. The shrinkage in wall thickness in immature culm is high than mature culm, and the shrinkage of round culm is controlled by the shrinkage of outermost wall layer (Tewari 1992). The percentage of shrinkage from green to air-dry and green to oven-dry condition in immature and mature culms of *B. bambos* in the wall thickness is 9.61, 12.29 and 13.39, 16.41; and in culm diameter 7.11, 9.88 and 10.52, 13.40 respectively. Seasons without much degrade; mature culms dry rather slowly; very little cracking; thick-walled pieces collapsing slightly; thin-walled immature pieces developing deformation. The strength properties of *B. bambos* in green and air-dry conditions from different localities are mentioned below (Table 3.3).

3.2.2.2 Reproductive

Flowering Nature

B. bambos commonly flowers gregariously after a long interval of time. Occasionally flowers sporadically or isolatedly. The reported flowering dates (years) are arranged in locality wise under each country, and the gaps between the dates are calculated to

Table 3.3 Some physical and mechanical properties of *B. bambos* culm in three localities of India

	Culm source from 3 localities					
	UP (Haldwani)		West Bengal		Kerala	
Parameters	Green	Air-dry	Green	Air-dry	Green	Air-dry
Moisture content (%): green/air-dry	65.1	15.6	NR	14.3	104.1	16.5
Specific gravity	0.583	0.649		0.664	0.438	0.641
Fibre stress at elastic limit (kg/cm^2)	464	629		176	315	200
Modulus of rupture (MOR); (kg/cm^2)	751	975		482	461	566
Modulus of elasticity (MOE); (1000 kg/cm^2)	94.6	134.5		31.6	27.4	39.1

Source: Tewari (1992)
Note: *NR* not reported

estimate the interseeding period or flowering cycle (Table 3.4). The variation in flowering cycle in the species seems to be due to different genetic population of different localities. After analysing all the flowering records available from its natural zones of *B. bambos*, Blatter (1931) concluded that its flowering cycle is 30–34 years and also pointed out that at Dehradun where it had been planted it has a cycle of 45 years. Gaur (1987) mentions the period of flowering cycle as 30–45 years. The species in the coastal part of India reported to exhibit 30 years interseeding period while in northern India 45 ± 3 years (Banik 2000). In a complete-flowering clump, blooming is not continuous, rather it occurs in three successive flushes (flush period) with two non-blooming period (rest period) in between. A clump completes flowering within 1–2 years and then dies. Precocious flowering is also known at the seedling stage of this species (Banik 1980, 2000; Fig. 3.2e).

Inflorescence

The leafless floral shoots (5.0–48.0 cm long) directly produce both on the nodes of culms and branches usually in the month of December. The inflorescence is very large. The spikelets (10–25 mm long) alternately develop on the nodes of floral shoots forming loose clusters. Flowering starts in the first week of February and completes within April and then clump dies. The number of florets per pseudospikelet is 6–8, out of which 5–7 are fertile and 1 terminal sterile. Another yellow, filament long, free, hanging freely. Opening of flower and pollen discharge mostly occurs in the morning at 6 am to 9 am. *Style* short, stigma 3, plumose; *ovary* elliptic-oblong. The complete-flowering clumps do not produce any culms in the current year of flowering.

During January 2011, groves of *B. bambos* have blossomed en masse in Muthanga Wayanad Wildlife Sanctuary in Karnataka and also in Tamil Nadu, Southern India (Shaji 2011).

Table 3.4 Estimation of flowering cycle in *Bambusa bambos* from available flowering records

Country/locality	Flowering date (calendar year)	References of flowering dates	Estimated flowering cycle (year)
Myanmar: *Martaban*	1916	Anon. (1916)	
India (north)			
Utter Pradesh: *Dehradun*	1832, 1836	Nicholson (1945)	48 (1832, 1836–1880, 1881, 1882)
Kumaon Hills	1880, 1881, 1882 1926	Blatter (1929)	
Shivalik division	1992 (Greg.)	Mohan (1992)	
Pilibhit, Khiri, Baraich, Genda, *Gorakhpur*	1818 1865–1870, 1880	Wood (1881) Trevor (1929)	47–52 (1818–1865, 1870)
Bihar: Upper Wainganga valley Balaghat District Perhat Div.	1818 1865–1870 1922	Nicholls (1895) Nicholls (1895) Anon. (1922)	47 (1818–1865)
India (northeast)			
Assam: Nawagang Div.	1928	Adhikari (1928)	49 (1928–1977)
Assam valley (Greg.)	1974–1977	Gupta (1982)	
Nalbari	1997–1998	Tripathi (2002)	
Lowland forest of Bajali area, Indo-Burma hotspots region	2008	Sarma et al. (2010)	
Meghalaya, Garo Hills	1929	Trevor (1929)	
India (east)			
West Bengal: Malda (North Bengal)	1874	Gamble (1896)	
Midnapore (South Bengal)	1946	Anon. (1946)	
Orissa	1812	Gamble (1896)	40 (1929–1969)
Nabarangpur Div.	1929	Nicholls (1895)	
Chandka Range, Koraput	1969 (Greg.)	Das (1969)	
Angul and Puri Div.	1912, 1944		
Ganjam Dist.	1926		
Puri Div.	1922,		
South Orissa	1976 (Greg.)	Das (1976)	
India (central)			
Madhya Pradesh: Satpuras	1869	Tewari (1992)	32 (1869, 1870–1900, 1901)
Kymore	1870	Brandis (1899)	
Central Province: ??	1900, 1901	Smythies (1901)	
Balaghat	2014–2015	Author observed	31 (1983–2014, 2015)
Maharashtra: Melghat	1938	Sagereiya (1941)	37 (1901–1938)
Nagpur [geographically Melghat is located nearer to Nagpur and Satpuras of MP]	1983	Tatwawadi and Kali (1983)	44 (1938–1983)

Table 3.4 (continued)

Country/locality	Flowering date (calendar year)	References of flowering dates	Estimated flowering cycle (year)
India (south)			
Karnataka: Wynaad and Coorg	1804, 1836	Brandis (1906), Kurz (1876), Blatter (1931) Gadgil (1980)	32–36 (1804–1836, 1866)
South Kanara, Coorg	1866		
Coorg dist	1977–1979		
Muthanga (Wayanad) Wildlife Sanctuary	2011 (Greg.)	Shaji (2011)	34–37 (1977–2011, 2012, 2014)
Western Ghat	2014–2015	Author observed	
Bandipur, Coorg and Dandeli Madikeri Division, Kodagu Narbada/Narmada valley	2011 2011–2012 2011–2013 1980–1981 1878–1884 1905–1918 1981 (1 clump) 1870	Shaji (2011) Anon (2012) Anon (2014) Brandis (1899) Troup (1921) Troup (1921) Menon (1918) Brandis (1899)	
Andhra Pradesh:			
Nallamala Hills	1889, 1892	Gamble (1896)	1889–1988 = 99 or 32–33 × 3 years
Kurnool dist	1987–1988 (Greg.)	Rao (1988), Sitaram (1988)	
Kerala: *Travancore*			
Cochin	1901	Pillai (1901)	
Parambikulam forests	1918	Menon (1918)	
South Travancore	1974–1975	Kondas (1981)	
Coimbatore Walayar forest	1816, 1817 1869–1870 1896	Nicholls (1895) Brandis (1906) Tewari (1992) Gadgil and Prasad (1984)	
Sampaje Forest Range	1976–1980		
Malabar	1866 1982, 1992, 1996, 2001–2004	Seethalakshmi (2006)	1866–1982 = 116, Or 38–39 years × 3; so might have flowered in around 1905, 1944, 1982
Tamil Nadu: Kodayar Forest Divn Kanyakumari District	2006	Kiruba and Jeeva (2010)	30–35 (1976–2006, 2011)
Mudumalai National Park	2011	Shaji (2011)	

(continued)

Table 3.4 (continued)

Country/locality	Flowering date (calendar year)	References of flowering dates	Estimated flowering cycle (year)
Bangladesh			
Chittagong :BFRI Bambusetum	1976, 1988	Banik (2000)	
Dhalghat	1988	Banik (2000)	
Sylhet: Lawachara Silv. Res Station	1988	Banik (2000)	
Brazil (Introduced, probably 1804 seed from Coorg)	1804, 1836 1868, 1899	Dutra (1938)	32 (1804–1836) 32 (1836–1868) 31 (1868–1899)
Malaysia: *Penang*, Waterfall garden at the Residency. Clumps flowered, seeded and died	1937, 1938–1940, 1941	Holttum (1958)	

?? exact location of flowering was not mentioned

Seed

Like small wheat grains, covered with glumes (Banik 2000, Fig. 3.2f). Seed ripens during March to April. A 10 g of sample contains about 990 seeds. Seed size 0.83 cm (length), 0.19 cm (mid-width) and 0.17 cm (mid-breadth). Seed production per clump varies from 300 to 800 g in Bangladesh; while in a gregariously flowered area of Shahdol and Jabalpur (Madhya Pradesh, India) seed production varied from 0.9 to 1.5 ton per ha (Prasad 1986).

Seed Germination

Germination starts within 5–10 days of sowing and continues up to 14–33 days. The germination percentage is about 52–80. Seeds short lived, lose viability within 50 days. Seeds remain viable for 6 months when stored at 5 °C or when stored over $CaCl_2$ in airtight containers. Drying of seeds to a moisture content of about 12 % definitely increased longevity when stored over hydrated lime under refrigeration. McClure (1966) reported that, at the Federal Experimental Station, Puerto Rico, *Bambusa bambos* seeds collected in 1945 and stored in unsealed glass jars at room temperature 21.1–32.2 °C gave a germination of 46 % at 160 days and 0 % at 202 days. Seeds from the same lot stored in sealed glass jars containing calcium chloride, under the same temperature conditions, gave a germination of 80 % at 160 days and 76 % at 202 days. Seeds can be germinated in nursery beds, and seedlings having 4- to 6-leaf *stage* are pricked out from the bed; these are then transferred into polythene bags of size 18 cm (flat width) × 22 cm. For maximum survival and establishment, 1-year-old seedlings are planted during rainy season (June to August).

Seedling Character

Grass like, develop rhizome system below within 45–60 days of germination forming a miniature clump by producing 1–2 shoots (culms). During first 5–6 months, partial sunlight (60 % light) is essential for better growth of green foliage and shoots in the seedlings. A seedling may attain 1.78 m length even at the age of 11 months provided seed source is good and maintained under partial shade with proper care at nursery. Instead of nursery shed, seedlings may be placed below the crown of trees having small leaflets on compound leaf, thus allowing partial light below to the bamboo seedlings (Fig. 3.2g).

Time to time weeding, adding soils and FYM (30–45 g) in the polythene bags; may need application of fertilizer NPK [$N_{100}P_{50}$ K_{10}] through water both foliar and ground. After 5–6 months of age, seedlings need to be shifted from one bed to another so that developing rhizome system and roots fail to penetrate the ground below and neighbouring polythene bags.

During rainy season, profuse regeneration of bamboo seedlings starts from the fallen seeds on the ground. To ensure the success in establishing the regenerating seedlings, proper aiding and management are necessary. Protection from weeds, grazing and fire has been found essential for hastening the success of natural regeneration (Aided Natural Regeneration) into a bamboo forest again (Banik 1988b, 2000, 2015b).

The densely populated bamboo seedlings compete strongly in the wilderness for survival and should be thinned out to minimize competition (Banik 1988b). Two to six leaved stages of wild seedlings of different bamboos are sometimes collected and transplanted in the polythene bags and used as planting stocks. Immediately after collection, *wild seedlings* should be brought to the nursery and transplanted to polythene bags containing soil (sandy loam) mixed with FYM/cow dung (3:1). At the beginning, seedlings have to be properly watered or misted keeping under partial shade for 3–5 days for hardening. Sixty to seventy per cent light may be provided by placing overhead shed net for healthy growth compared to direct sunlight (Banik 1988b, 2000, 2015b). Seedlings need regular weeding and daily watering at the nursery stage. It was observed in the nursery that the seedlings raised from the seed source Western Ghat were mostly healthy,tall with good rhizome systems than those raised from Balaghat and TamilNadu seeds. In these areas seedling population had more morphometric variabilities like dwarfness, bushy and tall.

Vegetative Propagation Methods

The guidelines for producing culm and branch cutting, collection and bagging of wild bamboo seedling and multiplication through macroproliferation and their nursery management are provided in Appendix I, and for details, the reader may consult the INBAR Technical Report No. 6 (Banik 1995).

SL. No.	Multiplication method *of B. bambos*	Success rate (%) and comment
1	*Offset planting*: Collect 1–2-year-old material during March to May and nursed at sand bed nursery, planted in July to August	65–75, costly, large-scale plantation not possible for limited availability
2	*Culm cutting*: rooting improves with the treatment of hormone NAA (I-naphthalene acetic acid) solution (100 ppm)	50–60
3	*Branch cutting*: rooting improves with hormone treatment	45–65
4	*Seedling macroproliferation*: Both wild seedlings and seed raised seedlings need to be collected and maintained in nursery into polythene bags containing mixture of soil, sand and FYM at the ratio of 2:1:1 and after 5–6 months age multiplied through rhizome separation (macroproliferation technique)	70–85, cheap and very efficient method, seed availability is frequent throughout the region
5	*Macroproliferation of cuttings*: May be practised when a cutting has 4–5 shoots	Possible but rarely practised due to poor success (25–30 %)

Tissue Culture

Species (explant)	Medium and environments	Results	References
Bambusa bambos (embryo)	N_6 + 2,4-D(7 mg/l mg/l) N6 + 2,4-D(7) + BAP(1 mg/l)	Callus Somatic embryos	Mehta et al. (1982)
(Node from mature clump)	MS + CW(10 %) + BAP(0.5 mg/l) +Kn(0.2 mg/l)	Multiple shoots	Nadgir et al. (1984)
(Node from non-flowering clump) (Shoot)	MS + 22 μM BA + 88 μM sucrose + agar 6 g/l MS + 5.4 μM NAA The cultures maintained at 25 ± 2 °C, 16 h photoperiod	Multiple shoots Rooting	Prutpongse and Gavinlertvatana (1992)
(Node with bud)	(i) Liquid or gelled MS or B_5 + BAP 1.0 mg/l by repeated transfer of culture at 3–4 weeks (ii) In same media within 6–7 weeks after the proliferation of shoot	(i) Shoot proliferation, 20–75 shoots/culture (ii) Healthy roots produced	Banik et al. (1993)

Cytology

$2n = 70, 72$.

Diversities

The common representative of *Bambusa bambos* has neither much bushy or too tall clumps, they are 18–20 m tall, with medium diameter (5–8 cm) culms with very long branches full of 2–3 slightly curved thorns at each nodes (Fig. 3.2h). These clumps are usually seen throughout India and other parts of South Asian countries.

Two major growth forms have been recognized (Gamble 1896) on the basis of geographical locations:

(a) The tall (24–30 m) (Fig. 3.2i(a)), handsome clump (*Bambusa bambos* var. *gigantea* Bennet and Gaur), having large diameter culms (8–10 cm) are naturally grown and cultivated in South India mainly for construction, making fences in the rural farms and agricultural fields, and used as raw materials in pulp mills. Clumps of largest size are found in the hills of Circars, especially around the river Godavari, on the hill ranges of the eastern and southern scarps of the Mysore plateau and in the Nilgiris.
(b) Dwarf (6–10 m) comparatively, very thorny thickly interlaced branched and small-culm type (*Bambusa bambos* var. *spinosa* Camus, Monogr., 1935, pl.75, f.A.), often crooked and knotty (Fig. 3.2i(b_1), i(b_2)) is often found on the low hills of Orissa, lower Bengal, the south-eastern part of Bihar, Jharkhand, Chhattisgarh and UP and across to Myanmar and also in some part of Thailand. Farmers also cultivate this type as homestead fences in some of the rural farm lands and homesteads in all the areas including northwest parts of Bangladesh (district of Dinajpur, Saidpur) and locally known as *Beur*, *Bish kantha*. The densely interlacing thorny branches and branchlets make it a close, almost impenetrable hedge; thus, it has been largely planted in 3–4 rows at close spacing (3×3 m) in the periphery or in boundary as a protective fence. Gamble (1896) wrote 'against such a hedge nothing but explosives would be much effect'; in respect to this, he further added 'In Hyder Ali's time, the town of Bednore in north-west Mysore, was defended by a deep trench filled with clumps of this bamboo'.

Four different morphologic types at seedling stage—grassy, grassy erect, erect and very erect—have been also recognized (Banik 1980).

Conservation Status No conservation work has been reported in natural forests. Different morphological types are conserved by the communities in farms, homesteads and waste lands. It is urgent to collect and conserve a wide range of possible varieties available in nature and commonly cultivated.

Cultivation Potentiality The species is productive and has been extensively utilized in construction works in Southern India, including many new modern bamboo-based industries. Thus, the species has untapped potentiality for cultivation in most of the countries of South and Southeast Asia, except too dry and cooler areas.

3.2.3 Cultivation and Management

One of the earliest bamboo plantations raised in 1883–1884 from bamboo seeds is in India. Nicholls (1895) took seeds from an 1882 flowering of *Bambusa arundinacea* in Jubbulpore and raised young plants 'which were distributed in great part along the Great Indian Peninsula line of Railway'.

3.2.3.1 Site Selection and Preparation

The species can grow almost all parts of the region except too dry and cooler places.

3.2.3.2 Planting Space and Pit Making

The plantation pits of 45 cm×45 cm×45 cm are usually dug in the centre of cleaned strip and keep some more space (say, 15 cm) to fill with mixture of loose sandy soil and FYM/cow dung/humus around the outside of the planting material during planting. The spacing of 5 m×5 m~6 m×6 m between the pits is practised mainly for production of timber culm. However, spacing may be changed depending on the purpose of plantation. Always select healthy and vigorously growing seedlings from the seedling population as planting materials for raising commercial plantation (Banik 1994c, 2008).

Closer spacing (2 m×2 m~3 m×3 m, 1.5 m×1.5 m) is practised for quicker regreening the land, gully plantation, biomass production, control of soil erosion and conservation of water catchment area. About 22 km long private plantation of *B. bambos* was raised on the bank of a canal, Masalgaon in Harda/Khandwa district of Madhya Pradesh, and this have been successfully controlling the land erosion and improved the water conservation in the nearby localities. Every year, recurrence of flood in the Gangetic plain areas of north Bihar causes heavy losses and damages of properties, standing crops and lives of the people. A plantation of *Bambusa bambos* was raised during 1999–2000 along the embankment areas (Pusa-Birauli) of the Burhi Gandak River to stabilize and bind the ground against the flood water current and thus has reduced the intensity of flood, and as a result, the plantation saved the lives and property of these areas. The thorny *B. bambos* also discourages the cattle to graze and trample on the embankment (Banik 2015e).

3.2.3.3 Fertilization and Soil Preparation

A general dosage that may be followed is 870 kg of urea, 500 kg of SSP and 900 kg of MoP (muriate of potash) per acre per year. For the first year 50 %, second year 75 % and third year onwards, full dose may be given in two split doses.

3.2.3.4 Weeding and Vine Cutting

As discussed under *Bambusa balcooa*

3.2.3.5 Aftercare and Tending Operations

All operational activities like intercropping, loosening of soil, watering, mounding and mulching, etc., may be followed as mentioned under *B. balcooa*.

Pruning Pruning is no doubt beneficial, particularly in a thorny thicket of *B. bambos* and in homestead groves. The branches of *B. bambos* bear thorns in all the nodes. Because of the strong spreading of thorny branches, the base of the clumps gets congested making it difficult to enter and extract the culms. As a result, it becomes difficult to take care and manage the clumps by coming closer to the plants. Harvesting of culms from the clumps also becomes difficult and in some occasions impossible. Pruning of branches from the mid- to lower culm part has been found useful in management of clumps in the plantation (Fig. 3.2j). The 4–6-month-old branches of this bamboo are cut/pruned from the standing culms during the months of December to January, when the moisture content was found to be 33 and 37 % as against the 40–68 % in previous month. The pruned-out branches provide supplementary income for the farmers in Kerala and drier parts of Bengal, Jharkhand, Orissa and Bihar since these branches are being utilized to make fence at the boundary of homestead, farms and crop land. Such stout branches are also used to support for small climbing vegetable annuals and also fire woods. These branches are also being used as reinforced materials at the centre of mud wall for construction of houses.

Pest and Diseases and Their Control Disease like damping off, leaf blight, leaf spot, stem infection, rhizome bud rot, rhizome decay, culm rot, basal-culm decay, leaf rust and sooty mould by differed pathogens have been reported. Proper sanitation and use of fungicides like Bavistin and Dithane M45 is effective in controlling diseases. Animals like porcupines, rats, pigs, elephants, squirrels, deer, goats, etc. also cause considerable damages in the first few years of plantation. Once the clumps are full grown with many thorny branches at the lower side of the clumps, the herbivore animals rarely visit nearer to the plant. Proper fencing and protection is required around the plantation site especially during the seedling stage.

Growth and Yield The average annual recruitment of culm gradually increased and numbered at 4, 5, 7 and 9 per clump raised from tissue culture plant (TCP) at Kummittapuram and 2, 5, 8 and 11 at Chickally during 1st, 2nd, 3rd and 4th year, respectively. The height growth found to be 4–4.5 m at 4th year on both study areas in Tamil Nadu, India. Monsoon dependency of the plantation was more prominent, since maximum recruitment of culms seems to be monsoon dependents rather than intermittent showers.

Extraction of Bamboo Culms Either all the old culms, i.e. those more than 3 years old (6 years in the case of clumps regenerated from seedling), or a certain number of mature culms are removed annually. Older culms in the interior of the clumps should be removed in a horseshoe pattern (Banik 2000, 2015c).

3.2.4 *Specific Spot Characters for Bambusa bambos Field Identification*

1. A very densely tufted thick-walled, spiny moderate- to big-sized clump.
2. Culm bright grassy green, shining thick walled.
3. Stout branches develop from all culm nodes, from the base to upwards with 2–3 recurved spines; lower branches spreading horizontally with a few leaves.
4. Leaf small to normal size, highly variable.
5. The young emerging shoot, metallic purplish green, with blunt growing apex.

3.3 *Bambusa cacharensis* R. B. Majumder

[*Bambusa cacharensis* R.B. Majumder in Bull. Bot. Surv. India 25 (1–4), 1983 [1985]: 237, pl. III; type: Cachar, Lakhimpur, Majumder 74265A (CAL)]

3.3.1 *General Information*

3.3.1.1 Vernacular and Local Names

Bom bans, *Pechee* (West Tripura-India); *Bethua* (North Tripura, Cachar-India), *Ba* (Jaintia, Meghalaya-India), *Moral*, *Sonarati*, *Bethua* (Southern Meghalaya-India); *Bethua* (Sylhet-Bangladesh).

3.3.1.2 Natural Distribution and General Habitat

Bambusa cacharensis is an endemic bamboo species to northeast India and Bangladesh. The species has been described from the specimen collected from Cachar hills of Assam (Majumder 1985). Occurring and cultivated throughout the villages of Assam in Barak valley, Patharia and Cachar (Badarpur, Silchar, Karimganj) and in Tripura from Udaypur through Agartala to Dharmanagar and Kailashahar (Banik 2004b). Pure patches of clumps covering several kilometres can be seen along both sides of the road leading from Silchar to Meghalaya and also from Silchar to Mizoram. This bamboo is abundantly distributed within the Brahmaputra and Barak valley of Assam and is commonly cultivated at Pasadwar and Umkiang, Jaintia Hills and other states of northeast India. Among the bamboo growers, 74 and 90 %, respectively, of the home gardeners and bamboo grove owners in different villages of Assam in Barak valley have been growing *B. cacharensis* (Nath and Das 2008). *Bambusa cacharensis* is also extensively cultivated in the homesteads located at south-eastern part (Moulvibazar, Habiganj, Shaistaganj, Fenchuganj, etc.) of greater Sylhet district and sporadically in northern (Sylhet Sadar, Tilagarh, Sunamganj) Sylhet, Brahmanbaria, Comilla, Sarail and eastern Mymensingh of Bangladesh (Banik 2000).

3.3.1.3 Climatic Conditions

The species grows well in humid tropical climatic condition around river valleys in moist and well-drained condition within 20–300 m above mean sea level. Available in mixed moist deciduous to semi-evergreen forests on the flat ground.

3.3.1.4 Uses

A much utilized common bamboo species of lower Assam and northern Tripura. As the species has long internodes with smooth nodes, strips taken out from the culm wall are commonly used as *bet* (weaving strips) in making mats, animal cage, ceiling and partition wall including many other household works and easy to bend for tying and weaving things, thus also known as *Bethu* or *Bethua* bamboo in Assam, north Tripura and Sylhet. Due to the creamy colour and smooth surface, the bamboo is used for making chopsticks, spoon and toothpicks. Nowadays, this bamboo is substantially used for making incense sticks in Tripura. Recently, poles of this species are used to manufacture bamboo wood for flooring in a factory (Mutha Industries Pvt. Ltd.) located at Bodhjungnagar, Tripura.

3.3.1.5 Ethnic Utilization

Shoots edible and taken by tribal people, some local Bengali people takes it by mixing and frying with other vegetables. Shoots are also used to make pickle and fermented depending on choice. Leaves used as fodder.

3.3.2 Plant Data

3.3.2.1 Vegetative

Plant Habit

Arborescent, loosely tufted clumps; culms in the clump are not so congested rather somewhat open (Fig. 3.3a). Culms erect. *Young shoots* without white powdery excretions below the culm sheaths.

Culm Height

Tall 15–29 m.

Culm Diameter

Three to eleven centimeter.

Culm Thickness

From tip to base 0.2–1.5 cm and about 0.7 cm at mid-culm zones, so the species is comparatively thinwalled.

Culm Colour

Dull green, densely covered with light rusty brown or dark brownish black pubescence (Fig. 3.3b). Young shoots blackish green, blade erect to reflexed (horizontal) covered with blackish brown appressed hairs, tip stiff, auricle large, curled.

Internode Length

Forty-two to ninety-five centimeter, nodes slightly greater in diameter than internode; nodal line single but sometimes two in some lower nodes, horizontal, verticils of aerial roots extend to the fifth nodes.

Branching

Branches generally from upper nodes, few slender from lower nodes, mid one large and the laterals are subequal. Branch buds about 2 cm long, oval-ellipsoidal with small apex, slightly keeled towards apex, ciliated along the keels. In *Bambusa cacharensis*, culms from plants of mature size may lack buds and branches in the lower $^1/_2$ to $^2/_3$ or even $^3/_4$ part of their length (Fig. 3.3a). This is a desirable aesthetic feature in bamboos planted for the purpose of ornamentation. In culms used in handicrafts or in industry, the absence of knots that mark the insertion of branch complements makes for economy and ease in working the material.

Leaves

Blades linear-lanceolate, 10–21 cm long × 1.0–1.5 cm broad, adaxial surface whitish green, pubescent, abaxial surface dull green, glabrous; highly variable in size. In

Fig. 3.3 (**a**) Arborescent, loosely tufted clumps of *B. cacharensis*; culms may lack buds and branches in the lower $^1/_2$ to $^2/_3$ or even $^3/_4$ part of their length. (**b**) Culm sheath *auricles* wavy with thick rigid cilia on the margins and short dense hairs outside on the body. (**c**) Culms are extensively used for making furniture especially the top of the tables and seats of chairs. (**d**) *Anthers* light purple, exserted by long whitish stamens. (**e**) Wild seedlings (2–4 leaves stage) were picked up and brought to a community nursery for transplanting in polythene bags. (**f**) At 5 months of age, the seedlings were 75–90 cm tall and each contained 4–6 shoots and maintained in the nursery. (**g (1)**, **g(2)**) Seedlings further multiplied through rhizome separation (macroproliferation) technique. (**h**) Commonly 2-node culm segments are collected from the mid-portion of 1–2-year-old culms during summer to rainy season (mid-February to September) and placed in propagation beds having sand rooting medium as culm cuttings. Profuse rooting 50–55 % in 60 days. (**i(1)** and **i(2)**) Clumps are straight, very tall 25–29 m, diameter of 9–11 cm and elegant in look—type as *Jati Bom*

a
b
c
d
e
f

Fig. 3.3 (continued)

Meghalaya, high rain fall zone, the leaves are comparatively big, 16–20 cm long and 1.5–2.5 cm broad. *Leaf sheaths* glabrous, slightly keeled on the back.

Culm Sheath

Deciduous, generally the blade is detached from the sheath proper before falling the whole sheath; matured sheath shiny brown to rusty brown, sharp apicular hairs; sheath proper 12–17 cm long and 19–32 cm wide at base, attenuated upwards into a rounded top, lateral margins oblique, adaxial surface glabrous, abaxial surface covered with glossy blackish brown hairs; *auricle* two, unequal, falcate, fringed with brown curly bristle; *imperfect blades* deciduous, 4–8 cm long, 4–12 cm wide, above the base, abaxial surface sparsely hairy, adaxial surface covered with brown pubescence, base cordate, one side overlapping and the other side overlapped by the auricles, margin bristled towards the base, tip acuminate, cuspidate; *auricles* wavy with thick rigid cilia on the margins and short dense hairs outside the body (Fig. 3.3b), while hairs absent in auricles of *B. tulda.*

Vegetative Growth

The emergence of culm takes place from late May to June and continues up to October with peak in July and August. Average 4–7 culms are produced per clump annually.

Physical and Mechanical Properties of Culm

The bamboo has high values of modulus of rupture (Table 3.5). Culms are extensively used for making furniture especially the top of the tables and seats of chairs (Fig. 3.3c).

Table 3.5 Some physical and mechanical properties of *Bambusa cacharensis* culm

	Culm		
Parameter	Top	Middle	Bottom
Moisture content (%)	79	93	104
Specific gravity (based on green weight)	0.61	0.57	0.55
Specific gravity (based on oven-dry weight)	0.84	0.81	0.79
Shrinkage in wall thickness (%)	5.1	6.5	10.7
Shrinkage in diameter (%)	4.1	5.5	7.3
Compressive strength, green (%)	419	361	320
Compressive strength, air-dry (%)	534	512	452
Modulus of elasticity (MOE); green (1000 kg/cm^2)	82	65	61
Modulus of elasticity (MOE); air-dry (1000 kg/cm^2)	96	70	60
Modulus of rupture (MOR); green (kg/cm^2)	373	426	469
Modulus of rupture (MOR); air-dry (kg/cm^2)	414	468	556

Source: Sattar et al. (1992)

3.3.2.2 Reproductive

Flowering Nature

Bambusa cacharensis commonly flowers sporadically or in isolated clumps (Table 3.6). From 1995 to 2011, *B. cacharensis* has been seen to flower in isolated clumps of in eastern part (north to south) of Sylhet, throughout lower Assam and Cachar, and also in different parts of Tripura (Bankumari-Jogendranagar, Anandanagar bypass-Moheskhola, Jirania-Mohanpur, Gandhigram bazar, Rangutia-Gandhigram, Ushabazar, Nayani mura, Ramshankarpara-Subalsingh, Teliamura-Khowai road, Bibeknagar-Amtali, Shekarkote, Bishalgarh, Katalia, Bishramganj-Kashba road, Fatik Roy road-Kumarghat). Since 2002–2004 and again in 2009–2011 in all the above mentioned places of Tripura, a total of 18 clumps were seen in flowering stage, of which 11 clumps isolatedly far away from each other; 2 clumps in Gandhigram bazar, 2 clumps in Anandanagar-Nagichara and 3 clumps together in Bankumari. Out of all the 18 clumps, 6 clumps in 3 places (Anandanagar bypass-Moheshkhola, Shekarkote, Jirania-Mohanpur) produced seeds and also wild seedlings. Thus, it appears most of the flowering clumps (about 66 %) are sterile and only few are fertile (about 33 %). Very few seeds could be collected as birds ate most of those while in the crown, only seedlings below the crown were collected. These wild seedlings were brought and maintained in the nursery for field planting (Fig. 3.3e, f). Detail DNA finger print of fertile and sterile mother need to be studied. Singha et al. (2003) reported that *Bambusa cacharensis* has flowered without producing any viable seeds. Pollen grains studied for their sterility and viability had shown that more than 70 % were found to be fertile and viable. A short gynoecium and the unsticky nature of the stigma resulted in failure of fertilization

Table 3.6 Estimation of flowering cycle of *Bambusa cacharensis* from available flowering records

Country/locality	Flowering date (calendar year)	References of flowering dates	Estimated flowering cycle (year)
India *Meghalaya*: Pasadwar	2007 (Iso. clump)	Naithani (2007)	
Tripura: Tulakona and Madhupur (Sadar Forest Division)	2001 (2 Iso. clumps)	Records of the Government of Tripura	
Different localities west Tripura, mentioned in section "Flowering Nature"	2002–2004, 2008–2009 (Iso. and sporadic flower)	Banik (2004a) Banik and Sharma (2009)	
Mokam, Birmohan and Suryamaninagar (Mohanpur Dukli RD block)	2009–2010 (Spor. flower) 2010–2011 (Spor. flower)	Sunita et al. (2013)	May be 30 years
Bangladesh *Sylhet,* Habiganj, Srimanal	1997–1999 (a few isolated clumps)	Author observed	

Note: *Iso. clump* isolated clump flowered, *Spor.* sporadic flowering and production of seeds. However, Sunita et al. (2013) observed variation in pollen viability 20, 44 and 59 % in the pollen grains collected from three localities.

A clump completes flowering within 1–2 years and then dies. In some occasions, a few branches in the clump may flower partly every year for 2–3 years then stop flowering and remain alive, a flowering characteristic similar to *B. nutan* (Banik 2000). However, some clumps are seen to be *part flowering* in nature and continues to flower for 4–5 years part by part in branches and in all culms and finally die within 6–7 years of time.

Inflorescence

Clumps usually start to flower sporadically on the small floral shoots in the month of November. *Spikelets* aggregated at the nodes of ultimate branchlets, with 3–6 florets, 3–3.5 cm long, florets 10–15 mm long, glossy green, rachilla internodes articulate; *glumes* two, unequal, 5–8 mm long, many nerved; *lemmas* 10–14 mm long, many nerved. *Paleas* small, 2 keeled, keels scabrous. *Lodicules* 3, ciliate. *Stamens* 6. *Anthers* light purple, exserted by long whitish stamens (Fig. 3.3d).

Seed

Seed ripens at the end of April and may continue up to July, and many small birds were seen to visit the clump crown for eating seeds. Seeds look like small wheat grains. Sample collected at Tripura (India) showed that about 30 g contained 700 seeds.

Seed Germination

Seeds germinate within 3–5 days of sowing at a percentage of 50–70. Viviparous germination of seeds was also recorded infrequently in some of the flowering populations of *B. cacharensis* in Tripura (Sunita et al. 2013).

Seedling Character

Grass like and develop rhizome within 30–45 days of age. During May 2004, in Tripura, 324 wild seedlings (2–4 leaves stage) were picked up carefully from the ground below the two flowering mother clumps grown in the village Brajanagar, Mohanpur, Tripura; then the collected seedlings were lightly watered and wrapped with banana leaves and brought to a community nursery established at Paschim Nalchar Self Help Group, named Chetana (Fig. 3.3e). These seedlings were then transplanted into polythene bags containing mixture of soil, sand and FYM at the ratio of 2:1:1 and hardened for 10 days under partial shade condition with adequate watering, gradually transferred to open sky and maintained at the nursery. At 5 months of age, the seedlings were 75–90 cm tall and each contained 4–6 shoots

(Fig. 3.3f). During the end of August 2004, these seedlings were distributed among the farmers of Melaghar and Kathalia block for homestead and farm land cultivation. About 538 wild seedlings were also collected in May 2010 from the ground below the two clumps that flowered at Anandanagar village, Agartala, Tripura, and transplanted to polythene bags. These seedling stocks were also maintained in nursery and further multiplied through rhizome separation known as macroproliferation technique (Fig. 3.3g(1), g(2)). Finally, seedlings were distributed among the communities for homestead plantation and also raised block plantation in the forests by Tripura Forest Development & Plantation Corporation (*TFDPC*) Limited.

Vegetative Propagation Methods

The guidelines for producing culm and branch cutting, collection and bagging of wild bamboo seedling and multiplication through macroproliferation and their nursery management are provided in Appendix I, and for details, the reader may consult the INBAR Technical Report No. 6 (Banik 1995).

SL. No.	Multiplication method of *B. cacharensis*	Success rate (%) and comment
1	*Offset planting*: 1–2-year-old culms along with underground rhizomes are to be collected during summer (March to May). Retain basal 4–5 nodes of the culm portion and nursed at sand bed transit nursery, plant during July to August in the field	40–75, availability limited, expensive. A few days draught after field planting may kill many, and if continues for few weeks, all may die. So water the planted propagules
2	*Culm cutting*: Commonly 2-node culm segments are collected from the mid-portion of 1–2-year-old culms during summer to rainy season (mid-February to September) and placed in propagation beds having sand rooting medium. Rooting hormone IBA or NAA (250 ppm) in talc formulation (available in market) may be used at branch/bud base on nodes of segments for better rooting (Fig. 3.3h)	50–55, it takes 60 days to root. Intermittent misting or watering is essential for rooting
3	*Branch cutting*: Branches are usually collected after pre-monsoon shower (any time after May to June) and placed in propagation beds containing sand rooting medium. Rooting hormone IBA or NAA (250 ppm) in talc formulation (available in market) be used at the base of cutting for better rooting	45–55, takes 60–75 days to root
4	*Seedling macroproliferation*: Seedlings raised or wild seedlings are collected in April to June from the ground below the flowered clumps and transplanted to polythene bags containing mixture of soil, sand and FYM at the ratio of 2:1:1. These seedling stocks are maintained in nursery; after 4–6 months each seedlings produce 4–7 shoots then multiplied through rhizome separation	80–90, Very effective and cheap method to produce large-scale planting materials
5	*Macroproliferation of cutting*: When a cutting has 4–6 shoots, separate them for multiplication	40–50, limited numbers

Tissue Culture

Not known

Cytology

Not known

Diversity and Conservation Status

Apparently, three different clump forms of *B. cacharensis* are seen in the natural habitat and mostly in cultivation.

(i) The normal common type clumps have culms 18–24 m tall, diameter 5–8 cm with smaller size leaves always in the upper 1/3 of the height.
(ii) Clumps are comparatively less tall 10–16 m with somewhat compact clump with smaller diameter culms (3–5 cm) and farmer locally call them '*Pechee*' bom bash.
(iii) Clumps are straight, very tall 25–29 m and elegant in look. Culms having diameter of 9–11 cm and somewhat loosely grown in the clumps and less branching below. Local people call this type as *Jati Bom* (Fig. 3.3i(1), i(2)). In local language *Jati* means superior quality, so *Jati Bom* means superior quality Bom bash. This type has been centralized in Bambusetum and Bamboo Clone Garden at Nagichara, Anandanagar, Tripura, India.

This natural phenomenon of failure in fertilization has contributed to a reduction of the resource. The destroying of the clumps by the inhabitants of the region has further accelerated this reduction and is contributing to the extinction of the species (Singha et al. 2003).

3.3.3 Cultivation and Management

3.3.3.1 Site Selection and Preparation

Fertile lower slopes and valleys to flat lands are ideal place for plantation. The locality should have sufficient rainfall, but the land should be free from any waterlogging. The existing plantations are mostly raised by the communities in their homesteads and farms, rarely raised any big forest plantations. Among different homestead sites, farmers through their experience prefer to grow this bamboo on red soil with good humus content. Nath et al. (2015) found that farmer-selected sites having red soil were the most predominant soil type (40 %) in *B. cacharensis* bamboo-based agroforestry system. Increasing clump age was positively correlated with the soil fertility for 0–10 cm, 10–20 cm and 20–30 cm depth, respectively: soil organic carbon ($R^2=0.95^*$,

0.94*, 0.91*), total *N* (R^2=0.95*, 0.88*, 0.69**), available *P* (R^2=0.94*, 0.77**, 0.90*) and exchangeable K^+ (R^2=0.91*, 0.79*, 0.61**), suggesting significant impact of bamboo in improving soil fertility through profuse root system and onsite nutrient conservation (*significant at 0.001; **significant at 0.005). The data on scientific analyses are consistent with the traditional soil taxonomy classification used by the farmers. Soil carbon stock (26–35 Mg C ha^{-1}) and sequestration rate (0.28–0.59 Mg C ha^{-1} $year^{-1}$) to 30 cm soil depth of bamboo agroforestry suggest its potential role in soil carbon sink management.

Planting Space, Pit Making and Intercropping

Plant to plant 4 m×4 m or 5 m×5 m spacing are maintained.

Bambusa cacharensis has erect culms, comparatively shorter branches and narrow leaves than other local village-grown bamboo species like *Bambusa balcooa*, *B. vulgaris*, *B. tulda*, etc. This clump architecture does not provide much shade to the associate crops to grow together and thus attract the farmers in lower Assam, Tripura and Sylhet to cultivate in more number of this bamboo in the homestead gardens. Due to such clump character *B. cacharensis* is also a good species for road side plantation at a closer spacing. Small patches of annual and seasonal crops like lady's finger, tomatoes, chilis, pigeon pea, pineapple and some perennial crops like banana, papaya, *Citrus* spp., *Curcuma longa*, *Zingiber officinale*, etc. are commonly grown with the bamboo species. Other tree species like *Areca catechu*, *Aegle marmelos*, *Moringa oleifera*, *Bombax ceiba*, *Azadirachta indica*, *Melia azedarach*, *Gliricidia* sp., *Sesbania grandiflora,* and *Ziziphus mauritiana* are also grown in the home garden along with the clumps of *B. cacharensis* and other bamboo species at wider spacing (Banik et al. 2008).

All other silvicultural practices may be followed as described under *Bambusa balcooa.*

3.3.4 Specific Spot Characters for Bambusa cacharensis Field Identification

1. Loosely erect culmed clump having branches mostly in the upper part of culms (Fig. 3.3a).
2. Culm rusty brown or darkish brown pubescent with comparatively longer internodes.
3. Leaves comparatively small than other common *Bambusa* species, somewhat similar shape and size to *B. polymorpha*.
4. Culm sheath *auricles* wavy with thick rigid cilia on the margins and short dense hairs outside on the body (Fig. 3.3b).
5. Young shoot blackish green, blade erect to reflexed (horizontal), densely covered with blackish brown appressed hairs, tip stiff, auricles large, and curled.

3.4 *Bambusa nutans* Wallich ex Munro

[Synonym: *Bambusa teres* Ham. ex Munro; Alam M K in Bamboos of Bang: field identification manual. BFRI Chittagong 2001. *Bambusa nutans* subsp. *nutans* [autonym], Stapleton in Edinb. J. Bot. 51 (1).1994:17 Tharu bans (Nepali), middle hills of central and western Nepal, North India; *Bambusa nutans* subsp. *cupulata* Stapleton in Edinb. J. Bot. 51 (1).1994:17.]

3.4.1 General Information

3.4.1.1 Vernacular and Local Names

Mokal/Mallo/Kali (Arunachal Pradesh-India); *Deobans*, *Mokal*, *Jotia-makal* (Assam-India); *Makla*, *Makal* (Tripura-India); *Utang* (Manipur-India); *Seringjai* (Khasi-India); *Utang wa* (Imphal-Nagaland-India); *Rungazumi* (Nagaland-India); *Mahlu*, *Mehla* (Lepcha-Sikkim-India); *Badia Bauns* (Orissa-India); *Makor* (north Bihar, Samastipur-India); *Malabans* (Hindi-India); *Mal bans* (Nepal); *Mal bans*, *Jhushing* (Bhutan). In Bangladesh the species has several names in different parts of the country: *Makhla*, *Makal*, *Nitai* (Chittagong, Comilla); Aile (Dinajpur); *Peechle* (Sylhet); *Keyitta* (Mymensingh); *May Bongvarn* (Laos) [Latin. *nutans* means weaving].

3.4.1.2 Natural Distribution and General Habitat

Naturally occurring in Sub-Himalayan zones from the Jamuna river to West Bengal. Mostly occurs in west of Yamuna River. Commonly cultivated in the villages of Bangladesh, northeast India and other provinces of eastern and north India. Moist hill slopes and flat uplands in well-drained sandy loam to clayey loam soils at elevations from 700 to 1500 m. Prefers a pH in the range 6–7, tolerating 5.5–7.5. It is distributed from inner Terai to 1500 m in mid-hills of central, midwestern and western part of Nepal particularly abundant in private lands and protected areas. According to Poudyal (1991), Kathmandu valley possesses large number of *Bambusa nutans* subspecies *nutans* and is the commonest cultivated bamboo in the hills of Central Nepal (Dolakha district) at least as far as Palpa district in western Nepal, but apparently not known in the Terai. *Bambusa nutans* subspecies *nutans* is found only in Nepal. Also one of the commonest cultivated bamboos in the lower hills Sikkim and Bhutan from 300 to 1500 m. Further reported to occur in the natural teak forests in the north and middle part of Thailand.

3.4.1.3 Climatic Conditions

Grows well in high rainfall monsoon zones from lower hills to plains. It tolerates dry stony sites. It prefers a mean annual rainfall in the range of 2300–3000 mm, but tolerates 700–4500 mm.

Uses

Important commercial species. The culms are thick and strong, but inflexible and brittle so of less use for weaving. Much used as poles in construction works and also used as raw materials in pulp and paper industries and furniture making. It is also used for weaving of baskets and mats, as the branches are small and the poles split easily, so further used to make agricultural implements and big size container for storing the food grains. Leaves widely used as fodder for cattle and goats in the hills of North India during winter draught season (Bhandari et al. 2015). The species has been preferred for wide cultivation in the homesteads. It is an attractive and clean-looking graceful bamboo species with ornamental value.

The shoots are bitter and usually not eaten. However, at Jharkhand, the shoots are also rarely eaten and the shoot weight was found to vary from 828.1 to 1771.1 g, with about 63.1–74.0 % of edible part (Das 2014).

In agroforestry practices, the plant is sometimes used to provide shelter for tea plantations.

3.4.1.4 Ethnic Utilization

Poles are used for carrying the dead and shoots are never eaten.

3.4.2 Plant Data

3.4.2.1 Vegetative

Plant Habit

Bambusa nutans is a moderate-sized graceful bamboo, straight clump sparsely below (Fig. 3.4a). The clump is much resembled to *B. tulda*, but the culms are not so densely clustered. Rhizome is loosely pachymorph.

Culm Height

Tall 10–20 m.

Fig. 3.4 (**a**) A moderate-sized graceful bamboo, straight clump of *B. nutans* sparsely below located at Bilaspur Himachal Pradesh, India. (**b**) Dull green, velvety whitish band both above and below the nodes. (**c**) Young shoot has unequal caterpillar-like auricles, body dark brown, densely furnished with long curved reddish bristles on the culm sheath, apex green. Strongly cupped culm sheath blades. (**d**) Culm sheath deciduous, somewhat persistent in young culms. (**e**) Inflorescence a panicle with many spikelets, stamens 6 sometimes 7, anthers long yellow; style short hairy

a
b
c
d
e

Culm Diameter

Five to ten centimeter in diameter.

Culm Thickness

Comparatively thick walled, from tip to base 0.3–2.5 cm.

Culm Colour

Dull green, straight, smooth, not shining with floury deposits on the surface, velvety whitish band both above and below the nodes (Fig. 3.4b).

Internode Length

Usually 25–45 cm long, nodes slightly raised, often hairy, lower ones (2–3 nodes) may bear small rootlets in rings. Culm nodes glabrous or pubescent.

Branching

Much branched in the upper parts of culms. The branches comparatively small. Lateral branches dendroid.

Leaves

Linear-lanceolate, 15.0–25.0 long and wide 2.0–3.5 cm, acuminate at apex, rounded and usually oblique at base, upper surface dull green, lower surface glaucous; glabrous above except hairy on mid ribs, slightly hairy beneath; veins prominent and pale below; tips pointed and sometimes twisted; petiole 3–5 mm long.

Culm Sheath

Strongly cupped culm sheath blades, 10–23 cm long, with 30 cm wide at base, sheath covered with appressed black hairs on the back, base with soft deciduous hairs (Fig. 3.4c). Auricles small, broad wavy, one usually erect, the other decurrent, densely covered with long curved reddish bristles. Culm sheath deciduous, somewhat persistent in young culms (Fig. 3.4d). Ligule 1–2 mm tall, margin finely serrated. Leaf sheath with a few long bristles.

Vegetative Growth

Culm emergence starts from June and continues up to October with peak time in July. A clump may produce 3–7 culms per year. All emerging culms do not always develop into full-grown culms. The natural mortality of emerging culms was found to be 41–44 %. A full-grown clump may have 15–40 culms.

Physical and Mechanical Properties of Culm

Generally 3–4-year-old culms of *Bambusa nutans* are stronger than other age groups. Dries fairly rapidly but liable to crack and collapse; cracks occurring mostly at the nodes; cracks at the other places closing up late during drying; drying by passing hot air from the solar air heaters through the bamboo tube obtained by puncturing the nodal partitions improved the drying behaviour with reduced surface cracking (Jain et al. 1991). The culm wood properties are given below (Table 3.7).

Table 3.7 Physical and mechanical properties of *Bambusa nutans*

	Culm source from 3 localities		
Strength properties parameter	UP	Dehradun	West Bengal
Moisture content (%) green	88.3	95.1	Not Reported
Moisture content (%) air-dry	14.0	12.1	17.8
Average specific gravity (green)	0.603	0.631	–
Average specific gravity (air-dry)	0.673	0.694[b]	0.758
Fibre stress at elastic limit (kg/cm^2) green	295	334	–
Fibre stress at elastic limit (kg/cm^2) air-dry	296	492[b]	225
Modulus of elasticity (MOE); green (1000 kg/cm^2)	66.2	98.2	–
Modulus of elasticity (MOE); air-dry (1000 kg/cm^2)	107.2	124.4[b]	62.8
Modulus of rupture (MOR); green (kg/cm^2)	529	623	–
Modulus of rupture (MOR); air-dry (kg/cm^2)	524	869[b]	565
[a]C P to grain. Max Crsh Str; (green) (kg/cm^2)	456	453	–
[a]C P to grain. Max Crsh Str; (air-dry) (kg/cm^2)	579	719	537

Source: Tewari (1992)
Note:[a]*C P to grain. Max Crsh Str* Compression parallel to grain. Maximum Crushing stress, [b]Kiln dry value

3.4.2.2 Reproductive

Flowering Nature

Bambusa nutans flower sporadically after long interval, usually in most cases isolated clump flowered and part-flowering nature, rarely gregarious. From observation in North India lowering cycle seems to be about 35 years (Bahadur 1980). But

Table 3.8 Estimation of flowering cycle in *Bambusa nutans* from available flowering records

Country/locality	Flowering date (calendar year)	References of flowering dates	Estimated flowering cycle (year)
India *North India* *Dehra Dun* FRI, *Dehra Dun*	1840, 1875 1893, 1894–1896 1915, 1922–1923, 1925, 1927 1931–32, 1935, 1942, 1944, 1958, 1963, 1966, 1975 1979–1980 (Greg.)	Brandis (1899) Gamble (1896) Tewari (1992) Bahadur (1980)	35 (1840–1875) 36 (1893, 1896–1931, 1935) 36–35 (1942, 1944–1979, 1980)
Himachal Pradesh: Pehrwin in Kandraur block, Bilaspur Dist.	2006 (1 clump, partly flowered for 2 years, not died)	Author observed	
Assam: Dhubri, *Lakhimpur*	1932 1997–1998	De (1932) Tripathi (2002)	65 (1932–1997)
Meghalaya Garo Hills (area not mentioned)	1932 1985	De (1932) Gaur (1987)	53 (1932–1985)
Nepal *Pokhara*	1996 (1 clump, died)	Author observed	
Bangladesh *Chittagong*: BFRI Bambusetum	1978–1979 (1 clump, died) 1978–1995 (1 clump flowered partly and continuously 17 years, not died)	Banik (2000) Banik (2000)	
Comilla	1979 (1 clump, partly flowered)	Banik (1986, 1997b)	

it appeared from the flowering reports from northeast India that the species flowered after longer interval 55–65 years (Table 3.8). However, no concluding estimation on flowering cycle could be made for other parts of the region due to inadequate information on flowering records.

One mutilated clump in the BFRI Bambusetum, Chittagong, flowered completely within 2 years (1978–1979) and then died, while another clump flowered partly every year for 17 years (1978–1995) and then stopped flowering and did not die. Every year, 0.2–8.0 % branches in the clump, mostly in lower mid-culm position, flowered partly (Banik 2000). This nature of flowering exhibited by a clump has been termed as *part-flowering clump* (Banik 1986, 1997b, 2000).

Inflorescence

Like other *Bambusa* species, the inflorescence directly born on the leafy branches. Inflorescence a panicle with many spikelets (Fig. 3.4e). *Spikelets* many, 1.5–3.0 cm long; *florets* 6 rarely 16, few fertile; *lodicules* 3, long fimbriate, *stamens* 6 sometimes 7, *anthers* long yellow; *style* short hairy, *stigma* 2–3; *ovary* sub-obovate.

Seed

Seed production is low. Seeds are wheat-like, oblong, obtuse and slightly hairy. The weight of 40 seeds about 0.30 g.

Seed Germination

Germinate within 3–7 days with 56 % rate of germination. Normally, viability remains for 25 days.

Seedling Character

Seedlings look like rice seedling. Seedlings start developing more than one shoot within 30–50 days of germination with rhizome development.

Vegetative Propagation Methods

The guidelines for producing culm and branch cutting, collection and bagging of wild bamboo seedling and multiplication through macroproliferation and their nursery management are provided in Appendix I, and for details, the reader may consult the INBAR Technical Report No. 6 (Banik 1995).

SL. No.	Multiplication method of *B. nutans*	Success rate (%) and comment
1	*Offset planting*: 1–2-year-old culms along with underground rhizomes are to be collected during summer (March to May). Retain basal 3 nodes of the culm portion and nursed at sand bed transit nursery, and plant during July to August in the field	45–65, costly, large-scale plantation not possible for limited availability
2	*Culm cutting*: commonly 2-node culm segments are collected from 1- to 2-year-old culms during summer to rainy season (mid-February to September) and placed in transit propagation beds containing sand rooting medium under mist. Rooting hormone IBA or NAA (250 ppm) in talc formulation (available in market) improves rooting	40–55, takes 6–8 weeks to root. Comparatively costly than branch cutting production
3	*Branch cutting*: Branches are usually collected after pre-monsoon shower (any time after May to June) and placed in propagation beds containing sand rooting medium. Rooting hormone IBA or NAA (250 ppm) in talc formulation (available in market) should be used for better rooting	40–45, profuse rooting within 7–8 weeks
4	*Seedling macroproliferation*: Flowers after different interval of time, so seedlings availability is rare. However may be used when seeds/seedlings available	Utilize when the seeds and seedlings are available

Tissue Culture

B. nutans (explant)	Medium and environments	Results	References
Explant node Nodal buds	(1) MS + BAP (1.0 mg/l) (2) MS + BAP (1.0–5.0 mg/l) (3) MS + ½ MS + NAA (1.0–3.0 mg/l) + IBA (1.0–5.0 mg l-l)	(1) Shoot bud initiation (2) Shoot multiplication (3) Mass propagation	Islam and Rahman (2005)
Nodal segments	(1) MS + BAP (2.22 μM) (2) MS+ IBA (49.0 μM) + Glucose	(1) Shoot bud initiation (2) In vitro rooting	Yasodha et al. (2008)

Cytology

Not known

Diversities

There may be considerable difference observed between populations, between plants of the same population growing in different sites, or even between plants of a single population growing together. Study on the intraspecific variation of this species has been carried out (Adhikari and Shrestha 2008) from six different stands (Kapan, Kirtipur, Sisneri, Naldum, Kafledi and Narayanghat) of Central Nepal ranging in altitude (from 256 to 1850 m) and climatic conditions. Altogether, 10 culms, from 3 to 4 clumps were selected, and 6 morphological characters having utmost importance in bamboo growth have been studied. Those characters were culm height, culm diameter at breast height, culm wall thickness, internode length, culm sheath length and culm sheath breadth. Culm wall thickness of bamboo showed highly significant difference among the population. It was concluded that significant differences were observed in phenotypic characters and argued the variation among the plants might be of genetic origin. *Bambusa nutans* subsp. *cupulata* has culm sheath with jet-black hairs; culm sheath blade is more cupped and promptly deciduous; usually found in Bhutan, Nepal, Bangladesh, Assam, northeast India, West Bengal on hills and plains and also in cultivation.

However, reciprocal transplant experiment and progeny testing should be conducted for further research. Provided information can be utilized to conserve the genetic resources and to exploit these resources scientifically and economically.

Conservation Status The species has been distributed from northern India to the east up to Vietnam; within this long distribution area, a number of spots need to be identified through exploration both for in situ and ex situ conservation of the genetic diversities of *B nutans*.

3.4.3 Cultivation and Management

Similar plantation techniques as *B. balcooa* and *B. cacharensis* may be followed.

3.4.4 *Specific Spot Character for Bambusa nutans Field Identification*

1. Somewhat similar to the clump of *B. tulda*, but culms are taller and straight, usually branching mostly in upper part.
2. Strongly cupped deciduous culm sheath blades (Fig. 3.4c) and black culm-sheath hairs.
3. Young shoot has unequal caterpillar-like auricles, body dark brown, densely furnished with long curved reddish bristles on the culm sheath, apex green.
4. White ring below the culm node (Fig. 3.4b).

3.5 *Bambusa polymorpha* Munro

3.5.1 *General Information*

3.5.1.1 Vernacular and Local Names

Jama betwa (Assam-India); *Faura* (Tripura-India), *Wakobor* (ADC Dukli, DebBarma-Tripura); *Kyanthung-wa* (Myanmar); in Bangladesh, the species is known as *Pharua* (Sylhet), *Mascumba*, *Burma*, *Rangoon* bans (Chittagong) and *Pai Hom* (Thailand).

3.5.1.2 Natural Distribution and General Habitat

Bambusa polymorpha has been growing naturally and cultivated in Eastern Central India, Bangladesh, Myanmar, south Yunnan, Indonesia and Thailand. Generally prefers moist shady habitat. However, in some parts of Kalenga, Adampur and Tilagarh forests of Sylhet (Bangladesh), this bamboo is also growing naturally as undergrowth in patches. Grows as a common bamboo in the Pegu Yoma forests of Myanmar, where usually associated with teak plants and abundant throughout the moist mixed deciduous forests.

Bambusa polymorpha, locally known as *Faura bansh*, is very commonly cultivated as homestead bamboo in Melaghar and Bishalgarh block of West Tripura district, Tripura (India) (Banik 2004b), as this bamboo has been used as raw material in the cottage industries for making many different excellent qualities of handicraft and novelty items.

3.5.1.3 Climatic Conditions

Grows well in moist and moderately high rainfall areas (3500–6000 mm), temperature 3–35 °C. The species is an indicator of deep, rich well-drained soil of the hill forests of Myanmar.

Uses

It has outstanding mechanical properties and durability, popular for house construction and agricultural implements in the sub-continent. The bamboo artisans in Tripura have been making many different finer toys, attractive handicrafts with the creamy colour of culm wood of the species and using the raw material mostly from their homestead cultivations. The Melaghar area especially the West Nalchar village of Tripura is very much well known for its skill in making bamboo crafts and novelty items and earned name and fame for the State. The people of this village are famous for their bamboo artisan activities and the Faura bamboo has been used as main raw material for such work. Due to increasing demand and State Rubber Board supported recent Rubber Plantation programme has alarmingly encroached the bamboo area in the villages. During 2000–2001, the price of a Faura bamboo pole was Rs 25–35, then in 2003–2004 the price of a pole went up to Rs 70–80, but in 2010–2011 it has been sold at Rs 120–160. This clearly shows the scarcity of this artisan bamboo species in the locality and as a result the artisans are compelled to use other alternative bamboo species (*B. tulda*, *D. longispathus*, etc.) which are not matched with the physical/mechanical properties of Faura bamboo suitable for making finer quality novelty items of bamboo. In recent time, the local bamboo artisans are losing interest in the age-old prestigious livelihood activities. So it is very important to replenish the Faura bamboo resource in this village to get back the lost interest of bamboo artisans and help them in livelihood earnings.

Also used as raw material for pulp and paper industries, for making attractive fibre board and sometimes sticks for agarbatti. The clump is attractive and suitable for landscaping.

Ethnic Utilization

The young shoots are edible and tasty. The young shoot of *B. polymorpha* is golden purple or yellowish green, sheath auricles biserrate, lower blade brown, other blades green and cup shaped. The weight of sheath cover and edible portion of a freshly collected shoot is 47.6 and 52.4 %, respectively. The shoot meat possesses creamy white colour (Fig. 3.5a) and taste is slight sweet to sweet with crisp to tender texture (Banik 1997a, 2000).

Fig. 3.5 (**a**) A newly emerged edible shoot, longitudinal section showing inner portion of the shoot. The shoot meat possesses creamy white colour. (**b**) A sympodial clump of *B. polymorpha*. (**c**) Young culms white scurfy and have appressed white deciduous hairs; after 1 year, they turn grey to greyish green. (**d**) Culm sheaths are arranged closely as crown at the apical part of the newly emerged and elongating shoot. Young shoot golden purple or brownish green. (**e**) Culm sheath purple or green strongly cup-shaped *imperfect blade* reniform, concave and much broader, reflexed, deciduous. (**f**) Branch cutting is cheap and efficient method. These cuttings were raised by a Self Help Group 'Chetana' of West Nalchar, Tripura at their community nursery. (**g**) Seedling layering may be used when seeds/seedlings available. (**h**) Dwarf variety about 25–35 cm tall with numerous grassy shoots

a
40
30
20
10
1M
90
b
c
d
e

Fig. 3.5 (continued)

3.5.2 Plant Data

3.5.2.1 Vegetative

Plant Habit

Bambusa polymorpha is a sympodial clump-forming (caespitose) species (Fig. 3.5b). The culms are erect, usually, in dense grouping or loosely oriented with branching in the upper half portion curving outward, with slightly pendulous tips. The pachymorph rhizome of each culm base is connected with other by a very short (5–10 cm) 'neck'. In few plants rhizome 'necks' comparatively elongated (10–25 cm) and thus form a somewhat open sympodial clump.

Culm Height

Tall, 15–25 m.

Culm Diameter

Five to fifteen centimeter.

Culm Thickness

The bamboo is smooth, wall thickness at culm tip 0.2–0.3 cm, mid-culm 0.6–0.9 cm, and basal positions 1.6–2.2 cm; so comparatively thin walled in nature.

Culm Colour

Young culms white scurfy and have appressed white deciduous hairs; after 1 year they turn grey to greyish green (Fig. 3.5c). Young shoot golden purple or green (Fig. 3.5d).

Internode Length

A 40–95 cm, nodes slightly thickened and lower ones fibrously rooted. *B. polymorpha* has more elongated internodes at the mid-culm zones, and therefore the internode length curve in respect of node numbers shows sharp peak in the median regions of the curves. Such culm architecture indicates that the species prefers to grow in association with crown of tall trees (Banik 2000, 2015a).

Branching

In *B. polymorpha* culms from plants of mature size may lack buds and branches in the lower $^{1}/_{2}$ to $^{2}/_{3}$ or even $^{3}/_{4}$ of their length. Branches are developed by breaking the buds from nodes in upper half of the culm. In most cases, buds below the mid-culm region are either poorly developed or remain underdeveloped and therefore no breaking of buds normally occurs.

Leaves

Small, lanceolate or linear-lanceolate, length 10–17 cm and width 1–2 cm, more or less sessile, unequally rounded at the base. Leaf-blade surface scaberulous, rough adaxially, pubescent, hairy on both sides. Leaf-blade apex acuminate. Leaf sheath compressed, slightly keeled at top, ligule very short (1–2 mm) with light brownish ciliated (1–2 mm) prominent auricles. The clumps of *B. polymorpha* are ordinarily evergreen in nature as its native places have high to moderate rainfall, but in some drier pockets and severe draught season, clumps of these species may exhibit deciduous behaviour (Banik 2015a).

Culm Sheath

Culm sheaths are arranged closely as crown at the apical part of the newly emerged and elongating shoot (Fig. 3.5a, d). These are thick, 15–35 cm long, 10–25 cm broad, persistent, covered on the back with densely and closely appressed brown to dark brown pubescence, *imperfect blade* reniform, concave about 7–8 cm long and much broader, reflexed, deciduous (Fig. 3.5e); Auricles 2, often one upwards, the other downwards. *Ligule* narrow and entire.

Vegetative Growth

The emergence of culm takes place from June and continues up to October with peak in July and August. In India at 3-year rotation, a yield of 22 air-dry ton per ha was obtained. Average 2–5 culms are produced per clump annually. However during 2010–12, two clumps were observed at west Nalchar Tripura producing 68–82 culms having 7.3–9.2 cm diameter annually, and thus selected as Candidate Plus Clumps (TR104C–1 & TR104C–2) and centralized at Bamboo Clone Garden, Nagichera Tripura. The culm life in a full-grown clump varies from 5 to 13 years. In a mature clump, production ratio of new culms to the old ones is about 1:4.

Physical and Mechanical Properties of Culm

Some important culm wood properties of the species are presented in Table 3.9.

Table 3.9 Some physical and mechanical properties of *Bambusa polymorpha*

Seasoning condition	Moisture %	Avg specific gravity	Fibre stress at elastic limit (kg/m^2)	Modulus of rupture (kg/cm^2)	Modulus of elasticity (1000 kg/cm^2)	[a]C P to grain. Max Crsh Str (kg/cm^2)
Green	95.1	0.619	136	283	31.3	321
Air-dry	13.9	0.659	161	355	44.0	–

Source: Tewari (1992)
Note: [a]*C P to grain. Max Crsh Str* compression parallel to grain. Maximum crushing stress

3.5.2.2 Reproductive

Flowering Nature

The species flowers at long intervals, and flowering is remarkably gregarious and occasionally sporadic. A clump takes 1–2 years to complete flowering, and then dies. The gregarious flowering of the species was mostly reported from Myanmar. However, it appears from the Table 3.10 that the estimated flowering cycle is about 60 years.

Some genotypes flower after longer interval, about 70±5 years. However, at BFRI Bambusetum, Chittagong, only a few twigs of a clump flowered for 2–3 years in 1983 to 1984 and then stopped flowering and growing again with full vigour, thus showing a *part-flowering* character. All clumps of this species were planted in 1972–1973 and planting materials (offsets) sourced from the northwest part of the country.

It appears there are a number of flowering genotypes (cohort) of the species that exist in nature. In 1895, the *Bambusa polymorpha* forests in the watershed of the Pegu River in Thaukyeghat District did not flower, but the cohort on the adjacent Yoma Range (watershed for the Irrawaddy and Hlaing rivers) did (Brandis 1899).

Inflorescence

A much branched panicle with spikes of frequent heads bearing few *spikelets* comprising 2–3 fertile *florets*; with diminished florets at the apex. *Spikelets* shining and brownish, 0.5–1 cm long, in lower heads 5–6, number gradually decreasing upwards of branched panicle. Empty glumes 1–3, then 2–3 fertile flowers. *Palea*

Table 3.10 Past flowering records and estimated flowering cycle in *Bambusa polymorpha*

Country/locality	Flowering date (calendar year)	References of flowering dates	Estimated flowering cycle (year)
India			
Dehradun	1881	Wood (1881)	
Calcutta Botanical Garden		Gamble (1896)	
Myanmar			
Prome (Pegu Yoma)	1852, 1853	Wathon (1903)	
Zamayi (upper Pegu), Bago Yoma	1859–1860 (Greg)	Troup (1921)	
Thaukyeghat Dist	1862, 1871	Brandis (1899)	
Yoma Range (watershed for the Irrawaddy and Hlaing rivers)	1895	Brandis (1899)	
Pegu forests	1899–1903	Troup (1921)	50 (1853–1903)
Tharrawaddy forests	1913 (Spor.)	Troup (1921)	
Minbu, Thayetmyo, Prome	1913 (Spor.)	Bodekar (1930)	
Prome Division	1914 (Greg.)	Bradley (1914)	61 (1853–1914)
Bago Yoma west of Taungoo and Oktwin district	1934–1936 (Greg.) 1981–1982 (Greg.)	Htun (1999) Htun (1999)	74 (1860–1934)
Pyinmana Divn. Yonbin and Palwe RF; Pyinmana Divn. Pegu Yoma	1929–1930 (Spor. and Greg.) 1934–1935	Bodekar (1930) Win (1951)	70 (1859, 1860–1929, 1930)
Bangladesh			
Kalenga	1982–1983	Banik (1987a)	
BFRI Bambusetum, *Chittagong*	1982–1983 (Spor.)	Banik (1987a)	

BFRI Bangladesh Forest Research Institute, Chittagong

keels not ciliate. Apical sterile florets resembling fertile though underdeveloped. *Anthers* 6; purple; anther tip smooth, or apiculate. *Stigmas* 3; pubescent; *ovary* obovate; pubescent on apex.

Seed

Caryopsis (seed) ovoid; 5 mm long; hairy at apex. Seed ripens in May to July. Seeds look like small wheat grains. Sample collected during 1914 from Prome forests of Myanmar showed that 1 oz. (31.3 g) contained 750 seeds (Troup 1921). Sample collected from Kalenga forests of Bangladesh in 1982–1983 showed 10 g contained about 1250 seeds. It appears the seeds are smaller in Bangladesh variety than those of Myanmar.

Seed Germination

Seeds germinate within 3–5 days of sowing at a percentage of 30–50.

Seedling Character

Grass like and develop rhizome within 30–45 days of age.

Vegetative Propagation Methods

The guidelines for producing culm and branch cutting, collection and bagging of wild bamboo seedling and multiplication through macroproliferation and their nursery management are provided in Appendix I, and for details, the reader may consult the INBAR Technical Report No. 6 (Banik 1995).

SL. No.	Multiplication method of *B. polymorpha*	Success rate (%) and comment
1	*Offset planting*: 1–2-year-old culms along with underground rhizomes are to be collected during summer (March to May). Retain basal 4–5 nodes of the culm portion and nursed at sand bed transit nursery, and plant during July to August in the field	45–65, costly, large-scale plantation not possible for limited availability
2	*Culm cutting*: commonly 2-node culm segments are collected from 1- to 2-year-old culms during summer to rainy season (mid-February to September) and placed in propagation beds containing sand rooting medium	45–55, comparatively costly than branch cutting production as one full culm has to be purchased
3	*Branch cutting*: Branches are usually collected after pre-monsoon shower (any time after May to June) and placed in propagation beds containing sand rooting medium. Rooting hormone IBA or NAA (250 ppm) in talc formulation (available in market) should be used for better rooting	60–75, profuse rooting within 50–55 days. Cheap and efficient method (Fig. 3.5f)

SL. No.	Multiplication method of *B. polymorpha*	Success rate (%) and comment
4	*Seedling macroproliferation*: Flowers after long interval so seedling availability is rare. However may be used when seeds/seedlings available	Uncertainty due to long interseeding period
5	*Cutting macroproliferation*	Not effective method
	Seedling layering: Seedlings in polythene bags are lied horizontally placing the shoot part inside the sand rooting medium under misting and in 60–70 days roots develop at the nodes, maximum at nodes on mid-portion, each rooted nodes are separated and used as planting materials (Fig. 3.5g)	

Tissue Culture

B. polymorpha (explant)	Medium and environments	Results	References
(Node from non-flowering clump)	(1) MS + 22 μM BA + 88 μM sucrose + 6 g/l agar + 5.4 μM NAA (2) MS + 5.4 μM NAA The cultures maintained at 25 ± 2 °C, 16 h photoperiod	(1) Multiple shoot (2) Rooting	Prutpongse and Gavinlertvatana (1992)
Nodal segment	(1) Explants first cultured on MS liquid basal medium with 1.0 mg/l BAP, 1.0 mg/l TDZ + 0.5 mg/l NAA in test tube. (2) Sprouts further multiplied on basal liquid medium containing 6.0 mg/l BAP, 0.5 mg/l Kn and 15 % CW. These cultures were put on a rotary shaker with 60–80 rpm (3) Shoots transferred to the rooting medium containing half strength of medium with 5.0 mg/l IBA and 0.5 mg/l NAA	(1) Sprouting of the dormant buds (2) Shoots multiplied sufficiently (3) Within 30 days of culture) shoots rooted 90 %, 7.5 ± 1.36 roots per shoot. After 7 days in vitro hardening TC plants transferred to soil	Haque (2010)

Cytology

$2n = 64, 72$ (aneuploid).

Diversities

A number of phenotypic variations, such as clumps with loosely grown culms, compact clumps, and colour variation in newly emerging shoots, etc., are observed. During 1988, a few dwarf varieties were found in a seedling population in BFRI nursery, about 25–35 cm tall with numerous grassy culm shoots (Fig. 3.5h), and remained in that stage for the last 27 years.

It appeared from the Table 3.10 that some population of the species flowered after 50 years interval while others flowered after 60 and in further long interval 70- 74 years. Clumps exhibited sporadic and part flowering behavior mostly flowered after short intervals and commonly seen in cultivation. Gregarious flowering population is mostly in natural forests and not so common.

Conservation Status Needs exploration, germplasm collections and conservation and needs exchange of information in the region. No planned conservation programme is yet observed. The local people have been conserving the species to some extent in their homestead and settled forest areas. The species is domesticated in the villages and grows wild in the forest; varieties found at low and high altitudes need to be properly identified and conserved.

3.5.3 Cultivation and Management

Similar plantation techniques as *Bambusa balcooa, B.cacharensis, and B. tulda* may be followed. Besides raising commercial and homestead plantation *Bambusa polymorpha* is one of the ideal species for road side plantation.

3.5.4 Specific Spot Characters for Bambusa polymorpha Field Identification

1. Clumps large, usually leaf shedding in dry season.
2. Culm grey to greyish green, white scurfy when young (Fig. 3.5b).
3. Persistent culm sheaths with falcate auricles.
4. Leaves comparatively small than other common *Bambusa* species.
5. Auricles on young shoots biserrate; lower blade brown; other blades green and cup shaped.
6. Culm sheaths are arranged closely as crown at the apical part of the newly emerged and elongating shoot. Young shoot golden purple or green (Fig. 3.5d).
7. Culm sheath with appressed white hairs on back; blade ovate triangular, purple or green strongly cup shaped (Fig. 3.5e).
8. The shoot meat possesses creamy white colour in a newly emerged shoot (Fig. 3.5a).

3.6 *Bambusa tulda* Roxb.

(Synonym: *Dendrocalamus tulda* (Roxb.) Voigt)

3.6.1 General Information

3.6.1.1 Vernacular and Local Names

Jati bah (Assam-India); *Mitinga*, *Talla* (Tripura-India); *Bon-Jai*, *Wandal* (Kokborok-Tripura); *Saneibi* (Meiteis Manipur-India); *Jhushing* (Darjeeling-India); *Kiranti*, *Matela* (Western Duars-West Bengal); *Telda*, *Mitenga*, *Jowa* (West Bengal); *Paoshiding ying* (Lepcha-Sikkim); *Taleda bauns* (Orissa-India); *Chawa* (north Bihar, Samastipur, Pusa-India); *Talla* (Kalinagar Uttarakhand- India); *Wati*, *Wago* (Garo, Meghalaya-India); *Ka shang*, *Siej Rhnnai*, *Rngai* (Khasi, Meghalaya-India); *Rawthing* (Mizoram-India); *Hada bans*, *Koraincho bans*, *Singhane bans*, *Chab bans* (Nepal); *Thaik-wa* (Myanmar); *Phai bong* (Thailand); *May Bongkhom* (Laos); *Trexim* (Vietnam); and *Bengal bamboo* (English). In different parts of Bangladesh, the species is locally known as *Mirtenga*, *Mitinga* (Sylhet); *Mitinga*, *Mita* (Chittagong); *Mahal*, *Bon* (Mymensingh, Comilla); *Makla* (Dinajpur); and *Talla* (Khulna).

3.6.1.2 Natural Distribution and General Habitat

The species is native to Assam, Bengal, Northern Cachar, Chittagong and Myanmar. *Bambusa tulda* is a native of Chittagong Hill Tracts, Sylhet of Bangladesh, and Myanmar and Assam, Mizoram, northern Cachar, Tripura, Manipur, Meghalaya, Nagaland, Arunachal Pradesh and North Bengal of India. This species occurs sporadically wild along the outer foothills of Nepal and Bhutan, being common around the Chitwan area of the Nepalese plains (Terai), Kathmandu valley and in Chirang district of southern Bhutan.

It is commonly found as undergrowth sporadically or in patches. Sometimes, it forms patches of pure to semi-pure stands. This bamboo frequently grows on the flat alluvial deposits along the streams in the mixed deciduous forests and also along the banks of dry to wet water courses (Banik 1994a). As regards its natural abundance, *B. tulda* is next to *M. baccifera* (*muli bansh*) bamboo in the forests of Chittagong Hill Tract (CHT), Sylhet, Tripura and Assam. The species is not so thin walled like *muli bansh* and the culm nodes are comparatively slightly raised. Sometimes it forms patches of pure to semi-pure vegetation as observed near Bachai Bari forest at Tripura and near Matamuhuri stream of Alikadam Range of CHT forest. In Adampur forest of Sylhet (Bangladesh), sometimes the species can be seen growing as pure stand. *Bambusa tulda* is an evergreen to semi-deciduous clump-forming bamboo, usually found in the mixed deciduous forests. It is also cultivated near the hilly areas and at the rural homesteads in many plain land districts, especially in the villages of Northeastern India; Madhya Pradesh and Andhra Pradesh; Orissa, West Bengal, Bangladesh to Myanmar, northern part of Thailand and Yunnan and Indonesia.

The species has been introduced in the gardens of Java (Indonesia) and the Philippines. The cultivated clumps of this species from southern Bengal have been successfully introduced to different villages (Kalinagar, Dineshpur and others) of Uttarakhand, India, by Bengali immigrants (from Khulna, East Bengal) during the time of partition of the Indian sub-continent.

3.6.1.3 Climatic Conditions

Bambusa tulda frequently found to grow as an undergrowth sporadically or in patches in the mixed semi-deciduous forests. Sometimes may form a pure to semi-pure vegetation. It grows well in moist and moderately high rainfall (4000–6500 mm) areas with temperature range from 4 to 37 °C and seen up to 1500 m altitude.

3.6.1.4 Uses

Bambusa tulda is a useful species for construction, owing to its strong culms. Generally used for housing, roofing and some construction works including in cottage industries for making toys, mats, screens, wall plates, wall hangers, hats, baskets and food grain containers, winnowing fan for rice and food, etc. The narrow culms and branches are used as support for tendril climber vegetables.

The species is also used as raw material for pulp and paper industries in Bangladesh and India.

The young shoots of *B. tulda* are usually not taken as food (Chittagong Hill Tracts) due to its slight bitter taste, however, sometimes used for making pickles in some states of India (Meghalaya, Manipur, Mizoram, Tripura) and also in Thailand. The weight of sheath cover and edible portion of a freshly collected shoot is 49.2 and 50.8 %, respectively. The shoot meat possesses creamy white colour and taste is slight bitter to bitter with tough to crisp texture (Banik 1997a, 2000). The young shoot has green surface usually with yellow stripes, calcareous band on one side of culm node; sheath asymmetric or oblique; auricles, at least one of them situated laterally.

3.6.1.5 Ethnic Utilization

Traditional milk-producing and marketing communities in northeast India and Bengal has been using internode as milk vessels (locally known as *Thunga* or *Chunga*), cups for measuring the milk amount during selling.

In Arunachal Pradesh, India local tribal priest has been using flute during *Dree* festival with the belief that the sound (locally called *eloo*) will keep the evil spirit away. The ethnic communities Dimasa (linguistically belongs to the Tibeto-Burman group) and Zeme Naga in Cachar Hill districts of Assam maintain sacred groves of *B. tulda* at the catchment areas of major rivers or rivulet or at the point of origin of perennial streams; it is evident that the sacred groves provide such important ecosystem services as conserving soil, protecting water sources and catchment areas and helping to maintain downstream water quality. The leaves and culms of the species are used in worshiping (Medhi and Borthakur 2013).

The *meitei* Manipuri have a rich traditional knowledge for utilization of *B. tulda* bamboo (Singh et al. 2010). In fact, it has got multipurpose use covering several aspects from religious to industrial. Because of its strength and durability, it is of great demand on the market. Its young shoots (*ushoi*) and the fermented young shoots (*soibum*) represent a significant vegetable for local Manipuri people (Singh and Singh 1994). Local medicine man uses this bamboo for healing properties. In Manipur, tender shoot decoction paste is applied in wounds and injuries; ash obtained after burning is used in the preparation of a traditional food item locally called *utee*. It forms a good raw material for various handicraft works, house building, paper industries, fencing, and several other useful equipments for day-to-day life.

In riverain Bangladesh, West Bengal, Assam and some part of Orissa, fishermen communities have been using this bamboo for making the fish trap and roofing of boats as the species is comparatively durable in water than many other local bamboos.

3.6.2 Plant Data

3.6.2.1 Vegetative

Plant Habit

Clump forming, evergreen, rarely deciduous, medium to large size (Fig.3.6a) with somewhat underground dorsiventrally flat pachymorph rhizome (Fig. 3.6b).

Culm Height

Tall 10–25 m.

Culm Diameter

Three to ten centimeter. (varying with the localities).

Culm Thickness

The natural forest grown plants are thin walled, 0.35–0.65 cm, while cultivated in the villages found to have comparatively thick walls (1.2–2.3 cm). The basal 1–2 internodes sometimes may be solid.

Culm Colour

The young culm is dark green with slight whitish bloom on the internode and comes off easily with finger. The culm grey green when older and sometimes streaked with yellow on the basal 2–4 nodes. With brownish hairs which is usually irritating, persistent on the basal nodes of the young culm usually up to the next February to March months.

The young emerging shoots, usually whitish green (Fig. 3.6c), may have yellow stripes (Fig. 3.6d) on green surface, calcareous band on one side of culm node sparsely pubescent with black hairs (Fig. 3.6c). A brown dark coloured wooly band (2–4 mm wide) is present on the 1–3 basal nodes especially in young culm.

Internode Length

A 25–70 cm long. A distinct white ring present below the node when node covered with culm sheath. The node slightly or not swollen, generally 5th and 6th the longest. Sometimes the lowermost node has little aerial roots. *B. tulda* has more elongated internodes at the mid-culm zones, and therefore the internode length curve in respect of node numbers shows peak in the median regions of the curves (Banik 2000, 2015a).

Fig. 3.6 (**a**) *Bambusa tulda* is a clump-forming, evergreen, rarely deciduous, medium- to large-size bamboo. (**b**) The clump possesses somewhat underground dorsiventrally flat pachymorph rhizome. (**c**) The young emerging shoots, usually whitish green; sheath *blade* leaf-like, broadly triangular, green cordate, cuspidate, erect. (**d**) On each culm node, usually 3 larger branches present along with numerous smaller branches (7–30 in number). In some clump yellow striations present on the 1–3 basal green culm internodes. (**e**) Leaves green glabrous above but glaucescent and puberulous beneath. (**f**) No floral shoots produced, inflorescences (pseudospikelets) develop directly on the leafy branches. The pseudospikelets are sessile. (**g**) *Florets* 3–7, *stamens* 6, slender, free and hanging. *Anther* yellow mottled with purple streaks and obtuse. (**h**) Seeds linear to oblong, covered with glumes. (**i**) A huge number of seedlings (wildlings) were produced below the flowering mother clumps and natural regeneration took place during 1978–1980 gregarious flowering in pure patches of *B. tulda* at Adampur and Lawachara forests of Sylhet. (**j**) Branch cutting is a cheap and efficient method, a rooted cutting with shoots in polythene bag. (**k** also in **d**) Some harvested poles of *B. tulda* having faint yellow striations on the lower culm internodes, staked at Chakmaghat collection depot, Tripura; yellow striations may also present on the 1–3 basal green culm internodes in some clump. (**l**) In some clump a few culms show bending of internode with one-side bulging at lower part

a
b
c
d
e
f

Fig. 3.6 (continued)

Branching

The growing culm is without branches, but branching begins either after the culm has ceased growth or in the following growth season. Simultaneously, many thin branches are also produced in assembly from each culm bud. Leaves start developing within 2–4 weeks from the developing branches. Branching throughout culms from tip towards the bottom except on few (2–3) basal nodes, lower branches leafless, rigid, horizontal. On each culm node, usually three larger branches are present (Fig. 3.6d) along with numerous smaller branches (7–30 in number). These smaller branches on the basal 1–4 nodes of older culm become dead and shed off keeping the black scars. *Branch buds* ovate, about 2 cm long, 2 keeled, slightly ciliate along the keels.

Leaves

Lanceolate to oblong, green glabrous above but glaucescent and puberulous beneath (Fig. 3.6e), 15–35 cm long, 2.5–5.0 cm wide. Leaf bases round and also truncated with pale green short stalk. Ligule distinctly brownish hairy, 1–2 cm, mostly on the older leaves, has prominent leaf sheath auricles with bristles. The tip of the leaves are sharply pointed, the ventral side is whitish faint green. Leaves are profuse in number in the upper branches. The amount of leaf fall from *B. tulda* is 5.8 ton per hectare.

Culm Sheath

A 15–25 cm × 15–25 cm, coriaceous, deciduous, adaxial surface smooth and often whitish powdered, abaxial surface sometimes covered with appressed brown hairs, slightly attenuate upwards and rounded or triangularly truncate. Sheath *blade* leaf-like, broadly triangular, green, reniform or cordate, cuspidate, erect (Fig. 3.6c), hairy within, imbricating, powdery above, 5–10 cm in diameter (varying with the localities). *Ligule* 1–3 mm tall, continuous with the sheath top, margin finely serrated or entire. *Auricles* 2, broad and long, unequal, larger one continuous with the blade and rounded with ciliate margin. *Culm sheath* at young stage is green. *Leaf sheath* has prominent auricles with bristles.

Vegetative Growth

A clump of *B. tulda* may start emergence of new culm at any time in the month of May to November with peak in July to September. All emerging culms do not always develop into full-grown culms. The natural mortality of emerging culms was found to be 29–37 %. A full-grown clump may have 40–60 culms. In *B. tulda*, all the buds on the culm nodes awaken at nearly the same time with the exception of a few at mid-culm or below. Most of the remaining buds start opening about age 1–2 years.

About 4–15 culms produced per year in an adult clump, normal life of a culm in a clump varies from 5 to 10 years (Banik 2000).

Physical and Mechanical Properties of Culm

The average fibre length of *Bambusa tulda* is 3.0 mm (Liese 1980). This bamboo has higher value for compressive strength and modulus of rupture (Table 3.11).

Table 3.11 Some physical and mechanical properties of *Bambusa tulda* culm

	Culm		
Observed parameter	Top	Middle	Bottom
Moisture content (%)	86	92	108
Specific gravity (based on green weight)	0.61	0.58	0.54
Specific gravity (based on oven-dry weight)	0.83	0.79	0.75
Shrinkage in wall thickness (%)	4.9	7.3	11.9
Shrinkage in diameter (%)	2.6	3.5	3.9
Compressive strength, green (%)	513	466	403
Compressive strength, air-dry (%)	620	596	529
Modulus of elasticity (MOE); green (1000 kg/cm^2)	147	138	105
Modulus of elasticity (MOE); air-dry (1000 kg/cm^2)	168	140	114
Modulus of rupture (MOR); green (kg/cm^2)	542	595	710
Modulus of rupture (MOR); air-dry (kg/cm^2)	671	745	883

Source: Sattar et al. (1992)

3.6.2.2 Reproductive

Flowering Nature

Clumps flower sporadically and occasionally gregariously. Flowering of isolated clumps is not rare, both sporadic and isolated flowering common in Bangladesh and India. Thus appear to have diversities in flowering habit. Clumps usually complete flowering within 1 year, sometimes few clumps may continue flowering and seeding for 2–6 years before dying. Records show that *Bambusa tulda*, in addition to its 20–30 years gregarious flowering, also exhibits frequent sporadic flowering. In lower Bengal, *B. tulda* flowered four to five occasions within 16–18 years of time during the period of 1866–1884 (Brandis 1899). Similarly, the species also flowered sporadically on nine occasions in Chittagong within 12 years (1978–1990) (Banik 2000).

In the Botanic garden, Singapore *B. tulda* has long been cultivation. Out of two old clumps of the species, one flowered in1940 and died, leaving a few seedlings in the adjacent area. The other flowered a little on leafy branches of all its culms over a period of at least 6 years and gradually died (Holttum 1958). Thus, one clump exhibited *complete flowering* in 1 year and other one was *part flowering* which continued for 6 years and then died.

Past flowering records show that the estimated flowering cycles are 10–12, 16–18, 19–20, 24–30, 40–42, 45, 48 and 57 years (Banik 2000; Table 3.12). Therefore, it appears that *B. tulda* has a number of flowering populations; some exhibit short and others long flowering cycles.

Precocious flowering has also been seen and recorded at the seedling stage of the species (Banik 1980).

Table 3.12 Estimation of flowering cycle in *Bambusa tulda* from available flowering records

Country/locality	Flowering date (calendar year)	References of flowering dates	Estimated flowering cycle (year)
Burma Tharrawaddy	About 1865	Troup (1921)	15 (1865–1880)
Prome	1880	Brandis (1899)	22 (1865–1892)
Gangaw	1892	Gamble (1896)	42–40
Toungoo, Zigon, Rangoon,	1903–1905	Troup (1921)	(1865–1903,
Pegu, Yames	1914	Troup (1921)	1905)
Shan States, Palwe and	1908	Troup (1921)	
Kaing ranges	1914, 1915	Troup (1921)	
Pyinmana	1911	Troup (1921)	
	1913, 1914, 1915	Prasad (1948)	
	1938–1939		
Singapore Botanic Garden	1940 (1 clump)	Holttum (1958)	
India	1889	Prasad (1948)	
Assam	1986–1987	Brandis (1899)	
BTC-Charaikhola	2007–2008	Naithani (2007)	
Guwahati-Moinakh	2007	DFO Kokrajhar[a] DFO Silviculture, Basistha	
Lowland forest Bajali Area	2010	Sarma et al. (2010)	
Arunachal Pradesh Along	2003 (Spor.)	Govt Record (T.K. Sharma)	
Mizoram	1785, 1833, 1881 1924, 1929 1977, 1994	Lalnuntluanga et al. (2003)	48 (1785–1833) 48 (1833–1881) 43–48 (1881–1924, 1929) 48, 53 (1924, 1929–1977)
Meghalaya	1866–1868 1872	Brandis (1899)	
Tripura	1994	Tripura Govt Record	
Ramdurg, Latia Chera	2003 (Spor.)	Author observed	
Bagafa	2008 (Iso. clump)	Author observed	
Brahmacharya Udaipur	2009 (Spor.)	Author observed	

(continued)

Table 3.12 (continued)

Country/locality	Flowering date (calendar year)	References of flowering dates	Estimated flowering cycle (year)
West Bengal			
Locality not stated	1867–1868	Troup (1921)	16–18 (1866, 1868–1884)
Malda district	1884 (Spor.)	Brandis (1899)	
Dighra, Hooghly	2003 (Spor.)	Bhattacharya et al. (2006)	
Kalyani, Nadia	2006–2008 (Spor.)	Author observed	
Jharkhand Ranchi, Mandar	2009–2010	Suraj and Nath (2011)	
Uttarakhand Motipur Ward No. 2, Dineshpur block of District US Nagar	2007 (Spor.)	Author observed	
Bangladesh			
Chittagong Hill Tracts (**CHT**)	1912–1920	Prasad (1948)	
Rangamati	1927, 1929 (Spor.) 1930 (Greg.)	Trevor (1927, 1929) Nath (1930)	
Kassalong Resv.	1960 (Greg.)	Hasan (1973)	30 (15 × 2) (1930–1960)
Shishak forest	1977 (Greg.)	Banik (1987a)	
Publakhali	1978	Banik (1986)	
Rangamati (Ghagra)	1979 (Iso. clump)	Banik (1986)	
Kaptai	1981 (Iso. clump)	Banik (1986)	
Matamuhuri Resv.	1984 (Greg.)	Banik (1986)	
Bandarban Chimbuk Hill (Remarkripransa)	1985 1988 (Spor.)	Banik (2000) Banik (2000)	
Chittagong Patiya Range	1876, 1988 (Greg.)	Gamble (1896)	12 (1876–1888) 57 (1888–1931)
Hazarikhil forest	1931 (Greg.) 1936	Coffey (1931) Prasad (1948)	
BFRI Bambusetum	1978 (Spor.) 1980, 1983–1984 (Spor.)	Banik (1986)	
Hathazari	1983, 1984	Banik (2000)	
Ichamati	1984 (Spor.)	Banik (2000)	
Hathazari, BFRI Field Station	1985 (Iso. clump)	Banik (2000)	
Nazir Hat	1987 (Iso. clump)	Banik (2000)	
BFRI Bambusetum	1988 (Iso. clump)	Banik (2000)	
Keochia	1989 (Iso. clump)	Banik (2000)	
Keochia	1990 (Iso. clump)	Banik (2000)	
Sylhet (Forest area of south)	1889 (Greg.) 1910 (Greg.) 1929–1930 (Greg.)	Gamble (1896) Prasad (1948) Baruah (1930) Prasad (1948)	19 (1889–1910) 20 (1889–1929)

Table 3.12 (continued)

Country/locality	Flowering date (calendar year)	References of flowering dates	Estimated flowering cycle (year)
Adampur forest	1978 (Greg.)	Banik (1986)	48 (25 × 2) (1929–1978)
Lawachara forest	1979, 1980 (Greg.)	Banik (1986)	
Sreemangal	1989 (Iso. clump)	Banik (2000)	
Akhaura, B. Baria	1988 (Iso. clump)		
Sirajganj	1988 (Iso. clump)	Banik (2000)	

Note: [a]Personal communication, *DFO* Kokrajhar, and *DFO* Basistha, *Spor.* sporadic flowering, *Greg.* gregariously flowering, *Iso. clump* isolated clump flowered

Inflorescence

No floral shoots produced; inflorescences (pseudospikelets) develop directly on the leafy branches (Fig. 3.6f). The inflorescences are initiated in December and within 4–5 weeks blooming takes place; pseudospikelets comparatively bigger in size, alternate, loosely arranged. After about 1–2 months of first blooming, the vegetative branches rapidly shed leaves, and elongated pseudospikelets developed from the buds on the branches. The pseudospikelets are 12 mm long, 5–8 mm broad, sessile; palea boat shaped, 5–7 veined between the keels and ciliated along keels. Lodicules three fimbricate. *Florets* 3–7, of which 3–6 fertile and terminal 1 sterile. *Stamens* 6, drooping, filament 8–12 mm long, slender, free and hanging. *Anther* yellow mottled with purple streaks and obtuse (Fig. 3.6g). Opening of flower and pollen discharge mostly occur in the morning 6 am to 9 am. *Style* short, stigma 3 and plumose. Ovary obovate, obtusely 3 sided and pubescent. Anthesis takes place in the morning 6–10 am. It was observed in 1979 at BFRI nursery Chittagong that out of 5000 seedlings of *B. tulda*, one seedling flowered at age of 18 months and produced fertile seeds, from these seeds 37 seedlings were raised, and of these 16 flowered at 12 months and also produced fertile seeds (Banik 1980). Such flowering behaviour appears to be genetically controlled and the responsible gene(s) is (are) segregated and expressed in each subsequent generation (Banik 1997b, 2000; Watanabe and Hamada 1981).

Seed

Like wheat grain, seed collected from CHT and Sylhet (Bangladesh) forest in 1978–1979 showed that 10 g weight contained on an average 150 seeds. So 1.0 kg may contain about 15,000 seeds (Banik 2000). A glumed seed 2.5 cm in length, 0.35 mid-width and 0.30 cm mid-breadth. Seeds ripen in April to August. Seeds linear to oblong, covered with glumes (Fig. 3.6h). One side of the deglumed seed smooth and other side furrowed longitudinally, seed tip carries small persistent style base. However, about 23696 seeds per kg were also reported from Ranchi, Jharkhand,

India (Suraj and Nath 2011). Thus, it appears seeds of Rachi source are lighter than those of CHT and Sylhet sources.

Seed Germination

Fresh seeds germinate within 5–7 days of sowing and continue up to 20–25 days. Seeds germinate better (58–65 %) under partial shade than direct sunlight (30–34 %). Seed longevity period of *B. tulda* at room temperature is 35 days only and can be retained up to 18 months by storing over silica gel in desiccators (Banik 1987b). However, Sarma et al. (2010) reported 70 % viability of *B. tulda* seeds from Bajali area of Assam.

Seedling Character

Seedlings 1–5 cm long, thin (1–2 mm), wiry grass like. Plumules pointed cylindrical, usually single, rarely two. Seed remains attached to seedling for about a month. After 30–40 days of germination, secondary shoots develop from base of the primary shoot forming a rhizome system. Huge number of seedlings (wildlings) were produced below the flowering mother clumps and natural regeneration took place during 1978–1980 gregarious flowering in pure patches of *B. tulda* at Adampur and Lawachara forests of Sylhet (Fig. 3.6i). The wild seedlings are thinned out from the forest floor and transplanted in polythene bags at nursery and finally planted in the field in raising plantation. Thinning of seedlings minimize the competition and assisted regeneration (Banik 1988b).

Vegetative Propagation Methods

The guidelines for producing culm and branch cutting, collection and bagging of wild bamboo seedling and multiplication through macroproliferation and their nursery management are provided in Appendix I, and for details, the reader may consult the INBAR Technical Report No. 6 (Banik 1995).

SL. No.	Multiplication method of *B. tulda*	Success rate (%) and comment
1	*Offset planting*: 1–2-year-old culms along with underground rhizomes are to be collected during summer (March to May). Retain basal 4–5 nodes of the culm portion and nursed at sand bed transit nursery, and plant during July to August in the field	45–65, costly, large-scale plantation not possible for limited availability
2	*Culm cutting*: commonly 2-node culm segments are collected from 1- to 2-year-old culms during summer to rainy season (mid-February to September) and placed in propagation beds containing sand rooting medium. Rooting hormone IBA or NAA (250 ppm) in talc formulation (available in market) should be used for better rooting	45–55, rooting comparatively costly than branch cutting production as one full culm has to be purchased

SL. No.	Multiplication method of *B. tulda*	Success rate (%) and comment
3	*Branch cutting*: Branches are usually collected after pre-monsoon shower (any time after May to June) and placed in propagation beds containing sand rooting medium. Rooting hormone IBA or NAA (250 ppm) in talc formulation (available in market) should be used for better rooting	50–65, profuse rooting within 60–65 days. Cheap and efficient method (Fig. 3.6j)
4	*Seedling macroproliferation*: Frequently flowers sporadically, also occasionally gregariously. So seeds and seedlings are often available	Often the planting materials may be produced. So efficiently this technique can be utilized
5	*Cutting macroproliferation*	Not effective method

Tissue Culture

Species (explant)	Medium (mg/l) and environments	Results	References
Bambusa tulda (surface-sterilized seeds)	(i) Placed on filter paper bridge in liquid MS (ii) Aseptic shoots from 3-week-old seedling cultured in liquid MS + 2 % sucrose BAP (8×10^{-6}M), Kn (4×10^{-6}M) (iii) Multiplied shoot on modified MS (½ strength NH_4NO_3) + IBA (1×10^{-5}M) + coumarin (6.8×10^{-5}M)	(i) Germinated (ii) Auxiliary shoot proliferation four- to fivefold in every 3 weeks and maintained over 2 years (iii) Shoot rooted 92 % After in vitro hardening, plants transferred to field, 80 % success	Saxena (1990)
(Single-node segment of mature clump)	(i) MS+ BAP (1.2×10^{-5} M) and Kn (4×10^{-6}M) (ii) Excised shoot cultured on MS + BAP (1.5×10^{-5}M) + IBA (3×10^{-6}M)	(i) Bud break within 5–6 days in 90 % cuttings. In a week, multiple (5–8) shoots developed (ii) Fourfold multiplication in 3 weeks	Saxena and Bhojwani (1991)
Nodal explant	MS + BAP (12.0 μM) + IAA (0.1 μM) + glutamine (100 μM) + coumarin (40.0 μM)	Organogenesis Micropropagation	Mishra et al. (2008)

Cytology

$2n = 70$, 72 aneuploid.

Diversities and Conservation

Bambusa tulda is one of the widespread common forest bamboo species of the region, and since the unknown past, the species has also been cultivated (domesticated) as one of the most useful homestead bamboos in eastern and northeastern India and Bangladesh. Such wide distribution in the natural forests and cultivation has evolved a number of phenotypic variations in the clumps of *B. tulda*. The species has been cultivated mostly in the villages of northern and southern Bangladesh, northeast India, West Bengal, Orissa, Bihar, Pantnagar (Uttarakhand). The cultivated ones are comparatively tall and have fewer branches at the lower parts of the culms. The village-grown clumps of *B. tulda* are taller with big size culms and have comparatively thick walls. The village groves were rarely seen to flower gregariously, rather flower partly or isolatedly. Side by side in nearby natural forests, the clumps are comparatively short and thin walled. Farmers through generations of experience have selected such elite type for cultivation in the homesteads and farms to get the better yield with less fear of clump death due to at a time flowering.

Singh (1993) reported nineteen representative genotypes of *B. tulda* from northeast India and concluded that intermodal length and girth of the culm contribute the most towards total genetic diversity. Also, number of culms per clump, length of internode and girth of the clump contributed as much as 75 %. Banik (1994a, 1997b) recognized three growth variations in this species as (a) normal, (b) large with thicker culms and (c) medium with large cavity and thin wall. Towards the western end of its range in the Terai of Lumbini District in West Nepal (where it is known locally as *chab bans*), *B. tulda* becomes progressively shorter with heavier branching, smaller leaves and more crooked culms (Stapleton 1994b). The culm of *B. tulda* is not known to have swollen nodes (Gamble 1896). However, swollen nodes were observed in most of the populations at 17 ecogeographical locations in different districts of West Bengal, India, except the populations at Memari, Singur and Sibpur (Bhattacharya et al. 2006).

Besides in some clumps of *B. tulda,* frequent presence of yellow striations on the 1–3 basal green culm internodes (Fig. 3.6d, k) are observed (Banik 2000) in small population of clumps naturally grown in the Khowai/Subalsingh forests of Tripura (India) and southern part of Sylhet forest (Bangladesh), but such striation is rare in the population of Chittagong Hill Tracts (CHTs). The seed fertility was found (Banik 1987b, 1994a) to be higher (48 %) from Sylhet source than that of CHTs (26.4 %).The Sylhet source is likely to originate from Khasia Jainta Hills, and the Chittagong Hill Tracts is a continuation of Lushai and Arakan Hills. It is also seen that the clumps in Chittagong Hill Tracts are more healthy and productive than those of Sylhet origin. In the Sylhet source, yellow stripe are frequently visible on the basal internodes of the culms. Yellow stripes are comparatively rare in Chittagong Hill Tracts bamboos.

Bending of internode with one-side bulging (probably due to rapid expansion of one side) (Fig. 3.6l) in few clumps of *B. tulda* is observed in village-grown cultivar in Kokrajhar Assam, Tripura and West Bengal (Kalayani) of India and northeastern districts of Bangladesh.

Selection strategy needs to be concentrated on such characters required for different purpose of utilization. Although there is no formal conservation plan, some of the local people are trying to conserve the species in their area through homestead cultivation. Emphasis has to be given on exploration, identification and collation of as many as possible variability of the bamboo species in bambusetum and clone gardens/gene banks to save them from getting eroded due to deforestation.

Continuous exploration of various germplasm and simultaneous investigation on physical and mechanical properties should go on for optimum utilization of the species.

No planned *conservation programme* is yet observed. Only the local people have been conserving the species to some extent in their homestead-settled forest areas and also as sacred conservation plots. The species is domesticated in the villages and grows wild in the forest; varieties found at low and high altitudes need to be properly identified and conserved.

Cytology

$2n = 70$, 72 aneuploid.

Morphological characters for culm-age determination of *Bambusa tulda*

Age up to	Morphological description
1st year	*Culm sheath*: Usually not present on the culm, may be present on the basal 1–2 nodes and tightly fitted *Bud break and branches*: The branch bud on the culm node breaks acropetally producing stout branch, while 3–8 buds in lower mid-culm zone remain dormant. Sometimes 3–5 branch buds may also remain dormant on the upper portion of the culm. Leafy branches are mostly confined to the top portion of the culm; smaller thin branches are present on the lower nodes. Branch bases and internodes cover with the straw colour sheaths *Culm*: Dark green colour with slight whitish bloom on the internode and comes off easily with finger. Basal 1–2 nodes may have a ring of small adventitious roots. A brown to dark colour wooly band present just above the 1–3 basal nodes
2nd year	*Culm sheath*: Absent *Bud break and branches*: Almost all the branch buds including dormant one of the first year on the lower midzone and upper portion of the culms become active and produce branches *Culm*: Whitish bloom may be slightly present, comes off with rubbing
3rd year	*Culm sheath*: Absent *Bud break and branches*: Generally, branchy sprouts in a branch complement on the basal 1–4 nodes of culm become dead and shed off keeping the black scars. As a result, the number of thin branches at the basal portion of the culm get reduced *Culm*: Green throughout, except yellowish stains present at the lower part of the internodes. Basal 1–3 internodes may not have this yellowish stain
4th year	*Culm sheath*: Absent *Bud break and branches*: Branches on the basal 3–7 nodes and in some cases on the upper portion the culms (2–4 nodes) become dead and shed off keeping the black scars. More number of thin branches are dead *Culm*: Dark green. Yellowish stain at lower part of the internode is comparatively prominent

3.6.3 Cultivation and Management

Planting of *Bambusa tulda* on farmland, borders of home garden, foothills, riversides, sides of ponds has been a traditional plantation practice in northeast India (Manipur, Assam, Tripura, Meghalaya), eastern India, moist part of north India and in all the districts of Bangladesh.

Similar plantation techniques as *B. balcooa* and *B. cacharensis* may be followed. Lower slopes and valleys are best sites for plantation. Spacing may be maintained as 4×4 m or 5×5 m. The species needs moist condition and occasional watering during draught at early stage of plantation.

3.6.4 Specific Spot Characters for Bambusa tulda Field Identification

1. A tufted bamboo with somewhat dorsiventral pachymorph rhizome (Fig. 3.6a).
2. Culm sheath blade leaf-like, green and somewhat open (Fig. 3.6c).
3. Branches from all nodes, numerous, usually three larger ones at each node (Fig. 3.6d).
4. Leaves green glabrous above, glaucescent and puberulous beneath (Fig. 3.6e).
5. White scurfy when culms young, green to grey green when culm old, sometimes faint yellow striations present on the lower internodes (Fig. 3.6d, k).
6. The abaxial surface of culm sheath sometimes covered with appressed brown hairs.
7. In the young emerging shoots, usually whitish green, may have yellow stripes on green surface, calcareous band on one side of culm node sparsely pubescent with black hairs; *blades* imbricating, powdery above, 5–10 cm in diameter (varying with the localities).

3.7 *Bambusa vulgaris* Schrad. ex Wendl.

[*Bambusa vulgaris* was first described and published in *Collectio Plantarum* 2: 26, pl. 47. 1808. 'Name: *Bambusa vulgaris* Schrad. *ex* J.C.Wendl.' (Author: Schrader, Heinrich Adolph; Ex author: Wendland, Johann Christoph); *Tropicos*. Saint Louis, Missouri: Missouri Botanical Garden. Retrieved June 17, 2011.]

3.7.1 General Information

3.7.1.1 Vernacular and Local Name

Different varieties and growth form has specific local names. These are mentioned variety wise under section "Plant Habit".

3.7.1.2 Natural Distribution and General Habitat

Bambusa vulgaris is a pantropical species. It was regarded as indigenous in Java and also native of Sri Lanka. *Bambusa vulgaris* may have travelled along the several ancient maritime spice routes between China, Malaya, Indonesia, Sri Lanka and India. This species has been known only in cultivation. Man has certainly carried propagules of *Bambusa vulgaris* throughout the tropics; its distribution is now so general that one cannot establish with certainty where it is native (McClure 1966; Soderstrom and Calderon 1979). It is widely cultivated in tropics covering South and Southeast Asia, Central and South America (Jamaica, Costa Rica, El Salvador, Puerto Rico) and east Africa and Madagascar, from sea level to 1200 m high altitude. In the continental USA, the plant is common only in frost-free or nearly frost-free parts of Florida and California, but it abounds in the populated parts of Puerto Rico where it has been an important source of building materials for temporary use (McClure 1966). Origin of the species is unknown but commonly cultivated everywhere in the tropics and subtropics.

It grows best under humid conditions, but can tolerate unfavourable conditions like low temperatures and drought. The species is also extensively cultivated in the rural areas of Bangladesh, Lower Assam, Tripura, West Bengal, Orissa and eastern India. In Bangladesh, it is a most common and priority bamboo species in the greater districts of Chittagong, Cox's Bazar, Khulna, Barisal, Faridpur, Noakhali and Comilla. It is also cultivated fairly in other parts of the sub-continent. In the unknown past the species might have been introduced towards the southern parts of Bangladesh through the sea, might be from Indonesia and or Sri Lanka (Banik 2000).

3.7.1.3 Climatic Condition

It grows best at low altitudes; above 1000 m altitude culms become smaller in length and diameter. It thrives under a wide range of moisture and soil conditions. Along rivers and lakes, it grows almost in permanently humid conditions, but it also grows in areas with a severe dry season where the plants become completely defoliated. In Southeast Asia, the green culm plants are widely naturalized on river banks, road sides, wastelands, open ground and homesteads. In Peninsular Malaysia, it even grows well on degraded soils containing tin.

3.7.1.4 Uses

B. vulgaris is the most used of all bamboos. Much used for construction works, props, scaffoldings, bridge making, fencing, boat masts and also as carrying poles. The species is widely used as wattle for crude temporary constructions such as out-houses, barns, shelters during harvest time, watch huts, fencing and props. It is often used for small bridges and as water pipes. The major demand for bamboo in housing is as scaffolding supports and props in the building sector within urban and new development areas.

Adewuyi et al. (2015) comparatively evaluated the flexural performance and deformation characteristics of concrete elements reinforced with bamboo (*Bambusa vulgaris*) and the twisted steel rebars. The Yield Strength of bamboo was 13 % of that of steel, while the ultimate tensile strength (UTS) was 16 % of that of steel. It has been concluded that the bamboo bars are suitable rebars for non-load-bearing and lightweight RC flexural structures.

In *Bambusa vulgaris*, the fibre morphology and chemical analysis show that bamboo portion location has significant effect on cell wall thickness, fibre length, slenderness ratio, holocellulose content and ash content. In this bamboo, the mean values for specific gravity, cell wall thickness, fibre length, Runkel ratio, slenderness ratio (fibre length/fibre diameter) and flexibility ratio (lumen diameter/cell diameter) were found to be 0.58 g per sq.cm, 0.0054 mm, 2.82 mm, 5.30, 218.48 and 18.60, respectively (Kasim and Jalil 1994). The flexibility ratio was found to be very low, but it was compensated by the high slender ratio which indicated its ability to form well-bonded papers. The base portion has longer fibre (3.50 mm), highest slenderness ratio (245.0), lowest holocellulose content (70.8 %) and ash content. Fibres with thin walls give compact, well-bonded sheets of paper, whereas thick-walled fibres give bulky and stiff paper of low strength.

Studies at Trinidad showed that over 4 tons of pure, dry cellulose pulp was obtained per acre per year on a 3-year cutting cycle; this could be substantially increased by the use of a longer cutting cycle. In fact, this bamboo has been extensively used as raw materials for many pulp industries in different countries of the region. The chemical composition of culm is given as holocellulose 67.8–60.6 %, alpha-cellulose 37.9–43.2 %, lignin 22.7–23.9 %, alcohol-benzene solubility 3.9–4.5 %, 1 % NaOH solubility 20.6–23.1 %, hot water 5.7–5.9 %, cold water 3.4–5.6 % and ash 1.8–2.1 % (Latif and Liese 1995).

In cottage industries, this bamboo is used for making toys and handicrafts. In many countries, split culms are used as walls either alone or plastered with mud for house construction.

The harvested culms of this bamboo are highly susceptible to powder-post beetle (*Dinoderus spp.*); this limits its value for many conventional construction works. Termite damage can also be serious, especially of harvested culms in contact with ground. Such attacks are prevented by treating the bamboos with preservative solutions of CCB (copper sulphate, sodium dichromate and boric acid mixed in 2:2:1 ratio) in water. Usually 2.0 kg mixture of CCB is dissolved in 10 litres of water (Banik 2000). The *sap displacement* or Boucherie method successfully treats the whole culm, while the soaking methods does bamboo strips and mats. However, according to McClure (1966), the reputation is redeemed by pulping studies that rate *B. vulgaris* very high among nearly 100 species selected in the Western Hemisphere.

Yellow and Buddha belly type are often planted as ornamentals and also as hedges.

Ethnic Utilization

The young shoot of *B. vulgaris* (green) is also edible, dark brown to greenish with apex green colour, auricle distinct, sheath blade somewhat triangular, acute, but the shoots of yellow type are very bitter and not edible. Water in which young shoots of the yellow culm form has been boiled and is used as a medicine to cure hepatitis. Leaves are used as fodder especially for goat.

The weights of sheath cover and edible portion of a freshly collected shoot are 30.5 and 69.5 %, respectively. The shoot meat possesses light white colour and taste is slight bitter to bitter with tough to crisp texture (Banik 1997a, 2000).

In Irian Java (Indonesia) culms are used to make traditional combs and penis gourds ('Koteka') in the phallocrypt tradition. This is also Papua New Guinea Traditional native costume for special ceremonies.

3.7.2 *Plant Data*

3.7.2.1 Vegetative

Plant Habit

Clump forming, culm erect or suberect, generally more or less curved, sinuous or slightly zigzag, evergreen, very rarely deciduous, medium to large size with strong underground sympodial rhizome system. Different growth forms (*diversities*) and cultivars are also reported.

(a) *The type form* (Green) *of the species*: Green culms (*B. vulgaris* var. *vulgaris*) variety is common in naturalized populations, and one of the major village-grown commercial and economic bamboo species in a number of countries of South Asia. This type has two growth forms:

- *Tall common type*: Usually 25–35 m tall, 9–12 cm in diameter at breast height, comparatively less branching at the basal portion of culm. Local names vary from locality to locality, and these are *Bari*/*Jai* (Tripura-India); *Wasul* (Kokborok-Tripura); *Jai* (Cachar-India); *Baruba*, *Rnaii Shilot* (Jaintia); *Wasshuti*, *Wamanna* (Garo, Meghalaya-India); *Bashini*, *Bakal* (Bengal, Manipur-India); *Sunderkania bauns* (Orissa-India); *Vairua* (Mizoram-India); *Shwe-wa* (Myanmar); T*elibans* (Bhutan and Nepal); and *Kola Una* (Sri Lanka). In Bangladesh, the names are *Baizya*, *Bariala* (Chittagong, CHT, Cox's Bazar), *Bashni* (Barisal, Khulna), *Jai bans* (Sylhet), and *Bangla bans* (Comilla).

  This green-type *B. vulgaris* (Fig. 3.7a), due to its various constructional uses, has been widely cultivated throughout Bangladesh, Tripura and lower Assam including Cachar; Dawki, Jaintia hills, Shella, East Khasi Hills, Gulpani area between Bagmara and Balpakram, South Garo hills and Phulbari area of West Garo Hills of Meghalaya; and West Bengal and Orissa. The species is also very sporadically cultivated in Sri Lanka and in some parts of Punjab of Pakistan.

- *Short and bushy type*: Culms are 7–15 m tall, 4–7 cm in diameter. Branching throughout the culm. At some nodes few thin auxiliary branches with or without leaves look thornlike structures develop usually from the mid- to basal-culm portion. Generally cultivated in south-eastern part of Bangladesh (Cox's Bazar, Chittagong) near the Bay of Bengal; local name is 'Kanta Bizzya Bans' (Banik 1994a). Clumps of 'bushy type' are rarely infected with blight disease and somewhat hardy to the cyclonic storm.

Another growth form has been also reported where *culms are green* at first then becoming *blotched with black* on ageing (McClure 1966).

(b) *Yellow type*: *Bambusa vulgaris* var. *vittata* A. Riviere; *B. vulgaris* var. *striata* (Lodd. ex Lindley) Gamble. The culms are yellow or in combination with green stripes. The commonest form with typically green-striped yellow culms (Fig. 3.7b). In some cases, the individual culms may be almost pure yellow or occasionally half green and half yellow. Kurz (1876) lists this type as varieties of *Bambusa vulgaris*.

Distribution: The green-striped yellow form thrives in areas where the dry season is so severe that the plants become completely defoliated. This yellow type is commonly cultivated all over the Indian sub-continent as an ornamental bamboo variety with yellow culms in the gardens, parks and campus of community buildings and also seen in China, Myanmar, Japan, Singapore, Thailand and Philippines under cultivation.

Vernacular name: *Golden bamboo* (English); *Painted Bamboo* (USA); *Sarna bans*, *Suna ban* (Bangladesh); *Rana Una* (Sri Lanka); *Bambu kuning* (Indonesia); *Buloh gading*, *Buloh kuning* (Malaysia); *Tamalang silau* (Sabah-Dusun); *yellow bamboo* (Tagalog-Philippines).

(c) Buddha's Belly bamboo (*pitcher type*): *Bambusa vulgaris* cultivar 'Wamin' Brandis ex Camus; 3 m tall, 8–12 cm in diameter, dark green shining and glabrous with short pitcher-shaped internode 4–10 cm long, lowest 3 or 4 nodes with rootlets (Fig. 3.7c).

Distribution: *B. vulgaris* cv. 'Wamin' is cultivated as an ornamental plant all over Indian sub-continent, China, Myanmar, Thailand, Singapore, Indonesia and other parts in the world. It may be assumed to have originated in cultivation, however, since the native home of *Bambusa vulgaris*, from which it clearly is derived, is still unknown (McClure 1966).

Fig. 3.7 (**a**) The common and green-type clump of *B. vulgaris*. (**b**) A clump with yellow culms green-striped type. (**c**) Buddha's Belly bamboo type clump, culms with short pitcher-shaped internodes. Clumps in this type are short. (**d**) In *B. vulgaris* (green type) the young emerging shoot is dark brown, apex green, auricles distinct. (**e**) A sulcation is present on the internode just above the bud or branch of *B. vulgaris*. (**f**) A culm internode of *B. vulgaris* used as a smoking pipe by tribal persons while taking rest after lunch at Chittagong Hill Tract. (**g**) Imperfect blade on culm sheath is somewhat triangular, greenish to bright yellow. (**h**) *B. vulgaris* rarely flowers, the pseudospikelets are alternate and loosely arranged; flowering sample collected in 1979 from Chittagong. (**i**) A long branch of *B. vulgaris* var. *striata* taken from mid-culm zone produced shoots and plenty roots at every node through ground layering in sand media. (**j**) Many profusely rooted branch cuttings of *B. vulgaris* (green type) taken out from the propagation bed for transplanting in polythene bags. (**k**) Planted on river bank for quick stabilizing the loose ground and erosion control

a
12 3 2005
b
c
1 3 2008
d
-50
-40
-30
-20
-10
-1M
90
80
e
f

Fig. 3.7 (continued)

Vernacular name: Buddha belly bamboo (English); *bambu blenduk* (Indonesia), Wamin (Myanmar); *ghati bansh* (Bengal).

It is believed that all these forms are developed through spontaneous mutation from green type. By using randomly amplified polymorphic DNA (RAPD) technology, *B. ventricosa* was found close to *B. vulgaris* var. *striata* (Nayak and Das 2003). Similarly, a high level genetic proximity (0.91) was obtained between *B. striata* and *B. vulgaris* (Das et al. 2007) that was in compliance with the proposition that *B. striata* is a somatic mutant of *B. vulgaris* (Bennet and Gaur 1990).

Culm Height

The green type—25 to 35 m (tall), 7–15 m (dwarf); the yellow type—15 to 25 m; the pitcher type—greatly shortened usually 4–8 m high.

Culm Diameter

A 9–12 cm (green, tall); 4–7 cm (green-bushy); 5–10 cm (yellow type); 1–3, 8–12 cm (pitcher-shaped type).

Culm Thickness

A culm may have wall thickness 6–17 mm (green type); 5–12 mm (yellow type); and 4–9 mm (pitcher type). At different height of a culm (20.6 m tall, green type) both diameter and corresponding wall thickness were measured and revealed that the basal internodes have thick walls but the diameter is slightly smaller than those internodes present within 3.5 m of culm height. The basal culm diameter was 11.3 cm with wall thickness (WT) 2.38 cm, at 3.4 m height culm diameter (CDia) 11.9 cm with WT 1.1 cm, at 6.8 m height CDia 11 cm with WT 0.8 cm, at10.7 m height CDia 8.5 cm with WT 0.65 cm, at 13.5 m height CDia 7.1 cm with WT 0.6 cm, at 16.9 m height CDia 5.0 cm with WT 0.5 cm and at 20.6 m height CDia 1.5 cm with WT 0.32 cm.

Culm Colour

Green to yellow (described under section "Plant Habit"). In green type, the young emerging shoot is dark brown, apex green, auricles distinct (Fig. 3.7d).

Internode Length

A 15–45 cm long, cylindrical (green and yellow type); *B. vulgaris* has a series internodes of more or less equal length at mid-culm region, and therefore the inter node length curve in respect of node numbers is broad and flat at the middle section, and this pattern of intermodal architecture indicates that the species is very light demanding and cannot satisfactorily grow under and any crown cover (Banik 2000, 2015a). A sulcation is present or can be felt by finger on the internode just above the bud or branch complement (Fig. 3.7e).

Ethnically some tribal people (*Mog*, *Murang*, *Lushai*) in Chittagong Hill Tract, Mizoram and in Meghalaya (*Khasi*) and Assam have been traditionally using a piece of culm internode of *B. vulgaris* as a smoking pipe (Fig. 3.7f). Smoke passes through the water stored in lower part of the smoking pipe filtering it and making a pleasant sound. In *pitcher type* and basally inflated internodes (pitcher shaped) in the lower part of the culms usually upper part of culms arching, internodes, 4–14 cm long, lower ones shorter, uniformly much swollen, upper ones longer, swollen at the basal part, gradually narrower upwards, matured pale yellowish; nodes constricted, branch nodes constricted.

Branching

Branches are thick and stout with light green colour. Except in 3–4 basal nodes, branching is throughout of the culm; 5–10 buds around the mid-culm zone remain dormant up to 3 years. Buds on the basal nodes may produce small, thin wiry leafy branches after 3 years of culm age.

In *pitcher type*, branching is on upper $^1/_3$ length of culm, single and the branch internode shaped similar to the main stem, branch internodes swollen at the basal part.

Leaves

Green type—oblong-lanceolate to linear-lanceolate, pale green. 10–25 cm long, 1.5–4.0 cm broad, round or attenuate at the base. Tip acuminate, glabrous on both surfaces, hairy beneath when young, margins and adjacent nerves scabrous, petiole 0.5 cm long; main vein narrow, pale, secondary veins 6–8 pairs, intermediate 8–9, frequent, pellucid glands giving the appearance of transverse veinlets on the under surface. Petiole 2–3 mm long. Leaf sheath has small auricles with bristles, short ligule and deciduous brown hairs.

In yellow type—15–32 cm long, 2–3.8 cm broad, linear-oblong or oblong-lanceolate, base obtuse and oblique, pale green, yellowish green along the veins, glabrous beneath, ending above in a long twisted scabrous point, margin scabrous, main vein narrow, secondary veins 5–7, intermediate 6–8; petiole-like base short, up to 5 mm long.

In Pitcher type—linear-lanceolate, 10–30 cm long, 1.4–3.5 cm broad, rounded or attenuate at the base, ending above in a long, twisted, scabrous point, glabrous on both surfaces, scabrous on the margins, main vein narrow, pale, secondary veins 6–8, intermediate 8–9; petiole-like base short, up to 5 mm long.

Culm Sheath

Fairly large 15–25 × 20–35 cm, felted, symmetric, broad, rounded and truncate at top, often beautifully streaked when young with green yellow, striate, adaxial surface densely covered with thick appressed brown black hairs, edges ciliate, covering a major part of the internode; *ligule* 5–8 mm tall, continuous with the top of the sheath; *auricle* 2, subequal, slightly pointed and curve backwards, 7–12 mm high, fringed by wavy stiff bristle; *imperfect blade* somewhat triangular, greenish to bright yellow (Fig. 3.7g).

In Pitcher type—*Culm sheaths* 10–20 cm long, 15–30 cm broad at base, rounded and covered with thick, appressed, dark brown hairs, margin ciliate; *imperfect blades* triangular, green when young, acute, 5–25 cm long, up to 11 cm broad at base, densely hairy within, glabrous or sparsely hairy beneath, lower edges ciliate; *auricles* 2, rounded, ciliate, subequal; *ligule* 3–5 mm high, dentate or sometimes entire.

The young shoot is dark brown, apex green, auricles distinct.

Vegetative Growth

In *B. vulgaris* a clump may start producing culms any time from the month of May and continue up to November; June to August is the optimum time for culm emergence. During the first few weeks of emergence, culm elongation is 5–10 cm per day; after that it may reach up to 65 cm per day, and a culm may complete its elongation within 65–80 days (Banik 2000). All emerging culms do not always develop into full-grown culms. A full-grown clump may produce 12–18 culms annually (Banik 1988a), but about 60–70 % of them may die at juvenile stage (Banik 1983a). Yielding only 4–7 fully elongated culms. The breaking of bud dormancy on the node of a culm was found to be in basipetal order, gradually moving towards the base and 5–10 buds around the middle of the culm remained dormant up to 3 years. After that, all opened except those in the 2–3 base nodes.

Clump girth rapidly increases every year and usually reaches up to 25 m after 7 years of planting. The natural life period of a culm in a clump is usually 7–10 years.

Physical and Mechanical Properties of Culm

The average fibre length in *B. vulgaris* is 2.3 mm (Liese 1980). In this species, moisture content varies from 79 to 118 %, oven-dry density from 0.27 to 0.57 g.cm^3 and radial shrinkage 6–11 %, and tangential shrinkage is 10–20 % (Latif and Liese 1995). Some strength properties of *B. vulgaris* are presented below (Table 3.13).

However, the average values of different mechanical properties data of *B. vulgaris* var. *striata* specimen collected from Bogor, Indonesia, showed that for MOE (kg/cm^2), MOR (kg/cm^2), compression strength (kg/cm^2) and tensile strength (kg/cm^2) were 76205.41, 1147.54, 455.02 and 1321.55, respectively (Elizabeth et al. 1987).

Table 3.13 Some strength properties of *B. vulgaris* (green type)

Modulus of rupture (MOR) MPa	Modulus of elasticity (MOE); MPa	Shear strength parallel to grain (MPa)	Compression strength parallel to grain (MPa)
62.3	6100	4.0	25.3

Source: Tewari (1992)

3.7.2.2 Reproductive

Flowering Nature

Never, since *B. vulgaris* (green type) was described in 1810, has it been observed to flower gregariously. Each occurrence of flowering has involved at most only a few plants (Table 3.14); the flowers so produced are reported not to have set seeds, and, therefore, no seedling progenies have been available. However, in spite of its not having rejuvenated in recent times by sexual reproduction, *B. vulgaris* remains to this day one of the most vigorous of all known bamboos (McClure 1966). Out of seven clumps of *B. vulgaris* (green type) studied, five flowered completely and died within 18 months without producing any seeds in Faridar Para Chittagong (Banik 1979). In the first year, these clumps flowered profusely up to September, with a pause from October to January, and then completed flowering within the next July through irregular flowering. The remaining two clumps were found to be *part flowering* in nature. Out of these two *part-flowering* clumps, one died after 3 years and remaining one stopped flowering and revived in vegetative state. It was learnt from old villagers that all the five *complete-flowering* and *part-flowering* clumps had been planted from two completely different stocks through offsets. Thus, it is likely that this species might have two genetic or physiological strains (Banik 2000), one was *part flowering* and the other was *complete flowering*. Thus, clumps may be either part flowering or complete flowering in nature. Complete-flowering clump usually dies within 1–2 years of flowering while part-flowering culms die after 3 years and sometimes may stop flowering and revived. The complete-flowering clumps flower in three successive flushes (*flush period*) with two non-blooming period (*rest period*) in between. Unlike other species, the flush periods in this species have long 45–75 days. The species flowers rarely and after long time, usually flowers sporadically or in isolated clumps. It appears from the available flowering records that the estimated flowering cycle in *B. vulgaris* (green) is more than 80 years (Table 3.14; Banik 1979, 2000).

The yellow variety has flowered in Europe in November 1873 (Naithani 2007). No flowering record has been found in pitcher type.

Table 3.14 Estimation of flowering cycle in *Bambusa vulgaris* (green) from available flowering records

Country/locality	Flowering date (calendar year)	References	Estimated flowering cycle (year)
Myanmar Locality not reported	1810	Wendland (1810), reported in McClure (1966)	
Locality not reported	1898	Moebius (1898) reported in McClure (1966)	88 (1810–1898)
Bangladesh *Chittagong* Faridarpara (Baddarhat)	1979 (7 clumps flowered)	Banik (1979)	81 (1898–1979)
India *Uttarakhand*	2005 (1 clump flowered)	Author observed	

Inflorescence

The floral shoots, 10–45 cm long, mainly develop from branch nodes during December to January and are not completely leafless. The shoot may also be very long 50–120 cm when the culms are injured. The pseudospikelets are alternate, loosely arranged 10–18 mm long and 11–19 in number (Fig. 3.7h). Compressed each with 6–11 floret, 5–10 fertile and 1 terminal sterile. The florets start blooming at the end of February (Banik 1979). *Palea* oblong, obtuse, 2–3 veined between the keels, ciliated along keels. *Lodicules* 3, fimbriate. *Anthers* are purple, slender, obtuse, 6 exserted and free. *Style* long, hairy, stigma 3 and plumose. Ovary narrowly oblong pubescent. *Anthers* do not produce viable *pollen* (Banik 1997b).

Seed

Caryopsis not known and described and also not seen by me during flowering.

Vegetative Propagation Methods

All different methods such as offset planting, branch cutting, culm cutting, ground layering and marcotting may be used for multiplication of the species. Any vegetative parts (culm segments, branches, branch nodes, etc.) of the species very easily produce roots and thus can be utilized as planting materials. Stakes freshly cut from culms of the living plants and used to support the yam vines, banana plants or placed for rituals in graveyard produce roots and leaves, may create extensive groves, even without any care. All the above mentioned three types of *B.vulgaris* are easy to propagate vegetatively (nodal culm-segment, branches, branch nodes, etc.). A long branch taken from mid-culm zone of *B. vulgaris* var. *striata* when placed horizontally (layered) under sand medium in propagation bed produced shoots and plenty of roots at every node only within 4–5 weeks (Fig. 3.7i). The guidelines for producing culm and branch cutting and their nursery management are provided in Appendix I, and for details, the reader may consult the INBAR Technical Report No. 6 (Banik 1995).

SL. No.	Multiplication method *of B. vulgaris*	Success rate (%) and with comments
1	*Offset planting*: Collect 1–2-year-old material during summer (March to May) and nursed at sand bed transit nursery, and plant during July to August in the field	65–75, costly, large-scale plantation not possible for limited availability
2	*Culm cutting*: Commonly 2-node culm segments are collected from 1- to 2-year-old culms during summer to rainy season (mid-February to September) and placed in propagation beds containing sand rooting medium	70–75, profuse rooting within 40–45 days
3	*Branch cutting*: Branches are usually collected after pre-monsoon shower (any time after May to June) and placed in propagation beds containing sand rooting medium	90–98, profuse rooting within 40–45 days. Cheap and efficient method (Fig. 3.7j)

SL. No.	Multiplication method *of B. vulgaris*	Success rate (%) and with comments
4	*Seedling macroproliferation*: No seeds and seedlings produced	Not possible
5	*Cutting macroproliferation*	60–65, limited numbers
6	*Ground layering*: During rainy season, 1–2-year-old culms are pulled and band down to ground level and buried in below the surface	40–50, cumbersome, need space near the mother clump

Tissue Culture

Species (explant)	Medium and environments	Results	References
B. vulgaris (node)	MS + CW (10 %) + BAP (0.5 mg/l) + Kn (0.2 mg/l)	Multiple shoots	Nadgir et al. (1984)
(Node from non-flowering clump)	(1) MS + 22 μM BA + 88 μM Sucrose + 6g/l agar (2) MS + 5.4 μM NAA The cultures maintained at 25 ± 2 °C, 16 h photoperiod	Multiple shoots, rooting	Prutpongse and Gavinlertvatana (1992)
Nodal buds (green type)	(1) MS + BAP (1.0 mg l-l), then (2) MS+ BAP (1.0–5.0 mg l-l) + ½ MS + NAA (1.0–3.0 mg l-l) + IBA (1.0–5.0 mg l-l)	(1) Multiple shoots (2) Mass propagation with profuse rooting	Islam and Rahman (2005)
B. vulgaris '*striata*' Axillary buds (single-node segments) from field-grown culm plant	(1) MS + BAP (4.0 mg l-l) (2) After 6 months of shoot proliferation cultured on BAP (4.0 mg l-l) + IBA (3.0 mg l-l) (3) Difficult-to-root in vitro shoots from adult field culms pretreated with 0.5 mg L^{-1} TDZ for 2–3 subcultures then placed in MS + IBA (3.0 mg/l) for rooting	(1) Shoots continuously proliferated (2) 40 % rooting in 27 days. In vitro shoots of 1-year-old TC plants rooted 83–92 % in 12 h photoperiod (3) Continuously illuminated shoots pretreated TDZ for 3 subcultures rooted 100 %	Ramanayake et al. (2006)
B. ventricosa shoot tips	(1) MS+ BAP (4.44 μM) (2) MS+ BAP (4.44 μM) + NAA (5.4 μM) + BAP (0.44 μM)	(1) Shoot bud initiation (2) Proliferation and rooting mass vitro propagation	Huang and Huang (1995)

Note: *BA* 6-benzyladenine, *TDZ* thidiazuron (1-phenyl-1-([1,2,3-thiadiazol-5-yl]) urea), *IBA* indole butyric acid

Cytology

Species	2*n*=Count	References
Bambusa vulgaris Schrad. ex J.C. Wendl.	70 (68, 70)	Chen, R. y., X. l. Li, W. q. Song, G. l. Liang, P. x. Zhang, R. s. Lin, W. x. Zong, C. b. Chen & H. l. Fung. 2003. Chromosome Atlas of Major Economic Plants Genome in China, Vol. 4, Chromosome Atlas of Various Bamboo Species. Science Press, Beijing.
Bambusa vulgaris Schrad. ex J.C. Wendl. cv. *Wamin*	64	Fang, W., J. r. Wu & Y. c. Sheng. 1991. Chromosome number of ten species of scattered bamboos. *Journal of. Zhejiang Forestry College* 8(1): 127–130.
Bambusa vulgaris Schrad. ex J.C. Wendl. 'Vittata'	52, 52, 67, 68, 70, 72	Li, X. l., R. s. Lin, H. l. Fung, Z. x. Qi, W. q. Song & R. y. Chen. 2001. Chromosome numbers of some caespitose bamboos native in or introduced to China. *Acta Phytotax. Sin.* 39(5): 433–442.
Bambusa vulgaris Schrad. ex J.C. Wendl. 'Womin'	56, 67, 68, 70	Li, X. l., R. s. Lin, H. l. Fung, Z. x. Qi, W. q. Song & R. y. Chen. 2001. Chromosome numbers of some caespitose bamboos native in or introduced to China. *Acta Phytotax. Sin.* 39(5): 433–442.

Diversity and Cultivars

Under sections "Plant Habit", "Culm Height", "Culm Diameter", and "Culm Thickness", variations in clump character are discussed. The more common cultivars are described below: [Laurence Hatch, *Cultivars of Woody Plants* (Volume I: A-G), section *Bambusa*, TCR Press, 2007, ISBN 978-0-9714465-0-2]

- Aureovariegata (*Bambusa vulgaris* var. *aureovariegata* Beadle ['*Bambusa vulgaris* Schrad.' *Plant List*. Kew, England: Kew Gardens. Retrieved 2011-01-31]: With rich golden yellow culms striped in green, sometimes in very thin lines, it is the most common variety of *B. vulgaris*. [Horace Freestone Clay, James C. Hubbard and Rick Golt, *Tropical Exotics*, page 10, University of Hawaii Press, 1987, ISBN 978-0-8248-1127-3]
- *Striata* (*Bambusa vulgaris* var. *striata* (Lodd. ex Lindl.) Gamble): A common variety, smaller in size than other varieties, with bright yellow internodes and random markings with longitudinal stripes in light and deep green.
- *Wamin* (*Bambusa vulgaris f. waminii* T.H.Wen): It is smaller in size than other varieties with short and flattened internodes. Likely to have originated in South China, Wamin Bamboo is spread throughout East Asia, Southeast Asia and South Asia. Basally inflated internodes give it a unique appearance.
- *Vittata* (*Bambusa vulgaris f. vittata* (Rivière & C.Rivière) McClure): A common variety that grows up to 12 m (39 ft) tall; it has barcode-like striping in green.
- *Kimmei*: Culms yellow, striped with green.
- *Maculata*: Green culms mottled with black, turning mostly black with ageing.
- *Wamin striata*: Grows up to 5 m (16 ft) tall. Light green striped in dark green, with swollen lower internodes.

3.7.3 Cultivation and Management

3.7.3.1 Site Selection and Preparation

B. vulgaris (green type) has been commonly grown in the homestead gardens, along road sides towards the villages, wastelands and open ground, generally in the low altitudes. It is a preferred species mostly on river and canal bank plantations for quick stabilizing the loose ground of the bank and erosion control (Fig. 3.7k). *Bambusa vulgaris* planted at certain strategic points along the course of the river, especially at points where the river curved, solved the problem effectively. A 2 km-long one to two rows of plantation of these bamboo species have been raised along the Tuichindrai Cherra in Teliamura and about 5 km-long plantation along the Rangapani canal bank at brick field near Bishramganj, Tripura. So bamboo should be strategically grown to reduce or prevent soil cutting from riverbanks, which is perhaps the major cause of siltation of riverbanks, resulting in reduced carrying capacity of rivers, and flooding during peak rainy season.

The yellow and Buddha belly types are commonly grown in the parks, gardens, near any historical building, hotels, etc. for ornamentation.

In Bangladesh, *B. vulgaris* (green type) is also being planted as an inner 1–2-line plantation behind coastal mangrove and casuarinas to shield the interior human habitation and farms from the effects of strong sea winds, cyclones and high inundation the Bay of Bengal. People also prefer to grow this bamboo species in coastal part of West Bengal and Orissa (India) This bamboo has been found most successful species against flood and can survive under flood water for 2–3 weeks (Banik 2000, 2015b).

In the villages of Bangladesh and Tripura and lower Assam of India, the most serious disease of *B. vulgaris* is bamboo blight. However, the disease is controlled successfully through regular sanitary cleaning and proper clump managements, soil works and controlled light firing at the clump base area (Banik 2000).

3.7.3.2 Planting Space and Pit Making

Planting pits of 60 cm cube are dug at 5 m × 5 m or 6 m × 6 m spacing for cultivating big size rhizomes and 45 cm cube for small size cuttings (branch and culm) of the bamboo species. Closer spacing 2.5 m × 2.5 m or 3 m × 3 m is required for quick stabilization of ground on newly dug canal bank and ponds. Pits should be dug and prepared 10–15 days before planting. All herbaceous and woody weeds within 1 m of the pit should be removed. All other practices may be followed as described under *B. balcooa* and *B. cacharensis*.

Morphological characters for culm-age determination of B. vulgaris [green type].

Bambusa vulgaris (source: Banik 1993b)	
Age up to	Morphological description
1st year	*Culm sheath*: Except basal 2–3 nodes no sheaths present *Bud break and branches*: One-third of culm top has leafy branches. Buds break producing central stout branches with small auxiliary branches throughout culm except 3–4 basal nodes *Culm*: Bright glossy green, basal 3–4 nodes have adventitious white root rings
2nd year	*Culm sheath*: Light dark colour sheath may be present at 1–2 basal nodes of culm *Bud break and branches*: More than half of the upper portion of the culm has thick stout leafy branches. Small auxiliary branches disappear *Culm*: Dull grassy green. Adventitious root rings on the basal nodes are drying out and turning black
3rd year	*Culm sheath*: Absent *Bud break and branches*: Branches are more thick and stout with light green colour; leaves are few. Bud breaks start at basal 3–4 nodes but forms small thin wiry leafy branches *Culm*: Slightly yellowish green, no adventitious root rings on the nodes
4th year	*Culm sheath*: Absent *Bud break and branches*: Thick stout branches are only on the upper one-third of the culm, turning yellowish with few leaves. Buds and branches on the basal nodes mostly dead leaving black scar *Culm*: Turning yellowish, smooth

3.7.4 *Specific Spot Characters for Bambusa vulgaris [Green Type] Field Identification*

1. Clump medium to large size.
2. Culm glossy green when young, light yellowish when old or exposed to the sun. Yellow and Buddha belly-type (pitcher type) internodes are also present in varieties.
3. The young emerging shoot is dark brown, apex green, auricles distinct (Fig. 3.7d).
4. A sulcation is present or can be felt by finger on the internode just above the bud or branch complement (Fig. 3.7e).
5. Slightly brown to creamy colour ringed 5–7 mm aerial roots usually present on the lower 1–4 nodes of young culms.
6. Culm sheath symmetric, broad and rounded at the top with triangular blade and two collar-shaped subequal auricles.

Chapter 4
Dendrocalamus Nees

The genus *Dendrocalamus* was first described by Nees in Linnaea 9: 476. 1834., under tribe *Bambuseae* and sub-tribe *Bambusinae*. Etymologically, *Dendrocalamus*: Greek word *dendron*, tree; *kalamos*, reed, tree-like reeds. This genus is closely related to genus *Bambusa*, but the number of keels on the inflorescence prophyll can distinguish between the two genera. Prophylls of *Bambusa* species have two keels, while those of *Dendrocalamus* species have only one (Stapleton 1994c). Most of the species of this genus can be recognized by its thick-walled culms, swollen nodes and aerial roots at the lower nodes. The species usually have white, blackish or light-brown hairs on the culm sheaths (Dransfield 1980). The genus is distributed from China (not in northern part), throughout India including Andaman Islands and Terai region of the Himalayas, Nepal, Bhutan, Bangladesh, Sri Lanka, Myanmar, Pakistan, Thailand, Laos, Kampuchea, Vietnam, mainly southern-central-China, Malaysia, Singapore, Indonesia, Philippines and Papua New Guinea.

There are 52 species known under the genus *Dendrocalamus* (Orhnberger 1999). Most of the species are restricted to high rainfall or montane areas. Out of these, the following seven bamboo species are the most popular to the local people of the South Asian countries due to their ecological importance, high demand in utilization and trading. Some of the species are naturally growing in the forests, others in cultivation or in both sources. In the next segments, these species are described in alphabetical order.

4.1 *Dendrocalamus asper* (J. H. Schultes) Backer ex K. Heyne

[Nutt. Pl. Ned. Ind. ed. 2, 1: 301. 1927. Handb. Fl. Jav. 2: 279]

[Synonyms: *Dendrocalamus flagellifer* Munro in *Trans. Linn. Soc*. 26:150. 1866. Gamble in Ann. R. Bot. Gard. Calc. 7: 91, pl. 80. 1896. Ridl., Flora 5: 265.

R.L. Banik, *Silviculture of South Asian Priority Bamboos*, Tropical Forestry, DOI 10.1007/978-981-10-0569-5_4

Gigantochloa aspera Kurz in Ind. Forester 1:221. 1876. McClure in Fieldiana, Botany, 24, pt. II: 141. 1955.

Basonym: *Bambusa aspera* Schult., Syst. Nat. 7:1352. 1830. Kurz in Journ. As. Soc. Beng. 39, pt.2: P 87.1870. *Dendrocalamus brandisii* Munro; Alam M K in Bamboos of Bang: field identification manual. BFRI Chittagong. 2001]

4.1.1 General Information

4.1.1.1 Vernacular and Local Names

Pai Tong (Thailand), *Bambu Betung*, *Sweet Bamboo*; common name: *Giant bamboo. Bambu Betung* (Indonesia); *Awi Bitung* (Sundanese); *Buloh Betong* (Malay); *Bukawe* (Philippines- Tagalong); *Manh tong* (Vietnamese); *Hok* (Lao).

4.1.1.2 Natural Distribution and General Habitat

Dendrocalamus asper is indigenous to south China. These species are commonly planted in Thailand, Vietnam, Malaysia (peninsular and east), Indonesia and the Philippines; a decade ago introduced in north-eastern part of India for its commercial importance in edible shoot production; also widely introduced elsewhere in tropical and subtropical botanic gardens, with origin somewhere in Southeast Asia. Among all the countries, it has been extensively cultivated in Thailand mainly for edible shoot. It is reported that the species was introduced to Thailand from south China any time around 1910–1915, about 100 years ago by the local farmer. It was primarily planted in Prachinburi province where it grows very vigorously and abundantly. The planting area rapidly increased with the popularity of edible shoots and development of the shoot-processing industry.

They are much planted throughout Malaysia for its edible shoots. About 30 km along the bank of Amandit river, from the city of Kandangan to Loksado, *D asper* forest has been growing vigorously in Indonesia. This stretch of bamboo grove, approximately 120 ha, is said to be planted by a local tribe of Dayak hundreds of years ago (Yudodibroto 1987).

4.1.1.3 Climatic Conditions

The species occurs naturally in tropical humid regions from the lowlands and found to grow in highlands 1000 m altitude (Toraja Sulawesi in Malaysia). The species thrives best at 400–500 m altitude in areas with average annual rainfall of about 2400 mm. It can tolerate 3 °C, requires much water with good drainage, commonly grows near water courses and prefers acidic soils (pH 4.5–5.5). It grows best in rich and heavy soils, but it also grows well in semidry areas in Thailand.

4.1.1.4 Uses

Its culms are also highly priced as building materials due to the strength of the culms which are relatively durable. It is also used in making good-quality furniture, musical instruments, containers, chopsticks, household utensils and handicrafts. Culm internodes used as containers. It is an attractive and clean-looking graceful bamboo species with ornamental value. Plantations for shoot production have been established in Thailand and other countries. During 2005-2010 edible shoots were also produced and marketed in some hotels/restaurants at Delhi from a Asper bamboo Farm located at Garh Mukteshwar, Hapur, Uttar Pradesh

4.1.1.5 Ethnic Utilization

The species is famous for its edible shoot. The young shoot is sweet and considered delicious. As an introduced plantation at Jharkhand, the shoot weight was found to vary from 916.7 to 1535.0 g, with about 52.9–66.5 % of edible part (Das 2014).

4.1.2 Plant Data

4.1.2.1 Vegetative

Plant Habit

This is a large bamboo with tall culms and lower nodes covered with a circle of rootlets (Fig. 4.1a).

Culm Height

A 20–30 m tall.

Culm Diameter

Usually 7.0–10.0 cm and near the culm base, the diameter is 15.0 cm or may be more.

Culm Thickness

Relatively thick walled (11–20 mm), but thinner towards the top of the plant.

Fig. 4.1 (**a**) *D. asper* clump with tall culms; rootlets on lower nodes. (**b**) Young culm covered with whitish brown hairs persistently near base, later become green. (**c**) A few smaller branches and the central healthy bud developed on the culm node, after 9–12 months the central bud produce one main stout branch. (**d**) Stout branching is from mid-culm to top. (**e**) Spikelets usually on long leafless short-hairy branches, in small nearly spherical and rather dense group. (**f**) The pre-rooted branches are excised from the node of a culm and have higher potentiality in rooting. (**g**) Most of the nodes on long thin branch produced rooted plants after layered in ground, collected and washed underground rooted portion for separating each of them, and each be a planting material. (**h**) In between the *D. asper* planting rows, vermicompost are cultured as an additional income. (**i**) A tissue culture plant of *D. asper* from old culture stock of narrow genetic base flowered precociously and finally died

Fig. 4.1 (continued)

Culm Colour

When young covered with closely appressed whitish-brown hairs persistently near base, later become green (Fig. 4.1b).

Internode Length

Commonly 20–45 cm long, sometimes 40–50 cm at middle of the culm; lower nodes usually bearing many roots in a circle.

Branching

In the first year of culm emergence, a few smaller branches and the central healthy bud develop on the culm node, then the central bud produce one main stout branch (Fig. 4.1c). Stout branching is from mid-culm to top (Fig. 4.1d). The branch tip is thin, long, and each node has leaves.

Leaves

A 30 cm long × 2.5 cm broad; base of lower ones unequally cuneate above a stalk 3–5 mm long.

Culm Sheath

Sheath appears large 30–40 cm long and broad even for its size. Top of sheath is rounded, auricles small. Upper surface covered with golden brown hairs. Very pale green when young, sparsely covered with loose pale hairs, but hairs darker at the base of the sheath. Under surface is not hairy. Sheaths fall off early. *Imperfect blade* is small, reflexed.

Vegetative Growth

Culm emergence starts any time during May to August. A culm may elongate at the rate of 58 cm in a day, an adult clump may produce 7–15 culms in a year. The culm life in a full-grown clump is usually within 10–13 years.

Culm Wood Properties

The chemical composition (% w/w) of *D. asper* culm wood is holocellulose (74.0), lignin (28.5), ash (1.5), cold-water solubility (6.4), hot-water solubility (9.2), 1 % NaOH solubility (24.7) and alcohol-benzene solubility (5.5) (Kamthai 2003).

Table 4.1 The mechanical properties of *D. asper* round culm wood

Properties	Culm condition	Average values (N/cm²)	
		On nodes	Without nodes
Modulus of rupture	Green	9472.9	8043.1
	Air-dry	11,953.7	9433.6
Maximum crushing stress	Green	2618.5	2743.2
	Air-dry	3432.9	3637.3
Shear strength parallel to grain	Green	757.1	591.7
	Air-dry	1050.4	876.0
Maximum tensile strength for whole bamboo	Green	28,421.1	
	Air-dry	29,215.6	

Source: Prawirohatmodjo (1988)

Some of the mechanical properties of this bamboo studied at Indonesia are presented in Table 4.1.

Cytology/Chromosome Number

$2n=72$, $n=12$

4.1.2.2 Reproductive

Flowering Nature

Clumps usually flower sporadically, also occasionally, flowers gregariously. During 1994–1995, *Dendrocalamus asper* flowered gregariously, produced seeds and died in an area of 40,000 ha in Thailand and caused a heavy loss in the bamboo shoot-processing industry (Tammincha 1996). No record of flowering cycle in the species has been seen. However, according to Bhodthipuks (1981), this bamboo takes 30–60 years to produce seeds. A clump blooms during November to January and yield ripen seeds in the following March to April.

Inflorescence

Spikelets usually on long leafless short-hairy branches (floral shoots), in small nearly spherical and rather dense groups (Fig. 4.1e), each spikelet is 6–9 mm long, flattened, 4 mm wide; empty glumes 1 or 2, florets 4 or 5, presence or absence of an imperfect terminal floret.

Seed

Grain like, oblong, covered with glumes. Seeds are short lived, usually 2–3 weeks.

Seed Germination

Fresh seeds of *D. asper* germinate 42–48 % within 3–7 days of sowing. *Dendrocalamus asper* seeds with moisture content of 3–8 % could maintain viability (36.5 % germination) up to 18 months when stored at 5 °C (Pukittayacamee 1999). The fresh seeds should be germinated in the seed beds for proper handling of the seedlings.

Seedling Character

Grass-like, produce more than one shoot after developing rhizome system within 45–55 days of age. One-year-old seedlings survive and establish better after field planting.

Vegetative Propagation Methods

The guidelines for producing culm and branch cutting, collection and bagging of wild bamboo seedling and multiplication through macroproliferation and their nursery management are provided in Appendix I, and for details, reader may consult the INBAR Technical Report No.6 (Banik 1995).

Offset Planting Like other big-sized bamboo species, *Dendrocalamus asper* may be propagated by offset planting for raising small-scale plantation and commonly used for homestead cultivation. Offsets are collected from the healthy mother in the month of February to May and brought to the nursery then placed in to the sand propagation beds for adequate rooting and shoot development. These are maintained in the nursery till transplanting in the field during rainy season (June to August). The final success rate is 70–80 % in the field.

Branch Cutting Propagation of *D. asper* by branch cuttings is a practical and effective method and can be used for raising commercial large-scale plantations. The species has aerial roots at the base of the lateral branch; these pre-rooted branches (Banik 1980, 1984, 1995) have higher potentiality in rooting (Fig. 4.1f). May to August are the best months for collecting the branches after a few shower from 1- to 2-year-old culms rooting in the nursery. Bigger branches have more potential for rooting than small ones. The cuttings from main branches (secondary branch cuttings are not growing that well) are efficient in rooting. These should be placed at nursery in sand propagation bed under misting when the bases are treated with rooting hormone. This induces enhanced rooting in bigger and clustered within 6–8 weeks. The rooting efficiency of each mother plant is different and depends on culm size and wall thickness as well as mother genotype. In general, thick-walled bamboos (both for culm cutting and branch cutting) posses potential for higher sprouting and rooting probably due to more food reserves in the culm and well-developed root primordial at the nodal position.

Branch Layering The long thin branch tip is brought down to the ground and buried inside the soil mixed with sand. It is done just before the rainy season, and within 12 months, most of the nodes produce roots with small braches (Fig. 4.1g). After removing the soil, the whole branch with rooted nodes is excavated carefully. Each of the rooted nodes is separated by secateurs and transplanted in polythene bag having mixture of soil, sand and FYM (2:2:1). The bags with rooted plants are hardened under fogging and shade net (60 % light) for 10–15 days and gradually shifted under the open-sky sun. These are planted in the field during rainy season.

Tissue Culture

Dendrocalamus asper (Explant)	Medium and environments	Results	References
Single nodal segments (3–4 cm) with unsprouted bud from secondary and tertiary branches. Explants swabbed with 70 % alcohol, washed with chlorhexidine gluconate soln, for 10 min.; bavistin (0.5 % w/v), streptocycline (0.1 % w/v) and rifampicin (0.1 % w/v) for 10 min, followed with streptocycline and rifampicin (0.5 % w/v) and ciprofloxacin (0.25 % w/v) for 2 min. Then surface sterilized with 70 % alcohol for 1 min followed by immersion in 0.1 % mercuric chloride for 5 min.	(i) Explants inoculated vertically to ½MS+ sucrose (3 % w/v) and agar (0.8 % w/v) for bud sprouting (ii) Transfer the sprouted nodal explants to solid MS + BAP 7.0 mgl^{-1}. Proliferating shoots maintained in BAP 7 mgl^{-1} using 3–5 shoots (iii) Transfer on medium with NAA 3 mgl^{-1} + IAA 1 mgl^{-1} Cultures kept at 35 °C ± 2 °C, 16 h light: 70 ± 5 μmol $m^{-2}s^{-1}$ and 8 h. dark	(i) The axillary buds sprouted in 10–14 days (ii) Maximum 97 shoots obtained within 12 weeks (iii) Roots emerged from shoots in 2–3 weeks on MS medium (iv) Most effective as roots in 80 % shoots with many long roots Plantlets raised transplanted to soil 80–90 % survival	Ali et al. (2009)
Nodal segments	(i) MS+ BAP (0.1–15.0 mg l^{-1}) (ii) MS+ BAP (1.0–10.0 mg l^{-1}) (iii) MS+ BAP (3.0 mg l^{-1}) (iv) MS+ NAA (3.0 mg l^{-1}) (v) MS + IBA (10.0 mgl^{-1})	(i) Shoot bud initiation (ii) Embryo germination (iii) Shoot multiplication (iv and v) Profuse rooting and large-scale plantlet production	Arya et al. (2002a, b)

(continued)

Dendrocalamus asper (Explant)	Medium and environments	Results	References
Methods and result summary			
(1) Seven axillary shoots were induced *in vitro* from each *excised tender node* (15–20 mm in length) containing single axillary bud when nodal segments were inoculated in semisolid MS medium fortified with 5 mg/l BAP. Maximum multiple shoot formation (14) observed when *in vitro*-generated axillary shoots were transferred to liquid MS + 5 mg/l BAP and 40 mg/l adenine sulphate. A maximum of 93.33 % shoots rooted when transferred to liquid MS + 1 mg/l IBA. A simple acclimatization primarily in cocopeat for 20 days and finally in a blend of sand, soil and farm yard manure (1:1:1 v/v), a very high survival rate within next 35 days. (2) After acclimatization, rooted plantlets were *further multiplied by splitting* of rhizomes, formed *in vivo* within 90 days of growth. After 90 and 180 days of acclimatization, plants were successfully transferred to the field and maintained in an unirrigated condition with the initial addition of farm yard manure (FYM) 10 kg/pit; about 85 % survivability with 25 culms per bush attaining an average height of 4.5 m was recorded up to 4 years			Banerjee et al. (2011)

Diversities and Conservation

In *Dendrocalamus asper,* six cultivars recognized [in INBAR's Priority bamboo species document, Rao et al. 1998], *D. asper*, cv. betung wulung, large black bamboo of Indonesia; cv. Thai green, suitable for plantations as used in Thailand, cv. Phai Tong Dam, less popular than Thai green, slightly black in colour. *Dendrocalamus asper* f. *niger* Hildebrand., culm internode blackish, distributed in Indonesia, is almost going to be extinct, common name *Betung Hitam* (Indonesian).

One variety is *Dendrocalamus asper* 'Thai Black'. Chiefly cultivated in Prachinburi, distributed throughout Thailand. Culm size: 8–10 m height, 6–12 cm diameter. Culm internodes dark green, covered with white powder when young; nodes nearly glabrous; foliage leaves large and thick, green to dark green; young shoots emerge black, flesh white, of exquisite taste, without splinter and weight 3–6 kg. *Uses*: Shoots for diet, culms for construction; frequently planted. Cultivation requirements: Easy growing in heavy moist soil with good drainage, exposed to full sun. The Thai name, phai tong dam (dam = black) refers to the surface colour of young shoots (when they are not taller than harvest height) and does not refer to the colour of the culm.

Conservation status is poor and needs priority attention, limited germplasm collection in Sumatra. There is clear potential for use of this species in agroforestry systems and also for use in manufacturing bamboo boards.

4.1.3 Cultivation and Management

4.1.3.1 Site Selection and Preparation

Comparatively, the drier sites are preferred by *D. asper* for edible shoot production in Thailand (Bhodthipuks 1981). Plantation for domestic use is rather in very small scale normally in small patches as in homesteads woodlots in the valleys and lower slopes. Generally, flat fertile land with good drainage is ideal for village plantation.

4.1.3.2 Planting Space, Pit Making and Intercropping

The wide spacing, 6×6 m to 8×8 m, is practiced as this large bamboo species ultimately need wider room to accommodate clump for healthy growth. During first and second year, different seasonal and annual crops (e.g. legumes, soyabeans, etc.) are grown wide space between the bamboo plants for additional income. In sub-montane and low-hill subtropical conditions of Himachal Pradesh, tulsi-wheat cropping system under the *D. asper* displayed maximum returns (Rs. 3,05,540–Rs. 4,86,419 per year). Whereas, under the mid-hill subhumid conditions, ginger crop intercropped under *D. asper* displayed better financial returns (Rs. 4,04,693 $year^{-1}$). Similar result was also obtained from a field trial at Kalyani, West Bengal, when ginger was intercropped with *D. asper.*

During draught period in summer from March to April, sometimes May, watering at least 2–3 times a week is needed to keep the plants alive and grow.

In addition, after loosening (hoeing), the soil at the clump base both organic manure (FYM, cow dung) and nitrogenous fertilizers has to be applied in later part of February to March before the rains for better growth and more shoot production during ensuing growing season. At Siriwat farm in Prachinburi province, Thailand farmers have been in success for more than 50 years together in *D. asper* plantation using farm tractors ploughing with manuring just before rainy season. Manure consumption is about 50–60 kg per clump (Bhodthipuks 1981). Weeding is required a few times just after rainy season because young bamboo shoots need no competitor especially for nutrient, light and moisture demands.

After 3 years of growth, a clump may produce three to six shoots annually. It is better to start felling older culms only after 5th year of clump age during non-growing season (November to February).

In between the planting rows, vermicompost may be prepared as an additional income (Fig. 4.1h).

A note on the performance of D. asper plantations in India raised through tissue culture plants

About 25 years back, *Dendrocalamus asper* was introduced to some of the countries of the region, and the success of introductory pilot plantations are not promising. In India, this species was introduced through importing a very few test tubes/vials (might be 4–5 number) containing tissue culture plantlets (TCP) from Thailand probably during 1991. Then these cultures (with very limited genetic base) have been further multiplied by thousands of numbers at least in two tissue culture laboratories of the country. Different public organizations and private owners have purchased several thousands of these TCP and raised plantations.

Some important observations are highlighted below:

(a) Two plantations of *D. asper* raised through these TCP are:

  (i) A 4 ha plantation raised in 2003 at Garh Mukteshwar Farm, Hapur, by the side of NH24 (National Highway), about 65 km away towards east from New Delhi. This plantation performed very well (Fig. 4.1h) with good growth, but after 8 years, 2011 showed some rots on the upper part of the culms and gradually it moved towards base and the whole culms started dying, and by 2013, most of the culms of 650 clumps died. Finally in 2014, all clumps were felled and sold by Mr Prakash Lohia, the owner of the Farm.

(ii) Another 8 ha plantation was raised at Paratia, Udaypur, Tripura (a state in northeast India), by the Tripura Forest Department during 2003–2004. At the beginning, clumps were bushy with many (25–45) thin culms. During 2007, selective thinning of thin culms was done from all the clumps and as a result emerged new culms with bigger size, and by the end of 2010, all the clumps became merchantable look. Within 2014–2015, many clumps started drying and showed culm rot and plantation deteriorated.

(b) Besides, during 2005–2012, all India basis multilocation trial of eight bamboo species started by National Bamboo Mission and Applications (NMBA) at nine sites distributed throughout the country from Assam at east to Rajasthan at west, Himachal at north to Kerala and Karnataka at south. Along with other bamboo species, TCP of *D. asper* were also distributed in all the nine sites. In most of sites, TCP of *D. asper* developed bushy clumps, some flowered precociously, and finally many infected and died (Fig. 4.1i).

Some Observations It appears TCP of only a few test tube source likely carried a narrow genetic base; original cultures of 1991 through repeated subculture might degenerated the quality and showed bushy clumps with many narrow culms and precocious flowering in *D. asper* and finally death of many clumps.

Plants in two farms showed good growth initially with financial and trading profit, thus appears the species can perform well in this region and an integrated approach of introduction of species needs to be taken. The healthy planting materials having wide genetic base such as seeds/seedlings and cuttings from different genetic sources may be collected and planted under proper silvicultural management practices.

4.1.4 Specific Spot Characters for Dendrocalamus asper Field Identification

1. Tall clump with culms when young covered with fine closely appressed persistent brown hairs, later green, near the base. Internodes of lower culm completely covered with velvety brown hairs; nodes of basal culm with abundant aerial roots
2. The pale green sparsely hairy sheaths on the upper parts of unbranched young culms
3. The lower culm nodes usually bearing many roots
4. Culms trumpet shaped with lowest diameter thicker than from breast height up
5. Culm sheath auricles large, bristly; culm sheath ligule with irregular coarse divisions or clefts; culm sheath blades dark purplish green when young

4.2 *Dendrocalamus giganteus* Wallich ex Munro

[*Dendrocalamus giganteus* Munro in Trans. Linn. Soc. 28: 150, 1868. Gamble 1896 in Ann. R. Bot. Gard. Calc 7.: 87, pl. 76, Ridl., Flora 5: 265. Backer. Handb. Fl. Jav. 2: 281, *Synonym*: *Bambusa gigantea* (Munro) Wallich ex A. & C. Rivière in Bull. Soc. Acclim. Ser. 3, 5, 1878: 685]

4.2.1 General Information

4.2.1.1 Vernacular and Local Names

Worra (Assam-India); *Hati bans* (Uttarakhand); *Dhungre bans*, *Rachhasi bans* (Nepal); *Budum bansh* (Bangladesh); *Yodha Una* (Sri Lanka); *Yak Nan* (Thailand), *Rawpui* (Mizo- India); *Wabo-gyi* (Myanmar); *Giant bamboo* (English).

4.2.1.2 Natural Distribution and General Habitat

Dendrocalamus giganteus indigenous to southern Myanmar and north west Thailand. The species is cultivated mostly for ornamental purposes in India, Thailand and southern China Yunnan at 580–1470 m altitude; also introduced and planted in many botanical gardens of Indonesia, Malaysia, the Philippines, Vietnam, south China and in Madagascar. This bamboo often cultivated in mid-country and in the montane zone of Sri Lanka and in Bangladesh, the species mostly cultivated in Buddhist temples of Chittagong and Chittagong Hill Tracts and Cox's Bazar as sacred plants (Banik 1998a). The tropical, subtropical to low hills and valleys having moisture and organic matter are natural habitat of *D. giganteus*.

4.2.1.3 Climatic Conditions

Dendrocalamus giganteus occurs naturally in humid tropical to subtropical highlands, up to 1100 m altitude, and can tolerate −2 °C. It can, however, be grown successfully in tropical low lands on rich alluvial soils. In northern Thailand, the species is found in natural forests with teak.

4.2.1.4 Uses

The big gigantic culm used for building purposes, house pillar, boat masts, vases, buckets and various other decorative purposes. Roof ceiling, mast of boats and small ships, floats for fishing net, matting, woven wares and also as props and scaffolding. The cylindrical big hollow culms are mostly used for building as structural timber, strong superior quality, buckets, boat masts and for making bamboo board; also used as water pipes. It is also useful for making pulp and household implements and furniture of very good quality. The big internodes are commonly used as storage containers, money (coin) saving box, wine storing vessel and also as flower vases and other decorative purposes. The culm walls are used for the production of bamboo boards and ideal decorative materials for interior applications such as walls, ceilings, floors, doors, shelves, etc. This bamboo is a suitable raw material for pulp and paper making.

Though *D. giganteus* produces big shoots (5–12 kg), they are bitter in taste and usually not eaten by the local people. On the basis of fresh weight, the outer sheath cover amounts to 58.2 % and the remaining 41.8 % contains edible inner meat portion of a shoot (Banik 1997a, 2000). However, the young shoots edible in Manipur are creamy and tender when cooked. They have a fair canning quality. The shoots contain cyanogenic compounds, including taxiphyllin, and give an irritant sensation in the mouth and throat. These compounds can be removed by cooking. Shoot residues (mainly sheaths and soft pieces of the stem) contain per 100 g dry matter: protein 13.1 g, fat 1.8 g, fibre 23.5 g, ash 6.4 g, Ca 53 mg, Mg 108 mg, P 261 mg, Fe 11 mg and Zn 5 mg. The hydrocyanic acid content is 213 mg per 100 g (Brink 2008). The residues can be used for fodder after removal of the hydrocyanic acid by boiling.

4.2.1.5 Ethnic Utilization

The large culm sheaths are sometimes used to make hats in Nepal, Thailand, China and Myanmar. The large leaves are used as fodder in Nepal.

Some of the Buddhist community are hill tribal people, such as Chakma, Marma and Barua, residing in the districts of Chittagong, Chittagong Hill Tracts and Cox's Bazar (Bangladesh). A number of Buddhist temples (locally called *Kiang*) are located in these areas for the religious activities of the community. It has been seen that almost every temple has a few clumps of *Dendrocalamus giganteus* and, in some case, *Bambusa polymorpha* in the same campuses. The community has a belief that these bamboo species are sacred plants (Banik 1998a). The species are used for the construction and repair of the temples and related works. The internodes of *D. giganteus* are mainly used as container for storing sacred religious water. Both these two species naturally occur in the forests of Myanmar. In the unknown past, these species might have been carried by man to the bordering districts (Chittagong Hill Tract, Chittagong and Cox'bazar) of Bangladesh and other nearby states of northeast India, since then conserved and maintained in the temple areas.

4.2.2 Plant Data

4.2.2.1 Vegetative

Plant Habit

A very large gigantic dense clump with sympodial rhizome (Fig. 4.2a).

Culm Height

Tall 30–36 m.

Culm Diameter

A 16–30 cm.

Culm Thickness

At the bottom of culm 2.5–3.5 cm, at middle 0.7–0.8 cm and at the top 0.2–0.1 cm. In comparison to culm diameter, the wall is not so thick, especially in the mid-culm zone. The basal culm diameter was 18.47 cm with wall thickness (WT) 3.42 cm, at 2.1 m height culm diameter (CDia) 15.5 cm with WT 1.55 cm, at 4.2 m height CDia 14.38 cm with WT 1.12 cm, at 8.4 m height CDia 13.56 cm with WT 0.98 cm, at 12.6 m height CDia 11.1 cm with WT 0.7 cm, at 16.8 m height CDia 9.2 cm with WT 0.7 cm, at 21.0 m height CDia 5.5 cm with WT 0.6 cm, at 23.3 m height CDia 1.1 cm with WT 0.35, and at 24.6 m height CDia 0.7 cm with WT 0.25 cm.

Culm Colour

Dendrocalamus giganteus culms closely grey green and the internode is covered with white waxy scurf when young (Fig. 4.2b).

Internode Length

A 20–70 cm, basal nodes marked with root scars, deep longitudinal groove present just above the nodal bud/branch in the upper part of the well-developed culm (Fig. 4.2c). *Dendrocalamus giganteus* being a tall and big-sized bamboo species has series of more or less equal length internodes in the mid-culm zones similar to other tall and large size of *Bambusa* species (*B. balcooa* and *B. vulgaris*). Bamboo species having culms with such nodal architecture suggests that they prefer to grow directly under open sky and not comfortable under or interplanting with the tall trees (Banik 2000, 2015a).

Branching

Branches usually slender, limited to the upper part, small and thin drooping branches commonly present on the lower nodes of 1-year-old culm where central prominent bud remains dormant. After 1 year, thin branches replaced by a stout branch developed from the central bud along with two to four axillary branches. The top half of the culm produces markedly trioclade branch complement and proceeds basipetally.

Leaves

Deep green, variable in size 25–45 cm long and 5.5–8.5 cm wide with truncated base; smooth above, hairy beneath; petiole short 4–6 mm; base of the blade and petiole purplish brown (Fig. 4.2d).

The amount of annual leaf fall from an adult clump of *Dendrocalamus giganteus* is about 7.0 ton per hectare.

Culm Sheath

Very large, 45–50 cm long and 40–55 cm broad at base, early deciduous, hard, glabrous and shining within, colour dull yellow, very thinly covered with golden stiff hairs. *Imperfect blades* 12–35 cm long and up to 10 cm broad, usually reflexed, broadly triangular on the lower culm nodes to narrowly triangular. *Ligule* 8–12 mm long, stiff. *Auricles* prominent, brown.

Vegetative Growth

Culm emergence starts any time during May to August. The daily (24 h) extension growth amounts to about 10–30 cm, but reaches 58 cm for *D. giganteus* (Osmaston 1918). All emerging culms do not always develop into full-grown culms. The natural mortality of emerging culms was found to be 46–67 %. An adult clump may produce 7–15 culms in a year. The culm life in a full-grown clump is usually within 10–13 years.

Fig. 4.2 (**a**) A 6-year-old large clump of *Dendrocalamus giganteus* raised in 2004 from branch cutting at Hapur, UP, India; observed in 2010. (**b**) (Courtesy: Thoop Nakasen, Thailand) A mature old clump in Thailand, culms closely grey green, the internode is covered with white waxy scurf when young. (**c**) Deep longitudinal groove present just above the nodal bud/branch in the upper part of the well-developed culm. (**d**) Base of the blade and petiole purplish brown. (**e**(**1**), **e**(**2**)) Part of the clump of *D. giganteus* flowered; stamens with long filaments, style also long, hairy ending in a feathery bright purple stigma visible from far away during February to March. (**f**) By January to February 2014, culms produced long pendulous leafless floral shoots on the nodes of leafy branches, pseudospikelets developed in cluster like a head on each node of the long pendulous leafless floral shoot. The spikelets were somewhat flattened ovate, acute and slightly sharp spiny. (**g**) In the second year by June 2015, remaining green part of the clump flowered but produced empty caryopses and died. (**h**) Seeds brownish, grain-like, oblong, 6–8 mm long, obtuse, hairy above and covered with glumes produced in April to June. (**i**(**1**), **i**(**2**)) Seedlings 40–45 days of age raised in polythene bags maintained in Hapur Garhmukteswar farm nursery. (**j**) Branches rooted profusely when collected after pre-monsoon shower, treated with 200–250 ppm rooting hormone and placed for 50–55 days in propagation beds containing sand-rooting medium

a
b
c
20
10
2M
90
d

Fig. 4.2 (continued)

Fig. 4.2 (continued)

Physical and Mechanical Properties

Mean value of fibre length in *Dendrocalamus giganteus* is 3.2 mm (Liese 1980). Papermaking studies have shown that African *D. giganteus* yields pulp suitable for paper with a high tearing strength. The average dimensions of the stem fibre are length 2.7 mm, diameter 26 μm, lumen width 19 μm and wall thickness 3.9 μm. Fibres from *Dendrocalamus giganteus* from Madagascar were on average 2.4 mm long with a diameter of 18 μm. The chemical composition of stems from Madagascar was cellulose 39.4 %, pentosans 18.4 %, lignin 25.3 %, ash 2.9 % and silica 0.4 %. The solubility in hot water was 5.1 %, in alcohol-benzene 6.5 % and in 1 % NaOH 24.4 % (Brink 2008). At a moisture content of 19 %, the density of the stem walls is about 0.9 g/cm^3. The modulus of rupture is 93–179 N/mm^2, modulus of elasticity about 14,000 N/mm^2, compression parallel to grain 39–62 N/mm^2 and shear about 4.5 N/mm^2 (Brink 2008). The stems are very susceptible to powder-post-beetle attack.

Some mechanical properties of air-dry culm (12 % moisture content) of *D. giganteus* at Myanmar were studied by Sint et al. (2008); the values are presented in Table 4.2.

Table 4.2 Mechanical properties of air-dried bamboo of *Dendrocalamus giganteus*

Static banding (N/mm^2)		Compression parallel to grain (N/mm^2)		Shear (N/mm^2)	
MS	MOE	MS	MOE	MS(N)	MS(WN)
32.1	6556	60.9	2117	13.3	14.4

Source: Sint et al. (2008)
Note: *MS* maximum stress, *MOE* modulus of elasticity, *N* specimen with node, *WN* specimen without node

Table 4.3 Comparison of some strength properties of *D. giganteus* to timber species

		Timber species	
Items	*Dendrocalamus giganteus*	*Gmelina arborea*	*Tectona grandis*
Moisture content of (%)	10	12.1	15
Volume weight (g/cm^2)	788.5	509.0	479.0
Horizontal hardiness (Mpa)	0.139	–	0.221
MOE-bending strength (kg/cm^2)	1405.0	1059.3	1032.0

Source: Yang and Xue (1999)

Yang and Xue (1999) have studied and compared some strength properties of *D. giganteus* to those of commonly used timber plants (*Gmelina arborea* and *Tectona grandis*), which are shown below (Table 4.3).

However, the average values of different mechanical properties of data of *D. giganteus* specimen collected from Bogor, Indonesia, showed that for MoE (kg/cm^2) and MoR (kg/cm^2), compression strength (kg/cm^2) and tensile strength (kg/cm^2) were 143,208.80, 1823.73, 627.02 and 1907.04, respectively (Elizabeth et al. 1987).

Cytology/Chromosome Number

$2n = 72$, hexaploid, $n = 12$

4.2.2.2 Reproductive

Flowering Nature

The clumps of *D. giganteus* usually flower sporadically. Sporadic flowering was recorded at Paschim Lataguri, Jalpaiguri district, West Bengal, on April 2007. Out of 25 culms in a clump, 22 have flowered; that is, the species also showed part-flowering character initially. In the next year (2008), almost all the clumps in that area have flowered and died. Since its introduction in Kurseong (India) during 1880–1888, there was no report of its flowering till 1974 when one clumps flowered completely, and thus, flowering cycle may be estimated as 86 ± 2.

Bahadur (1979) reported that a clump flowered at FRI Campus Dehradun in 1979. The planting year of that clump or age was not known. Considering that the plant might have introduced in the FRI Campus, Dehra Dun, India from Myanmar seeding in 1893, the flowering cycle also be estimated as 86±2, and which is of similar duration calculated for Kurseong (Table 4.4). Htun (1999) reported 80 years of flowering cycle for *D. giganteus* from Myanmar, but he did not mention the locality and years of flowering. However, the species is reported to have a life cycle of 75

Table 4.4 Estimation of flowering cycle in *Dendrocalamus giganteus* from the available flowering records

Country/locality	Flowering date (calendar year)	References of flowering dates	Estimation of flowering cycle (year)
Myanmar Tenasserim, Shan Hills, Southern Shan states ??	1892 1893 ??	Gamble (1896) Gamble (1896) Htun (1999)	 80
India *Dehra Dun*, FRI Campus UP: Hapur Garhmukteswar farm UK: Lalkuan, GBPUniv Campus	1979 2014 (1 cl prt flowr) 2015 (1 cl prt flowr)	Bahadur (1979) [a]Lohia [x]Rajesh	86±2 (1892, 1893–1979)
West Bengal Calcutta Bot. Garden Kurseong, Bamonpokhri 1 Clump Paschim Lataguri, Jalpaiguri	1860–1861, 1888 1880–1888 1974 (Spor.) 2007 (Spor) 2008	Gamble (1896) Gamble (1896) Lahiry (1974) [y]Mukhopadhyay	27 (1861–1888) 86±2 (1888–1974) 34 (1974–2007)
Nagaland Tuli-Wokha Road	1981 (Spor.)	Gupta (1982)	
Arunachal Lower Dibang valley, Lohit	2003–2005	[Z]Bhuyan	
SFRI Bambusetum Upper Siang, Tuting	1985–1986 (1-clp) 2003 (1-clp) 2004 (Spor.)	Sharma (2004)	
Bangladesh *Chittagong*, BFRI Bambusetum (1-clp. prt flowr)	1993–1994 (1-clp. prt flowr) 2005–2006 (3-clp)	Banik (2000) Author observed	
Sri Lanka (Introduced at *Royal Botanic Garden, Peradeniya* from Myanmar)	1831 (?) 1907	Janzen (1976) MacMillan (1907)	76 (1831–1907)
Kandy district 6 locations, 23 clumps flowered out of 111 clumps	1990–1996	Ramanayake and Yakandawala (1998)	
Malaysia FRI, Kepong, Selangor (1 clump flowered)	1951–1953	Holttum (1958)	

Note: *clp* Clump *prt flowr* Part flowered. Personal communication [a]Prakash Lohia, owner of the Bamboo farm Hapur; [x]Rajesh Kaushal Assoc Prof Agroforestry Divn, GBPAUniv; [Y]Arun Kumar Mukhopadhyay, IFS, DFO, NTFP Division, WB; [z]L R Bhuyan, Botanist, SFRI, Itanagar, Arunachal. *WB* West Bengal, *UK* Uttarakhand, *UP* Uttar Pradesh, India
[??]not reported

years by Janzen (1976), but Ramanayake (2006) stated 'our studies did not confirm it', because some clumps of this species in Royal Botanic Gardens (RBG), Peradeniya, Sri Lanka, did not flower at 75 years of age. The observation of Ramanayake and Yakandawala (1998) and documentation of past curators of the RBG, as reported by MacMillan (1907) that some of the clumps in the RBG flowered, set seeds and did not die at the end of flowering. Only the flowering culms died and there was subsequent regeneration from the rhizome, except for two clumps. Many of the flowering clumps resumed vegetative growth periodically. Similar to this report, one clump in the Bambusetum of Bangladesh Forest Research Institute, Chittagong, flowered partly in a few branches during 1993–1994 and continued for 2 years and after that stopped flowering and returned fully to vegetative state and such clumps are termed as part-flowering clumps (Banik 2000). Recently, 2014, flowering at Hapur, UP, also showed a part of the clump flowered (Fig. 4.2e(1), e(2), f), yielded seeds and seedlings and then dried up; the other part of the clump flowered in the next year 2015 and produced empty caryopses and died by June 2015 (Fig. 4.2g). Further, Ramanayake and Yakandawala (1998) reported that precocious flowering was seen in a 4-year-old seed-raised plant at Sri Lanka.

It may be inferred from the above discussion that *D. giganteus* often flowers in isolated clump, sporadically and gregariously after a long period of 85–90 years (Banik 2000), and might have more than one flowering populations—sporadic and isolated clump flowering. Besides, some clumps are part flowering and some are complete flowering in nature.

Inflorescence

Recently, during November 2013 in Hapur, GarhMukteshwar farm about 65 km away towards east from New Delhi on NH24, a clump of *Dendrocalamus giganteus* having 32 culms started producing long pendulous leafless floral shoots on the nodes of leafy branches (Fig. 4.2f). Interestingly, exact half number of total culm, i.e. only 16 culms, produced floral shoots and finally flowered. By January to February 2014 about 7–32 pseudospikelets developed in cluster like a head on each node of the long pendulous leafless floral shoot. The spikelets were somewhat flattened ovate, acute and slightly sharp spiny (Fig. 4.2f). In mutilated and broken culms, floral shoots were long and produced well-developed more number of pseudospikelets on the nodes. Florets were 3–6, all fertile except the uppermost one. Stamens with long filaments; anthers 0.5–1 cm long, acuminate; stigma feathery and brightly purple, palea bifid. Very soon within December to January, the leaves were shedded from the branches of flowering part of the clump (Fig. 4.2g). It is learnt from the farm owner Mr Prakash Lohia that the present flowering clump of *D. giganteus* was raised in 2004 by planting a cutting collected from Nagaland. It is also heard that the species has flowered during 2014 in some areas of Nagaland. The clump in Hapur started producing seeds in April and continued till September with the peak of production during May to July 2014. The ripe seeds from the mother fallen on the ground below germinated and produced seedlings simultaneously.

Seed

Grain-like, oblong, 6–8 mm long, hairy and covered with glumes (Fig. 4.2h), about 200 seeds per 10 g.

Seed Germination

A 3–7 days with 60–75 %.

Seedling Character

Grass-like, produce more than one shoot after developing rhizome system within 40–45 days of age (Fig. 4.2i(1), i(2)). One can also collect wild seedlings grown below the crown of flowering mother, transfer to polythene bags (having soil, sand and FYM at 2:1:1) and maintained in the nursery for 8–10 months. These are planted in the field during rainy season.

Vegetative Propagation Methods

The guidelines for producing culm and branch cutting, collection and bagging of wild bamboo seedling and multiplication through macroproliferation and their nursery management are provided in Appendix I, and for details, reader may consult the INBAR Technical Report No.6 (Banik 1995).

SL. No	Multiplication method of *D. giganteus*	Success rate (%) and comment
1	*Offset planting*: 1–2-year-old culms along with underground rhizomes are collected during summer (March to May). Retain basal 3–5 nodes of the culm portion with the rhizome, nurse at sand bed transit nursery, plant during July to August in the field	45–60, very costly, large scale plantation not possible due to limited availability and cost
2	Culm cutting commonly 2-node culm segments are collected from 1 to 2-year-old culms during summer to rainy season (mid-February to September) and placed in propagation beds containing sand-rooting medium	40–50, comparatively costly than branch cutting production, very difficult due to large culm diameter
3	*Branch cutting*: Branches are usually collected after pre-monsoon shower (any time after May to June) and placed in propagation beds containing sand-rooting medium. Rooting hormone IBA or NAA (250–300 ppm) in talc formulation used for better rooting	55–70, profuse rooting within 50–55 days. Cheap and efficient method (Fig. 4.2j)
4	Seedling macroproliferation: Flowers after long interval so seedling availability is rare. However may be used when seeds/seedlings available	Uncertainty

Tissue Culture

D. giganteus (explant)	Medium and environments	Results	References
Nodal segments collected from young shoot of mature culm	(i) MS + BAP (30.0 μM) (ii) BAP (20.0 μM) (iii) IBA (25.0 μM) + BAP (0.05 μM)	(i) Shoot bud initiation (ii) Shoot multiplication (iii) Rooting in the shoots Rapid multiplication	Arya et al. (2006)
Single-node segment from secondary branch of adult clump	MS+ BAP	A shooting and rooting percentage of 77 % was obtained in 3 or 4 weeks	Ramanayake and Yakandawala (1997)

Diversity, Selection and Conservation

Some clumps are part flowering and some are complete flowering in nature. The *Part-flowering* clumps produce seeds periodically but do not die completely and such clumps may be identified, heritability of part-flowering character studies in detail and conserved. Exploration should be made in the native habitat, especially in Myanmar and North West Thailand to identify clumps having superior characters, the productive and quality germplasm and centralized them; and conserve the population. Studies and proper efforts should be given to make expanded use of the species in the bamboo board and panel product industry.

The Sacred Groves, as being maintained in Buddist Temple (*Kiang*) of Chittagong Hilltract (Banik 1998a), have been playing an important role in conservation of diversity of the species as these places are experiencing least or without human interferences like *Jhum* cultivation, cutting of trees for fuel and/or for timber, etc. The sacred groves have been conserved through community participation.

4.2.3 Cultivation and Management

4.2.3.1 Site Selection and Preparation

Lower slope to Valleys are ideal sites for raising plantation. The species is generally cultivated around villages and farmlands and along waterways and 5–6 m away from the side boundary of highways.

4.2.3.2 Planting Space and Pit Making

Spacing between the planting pits 6 m×6 m to 8 m×8 m is preferred due to large size clump character. Gamble (1896) reported that an account of the plantations of *D. giganteus* at Myanoung on the Irawadi in Myanmar was given in *Indian Forester*

2 :311, where it was said that plantations have 15–20 clumps per acre and the good culms fetched British Indian Rupees (INR) 1–4 each (in the year 1877). As initial spacing is wide some seasonal/annual cash crop (legumes are preferred) may be grown in between the plants till the clumps crown form canopy and provide shade.

All other practices like, fertilization and soil preparation, weeding and vine cutting and aftercare and tending operations, etc. may be followed as discussed under *Bambusa balcooa*, *B. cacharensis* and *Dendrocalamus longispathus*.

4.2.4 Specific Spot Characters for Dendrocalamus giganteus Field Identification

1. A gigantic clump with close culms and slender branches on the upper part (Fig. 4.2a, b).
2. Culms large 20–36 m long, 15–30 cm in diameter, grey- green, covered with white waxy scurf when young (Fig. 4.2b).
3. A deep longitudinal groove present just above the nodal bud/branch in the upper part of the well developed culm (Fig. 4.2c).
4. The deep green leaf blade has purplish-brown stalk (Fig. 4.2d).
5. Culm sheath very large, blade broadly triangular, ligule serrate.
6. The young emerging shoot looks very big like a column and purplish in colour.

4.3 *Dendrocalamus hamiltonii* Nees & Arn. ex Munro

[Synonyms: *Bambusa monogyna* Griffith, *B. marima* Buch & Ham., *B. folconeri* Munro; *Bambusa monogyna* Griff. Notulae, 63. 1851 et Ic. Pl. 150. fig. 3. 1851. *non* Blanco (1837). *B. falconeri* Munro in Trans. Linn. Soc. Lond. 26:95. 1868. pp. *B. maxima* Buch.-Ham. in Wall. Cat. 5039. 1828 (*nom nud.*). *Dendrocalamus maximus* Kuntze, Rev. Gen. Pl., 2, 773. 1891].

4.3.1 General Information

4.3.1.1 Vernacular and Local Names

Pecha (Bangladesh, Tripura-India); *Tama*, *Choya*, *Ban bansh* (Nepal, Bhutan); *Kakon*, *Patsa* (Assam-India), *Kako* (Arunachal-India); *Maggor* (Himachal, Uttarakhand -India); *Mpei* (Rongmai sub-tribe of Naga tribe of Manipur-India); *Phulrua* (Mizo-India); *Wanoke*, *Wanok* (Garo Meghalaya-India); *Ka-sejlei*, *Siejbah* (Khasi Meghalaya-India); *Pao* (Lepcha-India); *Kaghsi bans* (Hindi-India); *Wabo-myetsangyi* (Myanmar); *Pai Hok*, *Paipor*, *Pai Nuan Yai* (Thailand); *Hoknar* (Laos).

4.3.1.2 Natural Distribution and General Habitat

Dendrocalamus hamiltonii is indigenous to the central and eastern sub-Himalayan region ascending to 1000 m, Nepal; Sikkim; Bhutan; Arunachal Pradesh, Assam, Manipur, Meghalaya, Mizoram, Nagaland and Tripura of India, Sylhet forest of Bangladesh; Bhamo, Katha and Rubimines districts of Myanmar; south-west China (Xishuangbanna); and north eastern part of Thailand, Laos and Vietnam. The distribution restricted to upland areas with moist fertile soil, especially in hilly semi-evergreen to evergreen forests.

By far the commonest Himalayan *Dendrocalamus* species is *D. hamiltonii*. This species is apparently found along the entire Himalayan range, where it grows in gullies in subtropical forest types, especially along the outer ranges of hills, such as the Siwaliks and Mahabharat Lekh. It is also found further into the foothills of Bhutan and Nepal where it is quite widely planted (Stapleton 1994a, b). This bamboo is also found throughout Meghalaya. It was found in West Khasi Hills. Large size culms are found in Nongkhyllem Reserve Forest, Ri Bhoi district and in West Khasi Hills.

Dendrocalamus hamiltonii has been reported to grow naturally in the Forests of Sylhet and Sitakunda Hills of Chittagong (Gamble 1896), during eighty's not a single clump of the species was found in the areas (Banik 2000). The species now mostly confined to Patharia Reserve of Assam and nearly Khadim Nagar forests of Sylhet (Bangladesh), especially in the tea gardens. Due to the destruction of forests the lands become mostly exposed that has decreased the humus and moisture content creating unfavorable conditions for this species. The clumps of *D. hamiltonii* are rarely found in the areas where human interference is not controlled. Tea gardens in the hilly areas being a protected area seem to be a better site for regeneration of the species. The species commonly occurs along banks of streams and in valleys, often bending over and forming dense thickets, in evergreen and moist forests. The species seems to be light demander, because clumps always occurred in the exposed condition rarely under the tree canopy. The labourer and management of Tea-gardens in some of those places have been cultivating *D. hamiltonii* for its utility value. It has been largely grown in north India, Himachal and Uttarakhand, found scatteredly on the hills of Garhwal and Jaunsar, but always cultivated in the lower hill slopes and plains near villages,

4.3.1.3 Climatic Conditions

It grows well in moist and moderately high rainfall areas of subtropical forests. The species seems to be light demander, because clumps rarely found under the tree canopy. Very much susceptible to biotic interference.

4.3.1.4 Uses

Used for walling, construction, basket making, mats, water and milk containers, large branches are used for weaving material and leaves used as fodder. Sometimes used as fuel and floats for timber rafts. Shoots are edible. Sheaths used for making hats. The rhizome with slight trimming and dressing, become an exact replica of a rhinoceros horn.

4.3.1.5 Ethnic Utilization

The species is used for walls of native huts, construction purposes, basket making, mats, water and milk vessels, fuel, floats for timber rafts. In Nepal *Dendrocalamus hamiltonii* is the favourite fodder plant for 77 % of farmers of Bariyarpur village (Bara district). Farmers in the Bagiung and Parbat district prefer fodder of *Dendrocalamus* (as the leaves do not contain any tannin) and *Bambusa* species. Both *Dendrocalamus* spp. and *Bambusa tulda* contain the highest crude protein content, 14.4 % and 14.05 % respectively, in the leaves of mid-culm zone (Anon 1995b). The tribals of Arunachal Pradesh use the tender shoot for preparation of 'hiyup' a sour pickle. Out of total available edible shoots of different bamboo species in the local Mizoram market about 12–15 % shoots are found from *Dendrocalamus hamiltonii* (Jha 2010). *Mesu* is a traditional fermented bamboo shoot pickle with a sour-acidic taste of Darjeeling hills and Sikkim. Locally grown young edible shoots *choya bans* (*D. hamiltonii*) are defoliated, chopped finely and pressed tightly into a green hollow bamboo stem. The tip of the vessel is covered tightly with leaves of bamboo or other wild plants and left to ferment under natural anaerobic conditions for 7–15 days. Completion of fermentation is indicated by the typical *mesu* flavour and taste. The dominant producer of *mesu* is the *Limboo* women of ethnic *Nepali* community. *Mesu* is eaten as a pickle. *Mesu*-pickle is mixed with edible oil, chilies and salt and is kept in a closed jar for several months without refrigeration. *Mesu* kept in a green bamboo vessel, loosely capped by leaves of fig plant tied by straw is commonly is sold during rainy season in local markets of Darjeeling hills and Sikkim by the *Limboo* women (Tamang and Tamang 2009).

The local pickle (*IROMBA*) made of *Mpei* (*D. hamiltonii*) bamboo shoot is a common delicious dish for both Naga and Manipuri tribes. The local people believe that *IROMBA* stimulates the blood circulation. The recipe of *IROMBA* are fermented (I month) bamboo shoots of *D. hamiltonii*, boiled potato, dry chili, fermented fish (collected from hill stream), coriander leaves, grounded garlic. The boiled potato is smashed and then mixed with smashed boiled fermented shoots. The fermented fishes are deboned and boiled or dried by hanging in air on the cooking furnace with slow heat. Finally the fish is thoroughly mixed with smash potatoes and bamboo shoots.

The *typical houses* in the villages of Tamenglong and Giribam (Manipur) are commonly made of bamboo *Mpei* (*D. hamiltonii*) for roofing and *Rieng* (*Melocanna baccifera*) for walling. The binding strips are made of culm skin of *Mpei* bamboo. These strips are very strong and usually 100–125 cm long and 1.0 cm wide. The villagers peel off the skin from the freshly cut one-year-old young culm of *Mpei* bamboo. Older culms are used for construction works.

The large leaves of *D. hamiltonii* make good fodder and wild elephants are very much fond of it.. Lopping of bamboo culms for browse for the cattle by local people and the migratory grazers is another mode of utilization of this bamboo plant. It was observed that in some villages located in the hills of Uttarakhand (India) during March to April (drier month), the culms of *Dendrocalamus hamiltonii* and *D. strictus* were lopped at the tops that induced more branches and foliage below, which then used as fodder. Local farmer thus control the culm height and the leaves are easily harvested to use as cattle fodder during draught period.

4.3.2 Plant Data

4.3.2.1 Vegetative

Plant Habit

A large tufted bamboo with strong sympodial rhizome system (Fig. 4.3a). The culms sometimes erect but often overhanging and long drooping tips, with large branches.

In the forests of Sylhet, Chittagong, south western Mizoram, and Myanmar, when there is no trees to support the clumps, the main culms bend over forming impenetrable thickets, and the lateral branches ascend vertically. In one village of Himachal it was observed that the culm tips of a few clumps were trained for making living green bamboo gate (Fig. 4.3i). Thus this character can be utilized for making ornamental green living structures in the park.

Culm Height

Tall 12–25 m.

Culm Diameter

A 9–19 cm.

Culm Wall Thickness

Thin walled, 0.8–1.2 cm thick, very flexible, good for weaving, but not strong enough for many constructional purposes.

Greyish-white when young with dense appressed pubescence, dull green when old.

Fig. 4.3 (**a**) A large tufted bamboo with strong sympodial rhizome system. (**b**) Culms usually unbranched below, much-branched above. Straw-creamy colour big branch buds on the culm nodes. (**c**) Branches are sometimes of similar diameter size of culms. (**d**) Purple-coloured anthers in florets arranged in heads on the long floral shoots. (**e**) Seeds are brownish, broadly ovoid, rounded at the base, per 10 g contains 264 glumed seeds, or 277 deglumed seeds. (**f**) Seedlings raised in seed bed and pricked up with roots and these bare-rooted seedling carried into a bowl for transplanting into polythene bags. (**g**) Pricked up bare-rooted seedlings are transplanted in polythene bags containing a mixture of sand, soil and FYM at 1: 2: 1 and maintained the nursery. (**h**) Cuttings in polythene bags raised from 2-node culm segments treated with rooting hormone. (**i**) The culm tips are thin and drooping nature and may be intermingled with each other usually observed in the high rainfall forests areas. Occasionally, such culms are made bend to form a living Bamboo Gate seen near the hills of Baijnath area at Palampur, Himachal, India

Fig. 4.3 (continued)

Internode Length

A 30–50 cm long, lower nodes marked with root scars.

Branching

Dendrocalamus hamiltonii culms usually unbranched below, much-branched above. Straw-creamy colour big branch buds on the culm nodes (Fig. 4.3b). Small, thin curving branchlets are on the younger culms. Strong, stout and long branches develop from the big buds on the nodes of older culms. Such branches are sometimes of similar diameter size of culms (Fig. 4.3c). Size correlation exists between bud size and branch size.

Leaves

Size variable, small on side branches, but large on main branches of new shoots reaching 37.5 × 3.75 cm, round at the base into a short thick petiole, broadly lanceolate; smooth above, rough beneath, finely serrate on the margins. The species produce big size leaves in warm humid (March–August) and comparatively small in the beginning of dry winter season (October–November). In *D. hamiltonii*, the leaves of young shoots and the end leaves of strong branches are usually very large, while those of medium branches are moderate in size and those of thin shoots from lower nodes are quite small. *Dendrocalamus hamiltonii* exhibits somewhat deciduous behavior in dry season and deciduousness of the species is compensated by faster production of a larger total leaf area in the early part of the growing season. Thus according to Rao and Ramakrishnan (1998) *D. hamiltonii* belongs to 'periodic growth-deciduous type'.

Leaf sheath covered with white appressed stiff hairs on outer side, no auricle or bristle. Ligule long, oblique and truncate.

Culm Sheath

Long and stiff, variable size, those of the lower part of large culms 35–45 × 20–22 cm, glabrous, shining within, rough with scanty patches of stiff brown hairs on outer side, truncate at top. *Imperfect blade* often 30 cm long, ovate-lanceolate, sides incurved, on outer surface glabrous, black hairs at the base of the inner surface. *Auricles* acute. *Ligule* smooth, entire.

Vegetative Growth

Emergence of new culm takes place during June to August. A full grown clump may produce 3–12 culms annually. Culm sheaths dislodge promptly while culm elongates.

Physical and Mechanical Properties of Culm

Some physical properties of the culm wood are mentioned below (Table 4.5).

The values of some mechanical properties of air-dry culm (12 % moisture content) of *D. hamiltonii* studies at Myanmar are also presented below (Table 4.6).

Table 4.5 Some physical properties of culm wood of *D. hamiltonii*

	Culm		
Some properties of *D. hamiltonii* culm (average value)	Top	Middle	Bottom
Moisture content (%)	77	82	101
Specific gravity (based on green weight)	0.61	0.57	0.52
Specific gravity (based on oven-dry weight)	0.80	0.75	0.64
Shrinkage in wall thickness (%) air-dry			
Shrinkage in diameter (%)	25.9	26.9	33.6

Table 4.6 Mechanical properties of air-dried bamboo of *Dendrocalamus hamiltonii*

Static banding (N/mm^2)		Compression parallel to grain (N/mm^2)		Shear (N/mm^2)	
MS	MOE	MS	MOE	MS (N)	MS(WN)
70.9	18,860	65.5	3404	15.3	15.1

Source: Sint et al. (2008)
Note: *MS* maximum stress, *MOE* modulus of elasticity, *N* specimen with node, *WN* specimen without node

4.3.2.2 Reproductive

Flowering nature: *Dendrocalamus hamiltonii* often flowers sporadically almost every year, may not be the same clump, but distributed in different clumps in northern India and the Himalayan region. As a result the clumps in flower may almost always be found. At the same time, like some other bamboo species, it also flowers gregariously after a long gap of time. Clumps flowered gregariously in Cachar, Assam (India), in 1912 and again in 1956 after 44 years of interval (Table 4.7). In 1968 it gregariously flowered in over about 15 km stretch along the Noagaon–Guwahati Road (Naithani 2007).

In 1996–1997 clumps *D. hamiltonii* flowered in Patharia reserves of Sylhet forest. Two clumps (Clump source was Patharia Reserve, Sylhet; border to Cachar hills) at BFRI bambusetum also exhibited flowering in November 1996, continued for 5 months and completed within April 1997 and died. The Patharia reserves of Sylhet forest is located near the border of Indian Cachar area of Assam, and may have similar interval of flowering cycle (after 43 ± 2 years) like Cachar hills. Khadimnagar forest reserve of Sylhet is the bordering area of Khasi hills and also flowered gregariously during 1999–2001, thus likely to have similar duration (45 years) of flowering cycle for *D. hamiltonii* (Table 4.7).

It is also reported that the species flowered in Sikkim and Dehra Dun in 1894 and again in 1992, the gap is after long 98 ± 2 years interval. However, the species in Sikkim and Dehra Dun might have flowered after 45 years during 1940s and may be remained unnoticed or unreported. Thus it may be estimated that the flowering cycle in *D. hamiltonii* may be about 45 ± 3 years, like that of Cachar hills of Assam (Table 4.7). With similar interval of time the species flowered in Khasi hills of Meghalaya.

During March 2002, while visiting from Imphal to Tamenglong (Manipur) through National Highway (NH) 53 starting from small market Longmai observed naturally growing many clumps of *Dendrocalamus hamiltonii* (locally called as ‘Mpei’ bamboo) on the lower hill slopes of both the side of the road flowered either partly or gregariously with characteristically purple coloured anthers in florets arranged in heads on the long floral shoots (Fig. 4.3d). Many healthy clumps of *D. hamiltonii* were also in flowering state on the hills of Phellong, Namtiram-II, Namtiram-I, Aziuram, Saramb and Thilon of Tamenglong, Later in 2004 almost all the clumps of the species covering this large tract of bamboo forest was looked as if burnt due to their large scale death. Finally this bamboo flowered gregariously and died within 1999–2006, covered 40 % land of total bamboo brakes (533 km). During mid of 1980–1995 species flowered gregariously and died in Arunachal and gradually spread

Table 4.7 Estimation of flowering cycle in *Dendrocalamus hamiltonii* from the available flowering records

Country/locality	Flowering date (calendar year)	References of flowering dates	Estimated flowering cycle (year)
India			
Dehra Dun	1894	Gamble (1896)	98 ± 2 (1894–1992)
Dehra Dun	1992 (Greg.)	Tewari (1992)	
Himachal, Kangra, Palampur	2003–2006 (Spor)	Author observed	
West Bengal (WB) Darjeeling	1900 1976 (Greg.)	Rogers (1900) Conservator of Forest WB (1976)	Continuous flowering ??
Central Province	1901 (2849 ha Greg flowered)	Smythies (1901)	
Assam Lakhimpur	1905	Cavendish (1905)	44 (1912–1955, 1956)
Cachar hills	1912		
North Cachar Hills	1955, 1956	Nath (1959)	
(Haflong-Nagaon road)	1997–1998 (Greg)	Tripathi (2002)	
Cachar area	2001	Yadava (2002)	
Nagaon-Guwahati Road, about 15 km	2007–2008 (Greg.) 1968 (Greg.)	Choudhury Rajen[a] Naithani (2007)	
Assam-Arunachal Border *Assam*-Nagaland Border	1996–1997 (Greg) 1994–1995 (Greg)	Tripathi (2002) Tripathi (2002)	
Meghalaya			
Northern Khasi Hills	1912	Troup (1921)	45 (1956–2001)
Khasi Hills	1956 (Greg.)	Rao and Ramakrishnan (1988)	
Northern Khasi Hills	2000–2001 (Greg.)	Tripathi (2002)	
??	1985	Gaur (1987)	
Sikkim	2006	Chaubey et al. (2013)	
Arunachal Pradesh East Kameng	1960 (Greg.)	Thakur (2005)	*Flowering wave* of 10–12 years (1983, 1984, 1990, 1993, 1994, 1995)
Western Kameng	1983–1984		
Kameng	1990	Anon. (1996b)	
Siang	1994		
Subansiri	1993–1994		
Siang and Papum Pare	1994–1995		
Namsai	2002–2004 (Greg.)	Banik (2004a)	45 (1960–2005)
Roing, Dibang Valley	2005 (Greg.)	Sharma (2005)	
All over state	2005 (Greg.)	Thakur (2005)	
Nagaland	2003–2004 (Spor.)	Banik (2004a)	
Mizoram	2001–2002 (Spor.) 2006–2008 (Greg.) 2009–2010 (Greg.)	Banik (2004a) Author observed Author observed	–

(continued)

Table 4.7 (continued)

Country/locality	Flowering date (calendar year)	References of flowering dates	Estimated flowering cycle (year)
Manipur, Tamenglong	2001–2004 (Greg.) 2005–2006	Banik (2004a)	
Bangladesh *Chittagong*: BFRI Bambusetum (Clump source Patharia Reserve., Sylhet border to Cachar hills)	1997–1998	Banik (1999a)	43 ± 5 (1955, 1956–1997, 1998)
Sylhet: Khadimnagar (nearer to north Khasi hills of *Meghalaya*, India, also flowered in 2000–2002)	1999–2000 (Greg.)	Banik (2000)	45
Myanmar Upper Burma, Ruby Mines district	1910, 1911, 1914	Troup (1921)	
Thailand (northern part) *Chiangmai*	2013–2014	Author observed	

Note: [a]personal communication, Mr Rajen Choudhury, IFS, DFO, Kokrajhar, BTC, Assam

in *waves* covering the whole area in 10–12 years within 2005 (Banik 2000); then in Manipur (Tamenglon): 2002–2004; and after that during 2007–2009 in Mizoram (Mammit area). Also it flowered earlier in Sylhet during 1998–2000. So flowering in this species is moving towards the east to Thailand now (2013–2014) maybe it is or will flower soon in Cambodia/Vietnam etc. area and nearby south China forest area.

Inflorescence

During flowering year 0.5–3.0 m long leafless pendulous floral shoots, sometimes much branched, nodes covered with boat shaped sheaths, are produced in the month of October–December. Floral shoots, at the beginning, develop directly on the younger culm nodes and as well as from the leafy branches. When floral shoot is branched it is usually long up to 5.0 m. The internodes (rachies-joints) on floral shoots are 2–7 cm, scabrous and white pruinose especially below the swollen nodes, furrowed on one side. These floral shoots start dislodging the nodal sheaths gradually from the base to the tip. But the buds on each node of floral shoot starts producing pseudospikelets from the tip to the base bearing half-verticillate semi-globular heads (inflorescence) of flowers with reddish purple anthers (Fig. 4.3d; Banik 1999a). Each inflorescence contained 45–139 spikelets. Spikelets fertile, 4–7 florets per spikelet; stamens 12–14 mm long, exserted, pendulous with white filaments; anther 5–7 mm, 4-lobed, tip acuminate, pollen smooth; stigma trifid-plumose exserted. First blooming starts from the upper part of culms at the end of December and continues through the next calendar year till the whole clump completes flowering. Almost all the culm tips (irrespective of young and old) flowered within first week of February during which leaves were also sheded off. Anthesis started in the morning at about 10 A M, usually pollen grains were discharged in the form of cloud after shaking. Blooming is not continuous rather occur in successive flushes (flush period) with non-blooming (rest period) periods in between. The seeds matured at the end of February (about 55–60 days after blooming) and attained milky stage during January to first week of February (i.e., in about 30–40 days). Squirrels and birds were found to eat softer part of inflorescence and milky to mature seeds respectively; thus they might have a role on pollination (Banik 1999a). The seed production in *D. hamiltonii* was very good where pollen was being actively collected by bees. To them the capitate inflorescences were vivid purple-red balls of anthers, and they would no doubt also be attracted to the very similar purple-red balls of stigmas that precede anthesis in this species (Stapleton 1982).

It is interesting to note that five new culms emerged during June to September, 1998 in the clump at BFRI Bambusetum that also started flowering in November- December.

Seed

Broadly ovoid, rounded at the base, per 10 g contains 264 glumed seeds, or 277 deglumed seeds (Fig. 4.3e).

Seed Germination

Starts within 3–10 days of sowing, 80–85 % germination of fresh seeds. At normal condition deglumed seed loses viability within 16 days of collection whereas glumed seed remains viable up to 25 days. It was possible to maintain the viability of seed for more than 1 year when kept in a paper packet covered with polythene bag and stored inside the deepfreeze.

Seedling Character

Grass like, starts forming a miniature clump by producing many shoots within 40–50 days (Fig. 4.3f). The bare rooted seedlings are carefully pricked up from seed germination bed (Fig. 4.3f), then transplanted in to polythene bags containing a mixture of sand, soil, and FYM at 1: 2: 1 or sandy loam soil and FYM at ratio 3:1 and maintained in the nursery (Fig. 4.3g) till out planting.

Vegetative Propagation Methods

The guide lines for producing culm and branch cutting, collection and bagging of wild bamboo seedling, and multiplication through macroproliferation and their nursery management are provided in Appendix I, and for details may consult the INBAR Technical Report No. 6 (Banik 1995).

SL. No	Multiplication method of *Dendrocalamus hamiltonii*	Success rate (%) and comment
1	*Offset Planting*: Collect 1–2 year old material during March-May and nursed at sand bed nursery, plant in July-August.	60–70, costly, heavy, large scale plantation not possible for limited availability of propagules
2	*Culm cutting*: 2-node segments, Rooting improves with the treatment of hormone NAA (I-naphthalene acetic acid) solution (100 ppm) (Fig. 4.3h).	70–80
3	*Branch cutting*: Collect pre-rooted branches during early rainy season and place in sand media propagation beds Rooting improves with hormone treatment,	70–75
4	*Seedling Macroproliferation*: Both wild seedlings and seed raised normal seedlings need to be collected and maintained in nursery in to polythene bags containing mixture of soil, sand, FYM at the ratio 2:1:1 and after 5–6 month age multiplied through rhizome separation (Macroproliferation technique)	80–90, cheap and very efficient method; only applicable when seeds and seedlings are available.
	Cutting Macroproliferation	30–35, rarely done.

Tissue Culture

Species (explant)	Medium and environments	Results	References
D. hamiltonii Seeds were disinfected with sodium hypochlorite (4 %) for 20 min.	(i) Cultured aseptically on semi-solid MS (agar 0.7 %) + 3 % sucrose supplemented with 5–45 μM BAP., pH 5.8. Cultures maintained at 25 °C ± 2 °C under a 16 h photoperiod of 2500 lux from white fluorescent tubes. (ii) The regenerated shoots excised (clusters of 3–4 shoots), sub-cultured on semi-solid MS + BAP 10 μM. These subcultured *in vitro* shoots were multiplied every 3 weeks. (iii) Shoots separated into propagules (cluster of 3–5 shoots), again subcultured on multiplication medium. (iv) *In vitro* shoots transferred to the liquid MS supplemented with 100 μM IBA	(i) In 3–4 weeks these shoots were further multiplied (ii) Shoot multiplication rate of 8–9 folds was obtained in 3 weeks along with sizeable shoot elongation (iii) Multiplication is maximum in 3 weeks. (iv) Rooting 93.93 % with 8–9 roots on an average per shoot. 4–5 week old plantlets with huge roots hardened and acclimatized prior to field transfer.	Arya et al. (2012)

Species (explant)	Medium and environments	Results	References
Nodal segments from mature clump sterilized with 0.1 % $HgCl_2$ for 10–15 min.	(i) Axillary shoots (3–4 shoots) proliferated within 10 days of culture on MS + BAP (1.0 mg/l), Initiated shoots excised and further multiplied on MS + Cytokinin (BAP). (ii) Subculture in every 3–4 weeks increased the rate of multiplication. (iii) Then excised propagules of 3–5 shoots inoculated on MS with high concentration of IBA for 7 days, shoots transferred to ½MS without auxin. (iv) Plantlets were hardened.	(i) Best shoot multiplication obtained on MS+ BAP (1.0–10 mg/l). (ii) & (iii) Rooted plants obtained within 19–15 days. (iv) Plantlets acclimatized and established in soil, where they exhibited normal growth.	Arya et al. (2009)

Diversity

The clumps of *D. hamiltonii* growing in northern India (Himachal and west Uttarakhand) generally produce culms with thicker wall than those from northeast-India (Mizoram) and Sylhet forest of Bangladesh. As for example; in September 2004 a clump was observed in Banuri grampanchayt (GPS reading: 32°5′0″N, 76°33′55″E, altitude 1322 m) village Molichak, about 5 km from Palampur of Himachal Pradesh produced 54 culms (8 in 2001, 13 in 2002, 12 in 2003 and 21 in 2004) with tall (17.4 – 19.7 m) and big size culms (diameter at breast height 9.86 cm –14.39 cm). The culms exhibited average wall thickness—3.04 mm (at node no. 2), 3.53 mm (at node no.5), 2.73 mm (at node no. 7) and 1.76 mm (at node no 22). The clump girth was 11.8 m. That clump of *D. hamiltonii* was selected after comparing with other nearby existed clumps as per criteria described by Banik (1993c, 1997b) and numbered as DH001/MI/HP and conserved at Bamboo Clone garden of GBP Agricultural University, Pantnagar. However, clumps growing in lower slopes and valleys on the hills of Mizoram generally produce taller (19.6–23.9 m) culms having longer internodes but with narrow diameter (7.8–11.7 cm) and less number of culms with fewer branches. Similarly longer internodes and thin walls characterize Naga Hills population and Tamenglong Hills of Manipur. The culm tips are thin and drooping nature and may intermingled with each other observed towards the high rainfall forests areas of Meghalaya, (India) and Sylhet (Bangladesh) (Banik and Das 1996). Such type was also seen near Bejnath area of Palampur, Himachal (Fig. 4.3i).

Across the six population in east Khashi hills of India the morphological variability in 120 sampled clumps was studied and observed that the within population variation was significant ($p \leq 0.05$) for two variables: culm internode length or in other words 'nodes per meter' and culm lumen diameter. High positive and significant ($p = 0.01$) correlations were found among variables culm length, culm diameter at breast height (dbh), number of nodes and culm lumen diameter (Pattanaik and Hall 2014).

The presence of such morphologic diversities also suggests there might exist a number of provenances in the species. As *Dendrocalamus hamiltonii* is growing naturally and also in cultivation in the vast region from the eastern Hindukush to Arakan ranges along the Terai and moist area covering about 4000 km (Himachal to Arunachal and also toward south to Mizoram and Sylhet) different provenances of the species might have evolved on the basis of site suitability. Multi-locational provenance trial of the species would help in identify the productive population of the species in relation to utilization objectives. Stapleton (1994b) also believes that there is substantial variation within this species and it has several distinct varieties—*Dendrocalamus hamiltonii* var. *hamiltonii* [autonym] Munro; [*Dendrocalamus hamiltonii* var. *edulis* Munro; *Dendrocalamus hamiltonii* var. *undulatus* Stapleton in Edinb J Bot 51,1994 (1):24].

It is also evident there are many number flowering genotypes (cohorts) growing throughout this region, which needs centralization and conservation for further studies.

4.3.3 Cultivation and Management

4.3.3.1 Site Selection and Preparation

Hilly, lower slopes and valleys near the water courses are ideal sites for raising plantation of *D. hamiltonii*. This bamboo can be planted along the bank of the water bodies inside the wild life game sanctuaries and protected forest reserves as food source for the elephant, as it is a favourite edible species for the animal. The species does not grow well in dry and climatically hot sites; so these sited should be avoided.

4.3.3.2 Planting Space Pit Making and Intercropping

Studies showed 4×4 m spacing is most productive for *D. hamiltonii* plantation. In humid and cooler places of Himachal when turmeric is intercropped with this bamboo the rhizome yield enhanced. The net returns were maximum (Rs. 54,400 per ha). Similar experiences were also obtained in Mizoram, Manipur, Meghalaya and Sylhet Teagarden forests as regards intercropping of turmeric with *D. hamiltonii.*

Other silvicultural practices may be followed as prescribed under *B. balcooa*, *B. cacharensis* and *Dendrocalamus longispathus*.

4.3.4 Specific Spot Characters for Dendrocalamus hamiltonii Field Identification

1. Persistent pale fur on the culms and long drooping culm tips.
2. Distinctly visible straw-creamy coloured big branch buds at the culm nodes (Fig. 4.3b).
3. Sometimes branches are very large and heavy look like a part of culm (Fig. 4.3c).
4. Leaf sheath with white hairs.

5. Very long (0.5–3.0 m) leafless pendulous floral shoots.
6. Reddish purple anthers (Fig. 4.3d).

4.4 *Dendrocalamus longispathus* (Kurz) Kurz

[*Bambusa longispatha* Kurz in J. Asiat. Soc. Bengal n,s, 42, 2, 1873:250.; *Dendrocalamus longispathus* (Kurz) Kurz, Prelim Rep. For. Veg. Pegu 1875, 94].

4.4.1 General Information

4.4.1.1 Vernacular and Local Names

Orah, *Khag bansh*, (Chittagong Hill Tract, Cox's bazaar, Sylhet- Bangladesh); *Rupai* (Tripura-India); *Wamlik* (Kokbarok and Reang-Tripura); *Rawnal* (Mizoram); *Siejlong* (Khasi, Meghalaya-India); *Orah* (Arracanese- Myanmar); *Wanet*, *Waya*, *Talagu* (Myanmar).

4.4.1.2 Natural Distribution and General Habitat

Dendrocalamus longispathus is native of Sylhet, Chittagong forests in Bangladesh, and Arakan and Tenasserim in Myanmar and distributed up to eastern part of India (Tripura, Nagaland, Mizoram, etc.). It is also found in north, central and southern Thailand. The species grows mostly along the streams in the most fertile loamy soil and partially shaded fringes of the forest covers. It is rarely seen on hill-tops, drier slopes and under the close canopy cover. Sometimes it forms pure patches at valleys and lower slopes as observed in Maharani Reserves and Attaramura area of Tripura. It is one of the preferred species by the tribal people for homestead plantation in the Chittagong Hill Tracts especially for edible shoot production (Banik 1998a).

In Tripura *D. longispathus*, are seen in a few pure to mixed patches at valleys and lower slopes in Attaramura and on both the side of the NH 44 Road from Teliamura, Chakmaghat, Ambassa, Kamalpur, Chamanu, Ganganagar, Manu, Lontarai R.F, Srinagar area (Banik 2015d). The species is also found in the forests of Karangichara, Chetrai, Laxmichara, Saidachhera, etc., mid-west part of Tripura and distributed towards north-west crossing the international border of Rajkandi R F of Sylhet (Bangladesh), where the species is named as *Khag bansh*.

Further from central to south Tripura this bamboo is more common in Debtamura RF catchments and Kalajhari Hill Range including purba Raima, Baluchhara, Sardong through central catchment area, Udaipur, Amarpur, Jatanbari, Shilachari, and distributed toward south-eastern forest area of Bagafa, Sabrum, Amlighat upto Feni River demarcating the international border to Ramgarh area under Chittagong Hill Tracts (CHTs) of Bangladesh (Banik 2015d). The species forms some pure patches towards the more humid part like Dighinala, Baghaichhari, Panchari, Khagrachari forests, Rainkhiang and Kassalong Forest Reserves of CHTs (Mog, Marma, and local Chakma tribal name of the species *Ora bansh*).. In Chittagong the species has been found in moist and low undulating land of Koila block of Karerhat forest. Further south of Chittagong the species is naturally occurring in Dulahazari, Alikadam, Sangu-Matamuhuri and Teknaf forests of Cox's bazar (Arrakanese name of the species *Ora*) (Banik 2000); distributed toward international border area and mid-west part of Myanmar (Rakhain, Bago, Thonze, Konbilin forests and Tungoo, etc.).

In Mizoram forests *D. longispathus* (local name *Rawnal*) is more commonly located throughout the Kolasib with the catchments of the rivers Serlui and Twlang. The River Serlui catchment has limited human settlements and thus has less biotic interferences and healthy stocks of the species are seen in such areas of exposed lower slopes and valleys, especially in Variengte, Mauchar, Bukpui, Serkham, Mualkhang, Hortoki and Bairabi. Some healthy patches of this bamboo are seen in the areas adjoining the river Twlang which flows in between Kolasib and Mamit districts of Mizoram; distribution extended to eastern border of Tripura state crossing the Longai River into the forests of Kanmun, Demcherra, Rahumchera, Panisagar, Bungthuam, Zawlnuam, Chailenga Reserves etc (Banik 2015d).

4.4.1.3 Climatic Condition

D. longispathus grows luxuriantly in the moist deciduous forests having annual rainfall varying from 2500 to 6000 mm, and average maximum Temperature 35.5 °C to minimum 11.4 °C. The clumps of *D. longispathus* found to grow mostly along the streams in the moist fertile loamy soil and partially shaded fringes of the forest covers in patches and also as undergrowth in the most mixed forests of *Albizzia* sp., *Aphanamixis polystachya*, *Anthocephalus chinensis*, *Artocarpus* sp., *Dipterocarpus* sp., *Duabanga* sp., *Gmelina arborea*, *Syzygium* sp., *Terminalia* sp., etc. in Tripura forests, CHTs, Sylhet and Cox's bazar and also sometimes with the members of *Lauraceae*; but not under the close canopy cover (Banik 2015d).

In Tripura, Sylhet, CHTs and Chittagong forests *D. longispathus* commonly forms scattered association with other common bamboo species *M. baccifera* and *Bambusa tulda*, rarely with *S. dullooa* and never seen with *Gigantochloa andamanica*; while in Mizoram forest the species forms common association with *D. hamiltonii*, *M. baccifera* and rarely with *B. tulda*.

4.4.1.4 Uses

The culms are commonly used for making baskets, furniture and food grain containers; also used as floats and rafts for timber transportation. Provides raw material for paper pulp industries.. It is a handsome species and also cultivated as an ornamental. Young shoots of *D. longispathus* are edible. On the basis of fresh weight the outer sheath cover amounts to 52.8 % and the remaining 47.2 % contains edible inner meat portion of a shoot (Banik 1997a, 2000). The shoot meat possesses creamy colour (Fig. 4.4i) and taste is somewhat sweet with crisp to tender texture (Banik 1997a). In local Mizoram market maximum amount (50–54 %) of bamboo shoots are available from *M. baccifera*, followed by *D. longispathus* (32–33 %), *D. hamiltonii* (12–15 %), and least from *B. tulda* (2 %) (Jha 2010).

4.4.1.5 Ethnic Utilization

According to local Marma tribe in CHTs *D. longispathus* is more durable. As the species is somewhat resistant to powder-post beetles (ghoon borers), the tribal people in CHTs use this bamboo to construct house posts and walls which are directly in touch with the soil (Banik and Islam 2005). According to the Marma tribal people it is more durable than *Melocanna baccifera*. The species is used for thatched wall, best among all the local bamboo species. The culms have somewhat long internodes usually each node has two ridges. The culms are also used for famous bamboo dance in CHT.

4.4.2 Plant Data

4.4.2.1 Vegetative

Plant Habit

Dendrocalamus longispathus is a clump-forming bamboo with sympodial rhizome system. It is a large tufted attractive bamboo with branches mostly in the upper part of the culms (Fig. 4.4a).

Culm Height

Tall 10–18 m.

Culm Diameter

A 4–10 cm in diameter.

Culm Thickness

From tip to base 0.2–1.8 cm.

Culm Colour

Glaucous-green when young, greyish green when old.

Internode Length

A 25–65 cm long with slightly swollen nodes, the internodes of mid-culm zone are distinctly longer (48–65 cm) than those in upper (6.5–16 cm) and lower (12.5–38.2 cm) part of the culms; and therefore the internode length curve in respect of node numbers shows a sharp peak in the median regions of the curves (Banik 2000, 2015a). The nodes usually have 2 close ridges of root rings (Fig. 4.4b), mostly on lower half of a culm, roots thick fleshy, 2–5 mm long on the nodes of young culms.

Branching

No branching in the lower $^{1}/_{2}$ to $^{1}/_{3}$ rd of the culm (Fig. 4.4a). Thin and small branches on the young culm, comparatively stout branch on the older culms.

Leaves

Green, 18–35 cm long and 2.0–4.5 cm wide, oblong-lanceolate to linear-lanceolate, tip acuminate and short stalked (2–3 mm) green with rough margin. The surface is glabrous. Main vein thick, prominent, yellow shining beneath, secondary veins 8–10 on either side and not prominent. Leaf sheath ligulate and margin ciliate, easily separable; ligule brownish hairy, small 1.0 mm. The leaf growth and defoliation were found to be a dynamic process where leaf growth started on the first year of culm emergence but there was no defoliation.

Culm Sheath

The long persistent fragile papery sheaths with dark-brown pubescence cover the internodes of young culms. Culm sheaths are long and cover the whole internodes or even may be longer than internodes (Fig. 4.4c). Some clumps have culm-sheaths

Fig. 4.4 (**a**) Large tufted attractive clump of *D. longispathus* with branches mostly in the upper part of the culms. (**b**) The culm nodes usually have 2 close ridges of root rings. (**c**) Culm sheaths are long and cover the whole internodes or even may be longer than internodes. (**d**) During flowering, 50–90 cm long leafless glaucous-green floral shoots produced in January usually on the culm tops. (**e**) Seeds grain-like, ovoid and somewhat oblique in shape and covered with glumes, seed diameter 1–2 mm; bottom 2 lines with glumed seeds and deglumed seeds are in the upper line. (**f**) Fresh seeds collected in 2011 from Amarpur, Tripura, are sown in lines in the seed germination bed for better germination and handling. (**g**) After having 4–6 leaved stage, seedlings (about 30–40 days old) are pricked up from seed germination bed and transplanted into polythene bags and maintained in the nursery bed (showing about 60–65-day-old seedlings in the nursery bed) till outplanting during rainy season. (**h**) In some cases, *Dendrocalamus longispathus* seedlings have purple colour at the basal part of new shoots, and colour may extend further below on the upper part of rhizomes. (**i**) The young emerging shoot is elongated, covered with papyraceous long blackish brown culm sheath, and the longitudinal section shows the internal edible creamy-coloured meat portion. On the basis of fresh weight, the outer sheath cover amounts to 52.8 % and the remaining 47.2 % contains edible inner meat portion of a shoot

a
b
c
70
60
50
CM
40
30
20
10
d
f
RUPAT
TYPE - SEED
e

Fig. 4.4 (continued)

shorter than the length of internodes. The sheaths usually 30–75 cm long, 15–25 cm broad, inner surface glabrous, outer surface clothed densely with patches of stiff dark-brown hairs. The Mog, Marma tribes call it as '*Wa-ya*' meaning ' stinging bamboo', for the hairs on the sheaths are especially irritating, The *ligule* is broad, much serrate or often long fimbriate, auricles usually absent, sometimes very small on one side. The blade is 20–45 long and 3.5–4.5 cm broad, lanceolate-acuminate and recurved.

Vegetative Growth

Optimum expression of the inherent character of *D. longispathus* is achieved in clumps about 10 years of age, but it is obvious that site and clump management plays an important role. In the third and fourth years of culm age leaf production both in number and area along with culm biomass growth are getting low and, subsequently, in the fifth and sixth years these sharply decline; therefore, more than 3–4-year-old culms are to be cut selectively to obtain sustainable yield (Banik and Islam 2005). In Chittagong and CHTs condition an adult clump may start producing culms any time within April to November with peak in June-August. Usually a clump may produce 4–9 culms per year (Banik 1988a). All emerging culms do not always develop into full-grown culms. The natural mortality of emerging culms was found to be 21–29 % (Banik 1983a). Normally the longevity of a culm in a clump is from 5 to 9 years.

Physical and Mechanical Properties of Culm

Some important properties of this bamboo were studied at BFRI Chittagong are described below (Table 4.8).

Table 4.8 Some strength properties of *D. longispathus* culm

	Culm positions		
Observed parameter Orah/Rupai	Top	Middle	Bottom
Moisture content (%)	60	80	90
Specific gravity (based on green weight)	0.72	0.64	0.60
Specific gravity (based on oven-dry weight)	0.80	0.76	0.71
Shrinkage in wall thickness (%)	5.4	6.5	7.3
Shrinkage in diameter (%)	2.8	4.8	5.7
Compressive strength, parallel to grain green (kg/cm^2)	552	438	426
Compressive strength, parallel to grain air-dry (kg/cm^2)	601	600	556
Modulus of elasticity (MOE); green (1000 kg/cm^2)	271	184	175
Modulus of elasticity (MOE); air-dry (1000 kg/cm^2)	277	196	199
Modulus of rupture (MOR); green (kg/cm^2)	695	533	732
Modulus of rupture (MOR); air-dry (kg/cm^2)	551	700	905

Source: Sattar et al. (1992)

4.4.2.2 Reproductive

Flowering Nature

The flowering in *D. longispathus* is often sporadic and occasionally also gregarious (Table 4.9). During 1977–1979 all seven clumps in Line 26 flowered at Bambusetum of BFRI Chittagong, of which four clumps flowered completely and died by shedding most of the leaves within 6–7 months and all by 15–18 months, thus designated as *complete-flowering* in nature (Banik 1986, 1997b, 2000).

Table 4.9 Estimation of flowering cycle in *D. longispathus* from the available flowering records

Country/locality	Flowering date (calendar year)	References of flowering dates	Estimated flowering cycle (year)
Myanmar: *Pegu*	1862 1871, 1875 1891	Brandis (1906) Gamble (1896) Brandis (1906)	29 years (1862–1891)
Tharrawaddy district (Thonze, Konbilin forests) Minhla, Mokka *Kadinbilin* and *Taungoo district*	1887 1912 (Greg.) 1913 (Greg.)	Troup (1921) Troup (1921)	26 years (1887–1913)
Bangladesh: *Cox's Bazar* [This locality nearer to Pegu, Myanmar]	1928	Trevor (1928)	35 ± 2 years (1891–1928) Cox's Bazar might flowered like Pegu in 1891, but not noticed and recorded
Chittagong (CTG) (localities not mentioned) Koila, Karerhat, Ramgarh	1876 (Spor.) 1879–80 (Greg.) 1885 (Greg.) 1930 1972 (Greg.)	Gamble (1896) Gamble (1896) Gamble (1896) Trigg (1930), Hasan (1973)	45 years (1885–1930) 42 year (1930–1972)
BFRI Bambusetum, (7 clumps) Chittagong Seedlings raised from the seeds of 1977–1979, *Bambusetum* and 1-year-old seedlings planted in 1978–1980	1977, 1977–1978, 1977–1979 Not flowered till to date 2015	Banik (1987b) Author observed	45 ± 2 years (1930–1977, 1978, 1979). >36–38 years (1977, 1978, 1979–2015)
Chittagong Hill Tracts (CHT) Locality not mentioned Shikshak (south) Shikshak (north) Raingkheong Forest Reserve Fatikchari Dabua Rubber Est. Khagrachari, Dighinala, Panchari, Baghaichari	1879 (Greg.) 1880 (Greg.) 1967 (Greg.) 1972 (Greg.) 1974 (Greg.) 1978–1979 (Iso cl) 1977 (Spor.) 2014 (Spor. ~ Greg.)	Troup (1921) Anon. (1881) Hasan (1973) Hasan (1973) Hasan (1979) Hasan (1979) Banik (1987b) Banik (2015d)	94–89 years (1879, 1880, –1967, 1972, 1974) or might flowered in between, around 1928–1930, at cycle of 45 ± 2 years, not noticed and recorded 42 years (1972–2014)

Table 4.9 (continued)

Country/locality	Flowering date (calendar year)	References of flowering dates	Estimated flowering cycle (year)
Sylhet (SYL) locality not mentioned, may be in Rajkandi R.F Rajkandi R.F Dholai Tea Estate	1889 (Greg.) 1973 (Spor.) 1974 (Greg.)	Gamble (1896) Hasan (1973) Hasan (1979)	85 years (1889–1974) or flowered 2 times in 85 years with gap of 42 years
India *Tripura*:			
Depacherra JFMC nursery, (JFM collected wild seedlings Nagichera, Bambusetum	2010 (Spor.) 2016	Banik (2015d) Author observed	No earlier records of flowering were found, so cycle could not be calculated. However, Tripura is close to CHT, SYL and CTG; so may have similar flowering cycle (FL Cy) 45 ± 2 years as estimated for these places
Amarpur (near to CHTs)	2011 (Spor.)	Author collected seeds	
Kalajhari Hills (near to CHTs) Damcherra (near Kolasib, *Mizoram*) Dhalai: Babusai TSR camp (2–3 ha), Chawmanu, Jawaharnagar (near to Rajkandi forests, Sylhet) Srirampur	2014 (Greg.) 2013 (Greg.) 2014 (Greg.) 2014 (Greg.) 2014 (Greg.)	Selim[a] Honnareddy (2014)	
Mizoram Bethlehem Vengthlang, *Aizawl*, Kolasib and Mamit districts	2014 (Spor.)	Sharma et al. (2014)	Mizoram is close to Tripura and CHT; FL Cy may be 45 ± 2 years.

Iso.cl isolated clump, *FL Cy* flowering cycle
Note: [a]personal communication: Selim Reza, TRIBAC, Tripura

The available first flowering records of *D. longispathus* were from 1862 to 1891 and most of the next flowering reports were after 1970, about 100–90 years gap from the first reports (Table 4.9). However, within this period (from 1891 to 1970) a few incidences of flowering in the species were reported (as second flowering record) in 1912, 1913 from Burma, 1928 from Cox's Bazar, and 1930 from Chittagong; and reveals that in those localities flowering occurred either 30, 35 ± 2 years (1891–1928) or 40 years (1891–1930) after the first incidences of flowering; and then the species might have flowered next in these localities around 1970 as happened in neighbouring Chittagong forests (Table 4.9) and not noticed and so not recorded.

During 2010–2014 *D. longispathus* exhibited sporadic and gregarious flowering in different forest areas of Tripura. No flowering records on *D. longispathus* are available from Tripura before 2010 and Mizoram before 2014; so not possible to estimate gap period between the past and present incidences of flowering in these 2 places. However, reports are there that the species flowered in nearby Bangladesh forest areas of CHTs, Chittagong (CTG) and Sylhet (SYL). The distribution of *D. longispathus* is continuous from Mizoram, Tripura to all these forest areas and not controlled by political boundary between India and Bangladesh. Thus it is likely that earlier the species also flowered in Mizoram and Tripura more or less same time to those of nearby forests of CHTs, CTG and SYL; but not noticed and remain unreported.

The estimated flowering cycle of *D. longispathus* seems to be 29–35 ± 2 years in Myanmar and nearby area of Cox's Bazar (Table 4.9); while towards north (north of Myanmar and Cox's Bazar) at Chittagong and CHT, Mizoram, Sylhet and Tripura the cycle appears to be little longer, about 45 ± 2 years. Thus, it appears, there exist distinctly 2 flowering genotypes (Cohort) in *D. longispathus* in the region covering from India, Bangladesh to Myanmar (Table 4.9).

Inflorescence

During initiation of flowering 50–90 cm long leafless glaucous-green floral shoots produced in January usually on the culm tops (Fig. 4.4d). Floral shoots usually take about 6 weeks (March) to start flowering. Pseudospikelets arranged alternately somewhat in cluster heads, 5–8 mm broad with few florets, sometimes 13–16 florets arranged in cluster. Lower florets fertile and 1-terminal floret sterile. Palea oval, truncated, faintly 2-keeled. Lodicules absent. *Stamens* 6, short, free, anther yellow, apiculate with black apex, style short with hairy stigma, ovary broadly ovoid and acute. Sequential flower opening from top to the bottom (basipetal) of the culms suggest that flowers in the clump pollinated by different groups of pollinators at different times.

Opening of flower and pollen discharge (anthesis) mostly occur in the morning from 6 am to 9 am. Anthesis takes place in the morning 6–10 am (Banik 1986).

Seed

Seed ripens in April-July, seeds grain like, ovoid and somewhat oblique in shape and covered with glumes, seed diameter 1–2 mm (Fig. 4.4e) and per 10 g contains 1350 seeds. Seed production per clump is 400–900 g.

Seed Germination

Fresh seeds are sown in lines in the seed germination bed for better germination and handling (Fig. 4.4f). Seeds germinate within 2–7 days of sowing. Germination rate more than 61 % under partial shade, 33 % under sun. Seeds collected from Amarpur,Tripura forests showed 76.4 % germination under shed net (50 % sunlight) (Banik 2015d). At normal condition seed viability remains up to 55 days after collection (Banik 1987a).

Seedling Character

Grass like, and look similar to other bamboo seedlings. After having 4–6 leaved stage seedlings (about 30–40 days old) are pricked up from seed germination bed and transplanted in to polythene bags and maintained in the nursery bed till out planting during rainy season (Fig. 4.4g).

Vegetative Propagation Methods

The guide lines for producing culm and branch cutting, collection and bagging of wild bamboo seedling, and multiplication through macroproliferation and their nursery management are provided in Appendix I, and for details may consult the INBAR Technical Report No.6 (Banik 1995).

Types of planting material of *D. longispathus*	Success (%), and comment
Seedlings (Fig. 4.4g)	75–85 in field
The Wild seedlings (*Wildlings*)	70–80 in field
Seedling Macroproliferation: Both wild and raised seedlings) are multiplied and April- June maintained in nursery in to polythene bags containing mixture of soil, sand, FYM at the ratio 2:1:1 and after 6 month age multiplied through rhizome separation.	70–85 after macropro liferation, survival is 60–65 at field planting.
The offsets are to be collected during mid-March to April and maintained in the Transit nursery bed (with sand as rooting media) under regular watering or intermittent mist up to middle of the June. Transplant in the field during rainy season (mid June– August).	Survival 60–75 in the field; only 10–15 when planted directly (without nursing) in the field.
Two-node Culm segments (*cutting*): collect during May-August and treat with IBA 200 ppm, place in sand medium propagation bed under intermittent mist.	Rooting 30–45 after 70–85 days in bed.
Branch cutting: collect during June-August and treat with IBA 200 ppm, place in sand medium propagation bed under intermittent mist.	Rooting 25–40 after 75–80 days in bed.

Tissue Culture

D. longispathus (Explant)	Medium and environments	Results	References
Single node segments from actively growing lateral branch, sterilized	(i) MS+ BAP (1.5×10^{-5}M) and Kn (3×10^{-6}M) (ii) MS+ BAP (1.5×10^{-5}M), IBA (1×10^{-6}M) and CM (10 %) (iii) MS + IAA (1.5×10^{-5}M), IBA (1×10^{-5}M), Coumarin 6.8×10^{-5}M)	(i) Maximum bud break (70 %), 3–5 shoot developed within 10–12 days (ii) Shoots multiplied 3 fold in 4 weeks. (iii) 68 % shoots rooted. Transfer to soil success 85 %	Saxena and Bhojwani (1991)
nodal buds of middle culms bamboo branch	(1) MS + 1.0 mg/l BAP + 1.0 mg/l TDZ + 0.5 mg/l NAA (2) MS + 3.0 mg/l BAP + 2.0 mg/l TDZ + 0.2 mg/l Kin + 4 % Sucrose + 10 % CW liquid cultr (3) ½MS + 6.0 mg/l IBA + 1.0 mg/l NAA + 1.0 mg/l BAP + 1.0 g/l AC + 2.2 g/l Gelrite	(1) Buds gradually broke, developed into shoots (2) Shoots further multiplied (3) Shoots rooted profusely regenerated plantlets 70–100 % successfully transferred to soil, plants grew vigorously.	Haque (2010)

Note: *CM* coconut milk, *CW* Coconut water, *AC* Activated Charcoal

4.4.3 Cultivation and Management

4.4.3.1 Site Selection and Preparation

The high humidity and warm climatic conditions with well drained moist soil, especially northern and western aspects of the hills, moist valley and lower slopes having sufficient organic matter and partially shaded fringes of the forest cover are *ideal sites* for the plantation of *D. longispathus*.. The species showed poor survival (30–45 % only) and stunted growth when planted on hill-tops, drier slopes and under the close canopy cover, so sensitive to drier edaphic and climatic condition.

4.4.3.2 Planting Space and Pit Making

The *planting pits* of 45–60 cm cube (depend on type of planting materials) are to be dug at 4 m × 4 m or 5 m × 5 m spacing and prepared 10–15 days before planting during rainy season (June to August); survival was maximum (75–85 %) under cloudy days and rains. The offsets of *D. longispathus* are very sensitive to direct sun light and draught condition. Therefore, north and west aspect of the lower slopes on the hill are suitable sites for offsets planting.

4.4.3.3 Weeding and Vine Cutting

First 3 years weeding and vine cutting have to be done.

4.4.3.4 Aftercare and Tending Operations

As the species is sensitive to draught, time to time watering need to be provided. Protect from fire and grazing. The newly emerged shoots are tender and wild animals like, porcupine, wildbore, monkeys etc. very favourite food and get damaged, and ultimately young (2–3 years old) clumps become weak and infected by insects and fungi through injured part. At present in northeast India, CHT and CTG there is acute scarcity of bamboo and edible shoots due to recent (2004–2012) large scale death of major bamboo species (*Melocanna baccifera*, *Dendrocalamus hamiltonii*, some populations of *D. longispathus* and *Gigantochloa andamanica* and *Schizostachyum dullooa*) on account of gregarious flowering. In such situation predation of wild animals on edible shoots is maximum, and newly raised bamboo plantation in the areas always remains in vulnerable situation. So raised plantation should be protected at least in early years of establishment.

Intercropping During first 3 year of plantation raised in the *Jhum land* soyabean (*Glycine max*), sesame (*Sesamum* sp.), pigeon pea (*Cajanus cajan*), bilati dhaniapata i.e., wild coriander leaf (*Erygium foetidum*), rice (Jhum variety), lady's finger (*Hibiscus esculentus*), local chillis, etc. are grown in between the lines (5 m×5 m to

6 m×6 m) of bamboo for immediate cash earnings. With the development of bamboo crown after 4–5 years turmeric (*Curcuma longa*), pineapple (*Ananus comosus*) and ginger (*Zingiber officinale*) are grown as intercrop keeping 1.5 m gap from either side of the bamboo clumps.

Within 5 years of plantation age one can start harvesting of culms.

4.4.4 Specific Spot Characters for Dendrocalamus longispathus Field Identification

1. Branching mostly on the upper part of culms.
2. Culms glaucous green when young, greyish-green when old, usually covered by long papery remnants of sheaths and dark-brown pubescence (Fig. 4.4a).
3. The internodes are usually covered by long papery culm sheaths densely clothed on the back with patches of stiff appressed irritant blackish brown hairs, persistent more than a year (Fig. 4.4c).
4. Often two ridges in the form of rings on a node (Fig. 4.4b).
5. The young emerging shoot is elongated, covered with papyraceous long blackish brown culm-sheath (Fig. 4.4i).

4.5 *Dendrocalamus membranaceus* Munro

[*Dendrocalamus membranaceus* Munro in *Trans. Linn. Soc. London 26*: *149*. 1868; Kurz., For. Fl. Brit. Burma 2:500. 1877; Gamble., Ann. Roy. Bot, Gard. Cal.7:81. 1896; Brandis., Indian Trees 876, 1906. Variation also reported (Ohrnberger 1999): *Dendrocalamus membranaceus* f.*striatus* Hsueh & D. Z. Li in J Bamb Res. 7 (4), 1988: 3; the distinctive character is culms with yellow stripes.; mainly distributed in China (Yunnan), cultivated the Tropical Botanical Garden, Mengiun, 580 m altitude.

Dendrocalamus membranaceus f.*pilosus* Hsueh & D. Z. Li in J Bamb Res. 7 (4), 1988: 3; the distinctive character is culms covered with brown hairs; mainly distributed in China: Yunnan, Jinghong Xian.

Dendrocalamus membranaceus f.*fimbriligulatus* Hsueh & D. Z. Li in J Bamb Res. 7 (4), 1988: 4; the distinctive character culm sheath ligule fimbriate.; mainly distributed in China: Yunnan.]

4.5.1 General Information

4.5.1.1 Vernacular and Local Names

Phai Saangnuan (Thailand); *Waphyu* (Myanmar).

4.5.1.2 Natural Distribution and General Habitat

Dendrocalamus membranaceus is one of the most frequently occurring, clump-forming woody bamboos (with pachymorph rhizomes) in Southeast Asia. The species is naturally distributed in Laos, Myanmar, Northern Vietnam and Northern Thailand in addition to China's Yunnan Province, especially along the Lancang-Mekong River Valley. Its natural habitat is a tropical mixed deciduous or monsoon forest below 1000 m elevation (Li and Hsueh 1988). This bamboo is indigenous to Upper Burma and Martaban (Troup 1921). Commonly seen in moist forests and low ground in Eastern Myanmar down to Tenasserim. The tropical, subtropical to medium and low hilly ranges (below 1000 m altitude), valleys having moisture and organic matter are natural habitat of *D. membranaceus*.

In 1995, Yunnan Province had approx. 70,000 hm^2 of natural *D. membranaceus* forest, which provided important support for the local ecosystem, including uses as a food resource (bamboo shoots and young culms) and as habitat for the wild Asian elephant (Xue et al. 1995). As a priority species, this bamboo has been crucial in efforts toward the protection of local soils and biodiversity in the southern and western portions of the province; however, because most of its native habitat does not exist in nature reserves, this species has long been overexploited in Yunnan. Due to its overexploitation, the habitat of *D. membranaceus* in Yunnan has been reduced, and the quality of the stand has declined. For example, during the past two decades, the development of tropical agriculture plantations for rubber (Ziegler et al. 2009) and tropical fruits has meant that the area traditionally comprising stands of *D. membranaceus* has been dramatically reduced, to less than 30,000 hm^2 in 2008, with obvious degradation of the quality of the remaining forest. This has raised great environmental questions about water loss, soil erosion, and a decline in biodiversity (Han-Qi et al. 2012).

The species has been introduced and planted at Dehra Dun, where it grows well (Troup 1921). According to Bose et al. (1987) plants and offsets of *M. membranaceus* were collected by A. Hoque from Martaban (Burma) during 1892 and planted in Indian Botanic Garden, Howrah, Calcutta. Lately during 2005, a good amount of seeds of the species was procured from Thailand by the Halduani Forestry Training Institute, Uttarakhand State and raised seedlings in number of Nurseries of the Forest Department, also started experimental plantation in northern India.

However, the species has been seen infrequently cultivated in the homestead of Chunati, Cox's Bazar, Chittagong, Bangladesh, bordering to Myanmar (Banik 2000). During 1991, some planting materials (offsets) were brought from the village Chunati and planted in the Line No 58 at BFRI (Bangladesh Forest Research Institute) Bambusetum, Chittagong of which three plants survived and growing, also simultaneously two planted and survived at Keochia Silvicultural Research Station (KSRS) located near Chittagong Hill Tract of the country. One plant at KSRS flowered in 1995 and died without any seed production.

4.5.1.3 Climatic Conditions

Grows luxuriantly in moist forests and low ground in eastern Myanmar down to Tenasserim.

4.5.1.4 Uses

D. membranaceus is commonly used in building, and also as raw material for furniture, bamboo board, construction, and industrial paper pulp. It is economically important as a vegetable crop. After studying 27 bamboos belonging to 10 genera for the edibility of shoots, *D. membranaceus* was considered to be excellent from processing point of view as the young shoots are smooth and easy to handle (Kennard and Freyre 1957). Its slightly bitter edible shoots are prized. The species produces superior timber. In Thailand the species is commonly planted for scaffolding, furniture-making, chopsticks, toothpicks, fruit-picks, but has ornamental value.

4.5.2 Plant Data

4.5.2.1 Vegetative

Plant Habit

Clump forming, very smooth straight erect lower branch-free culms followed by fine light green leaves (Fig. 4.5a). Its pale golden shoots are purple leaf tipped and blue tinged with powder.

Culm Height

Tall 18–25 m.

Culm Diameter

An adult culm may have diameter at breast height from 8 to 11 cm.

Culm Thickness

Thin walled, at top 0.8, middle 1.03, bottom 1.6 cm thick, very flexible.

Culm Colour

Green culms are white powder covered when young giving a misty blue appearance.

Internode Length

A 30–50 cm long.

Fig. 4.5 (**a**) *Dendrocalamus membranaceus* is a clump-forming bamboo having very smooth straight erect lower branch-free grassy green culms and light green leaves. (**b**) The deglumed seeds are broadly ovate, rounded at the base with a pointed end, 5.0–7.5 mm long and dark brown in colour, sown on moist filter paper inside the petri dishes. (**c**) *Dendrocalamus membranaceus* seedlings are raised directly on the ground at the forest nursery at Uttarakhand, India, and managed in the nursery beds as seedling bank. (**d**) Seedling of *D. membranaceus* taken out from the bed and multiplied through rhizomes separation as *macroproliferation* technique

Branching

Branches are present on upper half part of the culm, each node has stout branch with a few comparatively smaller branches. Upper branches are thin, drooping and leafy.

Leaves

Leaves are on slender branches; lanceolate, 18 cm long, 1.5 wide margin: sharp serrate, pointed tip, obtuse base. Leaf dark green and soft when young; pale green with small rusty spots when mature.

Culm Sheath

Sheath is bellshaped, Culm sheaths deciduous, initially orange-green, elliptical to oblong, usually longer than internodes, papery, margins ciliate; auricles small; oral setae short; ligule 8–10 mm, serrulate; blade reflexed, linear-lanceolate, 30–40 × 2–3 cm. Leaf sheaths initially sparsely hairy, becoming glabrous; ligule short, serrulate; blade lanceolate, 12.5–25 × 1.2–2 cm.

Vegetative Growth

D membranaceus is one of the highly productive bamboo species, production average more than 100 ton per ha at Yunan (Yang and Xue 1999).

Physical and Mechanical Properties of Culm

The average fibre length in *D. membranaceus* is 4.3 mm (Liese 1980). Following are some physical and mechanical properties of this bamboo (Narasimhamurthy Maya et al. 2013; Table 4.10).

Table 4.10 Some values of physical and mechanical properties of *D. membranaceus*

	Culm		
Parameter	Top	Middle	Bottom
Moisture content (%)	87.5	89.6	96.3
Basic density (oven-dry weight basis) (kg/m^3)	563	593	627
Compressive strength, green (Mpa)	64.7	69.7	66.7
Modulus of elasticity (MoE); air-dry (MPa)	6609	7458	6414
Modulus of rupture (MoR); air-dry (MPa)	102.3	97.3	127.7

4.5.2.2 Reproductive

Flowering Nature

Dendrocalamus membranaceus flowers gregariously over a large area and also sometime sporadically (Troup 1921; Rhind 1945). Gamble (1896) reported that the flowering materials were collected in Martaban by Wallich in 1827; in the Yonzalin valley

Table 4.11 Estimation of flowering cycle in *Dendrocalamus membranaceus* from the available flowering records

Country/locality	Flowering date (calendar year)	References of flowering dates	Estimated flowering cycle (year)
Myanmar			
Martaban	1827	Gamble (1896)	30 (1827–1857)
Yonzalin valley	1857, 1862		
Shweli forest, Ruby mines districts Shwegyin	1912 (Greg.) 1915 (Greg.)	Troup (1921) Troup (1921)	
India *Dehra Dun* FRI *Bambu setum* FRI and other area of DehraDun	 1944 1973 1991 (Greg.)	 Naithani and Biswas (1992) Lohani (1973) Naithani and Biswas (1992)	 29 (1944–1973) 46 (1944–1991)
Bangladesh Keochia Silv Res Station	1995	Banik (2000)	
Thailand			
Most places flowering started	1994–1995	Tammincha (1996)	
Kanchanaburi	1999 (Greg.)	Pattanavibool (1999)	

by Brandis in 1857 and 1862; in Tenasserim by Beddome in 1879; in the Yomas by Kurz in 1871; and in the Wuntho and Pinhwe forests by J.W. Oliver in 1890. As per herbarium specimen deposited at Forest Research Institute at Dehra Dun, India, the species flowered during March 1944 at Dehra Dun (Naithani and Biswas 1992). It appears the flowering cycle in the species is about 30 years (Table 4.11).

Inflorescence

A large compound panicle with distant globular heads, 1.4–2.5 cm in diameter, spinescent, Spikelets slightly compressed, glossy, nearly glabrous, 2–3 fertile flowers, empty glumes 2. *Stamens* exserted, filaments long, fine; *Anthers* yellow; Ovary ovoid, hairy above, produced in to a long hairy style ending in a purple plumose *stigma*.

Studies on the floral biology have indicated that *D. membranaceus* is likely anemophilous and prone to be an out-crosser (Du et al. 2000), which also was supported by the genetic differentiation ($GST=0.252$) that was similar to the average of out-crossing species (Han-Qi et al. 2012).

Seed

D. membranaceus can reproduce by seeds in the wild, although this phenomenon is rare, and the rate of seed set is low (Dransfield and Widjaja 1995). Caryopses are more elongated grooved. The de-glumed seeds are broadly ovate, rounded at the base with a pointed end, 5.0–7.5 mm long and dark brown in colour (Fig. 4.5b).

Seed Germination

Normally germination is very poor. Seeds germinated in 5–7 days after sowing and the germination period lasted 10–15 days (Xie et al. 2016). However, germination rate is found to be improved through aseptic culture. The seeds were sterilized by Hg Cl_2 (0.1 %) along with any sodium hypochlorite (NaOCl 15 %), soaked in Gibberellins solution (50 ppm GA_3) overnight dark under 30 °C temperature then sown (inoculated) on MS (Murashige & Skooge). 16 h dark/8 h light supplemented with BAP and Kinetin stimulated high percentage (70 ± 13 %) as compared to 63.13 % when used individually (Brar et al. 2012a).

Seedling Character

Grass-like and looks like other species of bamboo seedlings (Fig. 4.5c).

Vegetative Propagation Methods

The guidelines for producing culm and branch cutting, collection and bagging of wild bamboo seedling and multiplication through macroproliferation and their nursery management are provided in Appendix I, and for details, reader may consult the INBAR Technical Report No.6 (Banik 1995).

SL. No	Multiplication method of *D. membranaceus*	Success rate (%) and comment
1	*Offset planting*: Offset having 4–5 nodes of culm part should be collected from 18–24-month-old culm during April to May. Maintain them in sand bed before planting in rainy season	30–45, costly, not available in plenty. Suitable for homestead plantation
2	*Culm cutting*: Propagation by culm cuttings: 1–2-year-old culms cut before or after they produce new shoots	Rooting success 50–55
3	*Branch cutting*: Use branches of 12–18-month-old culm, treat with growth regulators such as NAA, 2,4-D (20–30 mg/l water); within 4 weeks, profuse root in sand propagation beds under misting	60–65, plant during rainy season, survival rate is very high
4	*Seedling macroproliferation*: Both wild seedlings and seed raised normal seedlings since April to June and maintained in nursery into polythene bags containing mixture of soil, sand and FYM at the ratio of 2:1:1, and after 5–6 months, age multiplied through rhizome separation (macroproliferation technique)	80–90 (Fig. 4.5d) but seedlings are not found plenty, as seed setting is poor
5	Cutting macroproliferation: May be done when cuttings have more than 3 shoots	55–65 %, rarely done

Tissue Culture

Seeds of the species have been procured by IHBT (Institute of Himalayan Bioresource Technology, Palampur, Himachal Pradesh, India) from M/s Allen Green Seeds, Wuyang Province, China (Brar et al. 2012a). *In vitro* propagation protocol was developed by using the collected seeds; viable plantlets were produced, and after hardening, those for 5–6 months were planted in the IHBT Campus experimentally.

D. membranaceus (explant)	Medium and environments	Results	References
Mature seeds collected from Kanchanaburi province surface sterilized with 75 % ethanol for 1 min, dipped into 10 % sodium hypochlorite solution for 15 min and then rinsed thoroughly in sterile water	Cultured on MS medium. Nodal segments of 7-day-old seedlings were cut to 2–2.5 cm and implanted on semi-solid MS media containing different concentrations of 2,4-di-chlorophenoxyacetic acid (2,4-D) ranging from 0 to 8×10^{-5}M and 6-benzylaminopurine (BAP) from 0 to 0.8×10^{-5}M. Callus formed in 4–6 weeks on medium containing $1–1.5 \times 10^{-5}$M of 2,4-D and 0.2×10^{-5}M of BAP. On media with higher concentrations of 2,4-D, the callus developed slowly with dark brown soft masses. After 6–7 months, the white and fresh callus was transferred to a medium with a lower concentration of 2,4-D ($0.5–1 \times 10^{-5}$M) Cultures kept under fluorescent light of 2000 lux, 12/12 h day/night regime and at a constant temperature of 25 ± 2 °C	Embryogenic callus formed within 3–4 months	Pattanavibool (1999)
Axillary buds from the field-grown plants were taken as explants	(i) MS+ 4.4 μM BAP + 1.16 μM KN (ii) A clump comprising of a minimum of 3 shoots inoculated onto ½ MS medium containing 5.4 μM NAA in addition to low concentrations of 4.4 μM BAP for root induction	(i) Induced multiple shoots (ii) About 65 % success in plant establishment after acclimatization in greenhouse	Brar et al. (2012b)

Note: N6-benzylaminopurine (BAP); kinetin (KN); α-naphthaleneacetic acid (NAA)

Cytogenetics

The chromosome number of *D. membranaceus* (collected from Yunnan) is $2n = 70 \pm 2$

Diversities and Conservation

A large portion of the genetic variation (78.95 %) resides among individuals within populations, whereas only 21.05 % exist among the populations. Considering that *D. membranaceus* is a monocarpic and has prominently low sexual reproduction, it has been suggested that the conservation area should be extended to its main natural habitats, the Lancang-Mekong River Valley (Han-Qi et al. 2012).

4.5.3 *Cultivation and Management*

4.5.3.1 Site Selection and Preparation

D. membranaceus is easy in cultivation in all tropical climates, prefers heavy soils and can also tolerate some drought. The natural and protected habitats are the good and ideal sites for planting this bamboo. This bamboo can be planted along the bank of the water bodies inside the wildlife game sanctuaries and protected forest reserves as food source for the elephant, as it is a favourite edible species for the animal.

Usually, 3×3 m or 4×4 m planting space is maintained, and 45–60 m^3 size pits are dug depending on the size of planting materials like seedlings, cuttings and offsets. The pit size is usually 45–60 m^3 depending on the size of planting material. Twelve- to eighteen-month-old seedlings are best planting materials; however, both culm and branch cuttings are also good when seeds/seedlings are not available. The offsets are conventionally used for small scale including homestead plantations.

All operational activities like weeding and vine cutting, intercropping, loosening of soil, watering, mounding and mulching, etc. may be followed as mentioned under *B. balcooa* and *B. bambos.*

4.5.4 *Specific Spot Characters for Dendrocalamus membranaceus Field Identification*

1. Thinner leaves, less rounded at the base.
2. Culm sheath dome shaped, felted, blade narrow, elongated tapering.
3. The sheath is light brown, mottled having linear blade on young shoot.
4. Culm sheaths deciduous, initially orange green, elliptical to oblong, usually longer than internodes, papery.

4.6 *Dendrocalamus strictus* (Roxb.) Nees

[Synonyms: *Bambos stricta* Roxb., Cor. Pl. 1, 80: 58 (1798); *Bambusa stricta* (Roxb.) Roxb., Fl. Ind. ed. 2, 2: 193 (1832). *Bambusa tanaea* [Buch.-Ham., Cat.: 118 (1822) *nom. nud.*; Wall., Cat.5038A (1831–2) *nom. nud.*]

4.6.1 General Information

4.6.1.1 Vernacular and Local Names

Male bamboo, Solid bamboo (English); *Latare, Katli, Lathi* (Pusa, Samastipur—Bihar); *Dominee* (Pusa-Samastipur, north Bihar); *Salia bauns* (Orissa); *Tursing* (Mizo—India); *Karka* (Balaghat, Madhya Pradesh—India); *Nakur bans, Kiri bidiru* (Gujarat); *Male bamboo, Narvel* (Maharashtra); *Kalmungil* (Tamilnadu); *Kallumula* (Kerala); *Sadanapa Veduru* (Telugu-Andhra); *Karali* (West Bengal); *Lathi bans* (Tripura); *Lathi, Karail* (Bengladesh); *Lathi, Karka* (India); *Lathi* (Nepal); *Hmyin-wa* (Myanmar); *Bambu batu* (Indonesia).

4.6.1.2 Natural Distribution and General Habitat

Dendrocalamus strictus is most common and widely distributed in the dry hills 400–1000 m, occasional to common in dry regions of India, Myanmar and Thailand and also scatteredly found in Jinghong and Mengla of Yunnan; southwards, it is said to extend to Singapore and Java. In Indian sub-continent, the species is found to the north in the Punjab, Salt Range, and extends down along the base of the Himalaya and in the Siwalik Range to Nepal, but does not occur naturally in Sikkim and in the Assam valley (Gamble 1896). The Dry Bamboo Brake forest type occurring in Madla range of Panna Tiger Reserve in Madhya Pradesh is characterized by *Dendrocalamus strictus*—*Tectona grandis*. In this forest type, *D. strictus* formed relatively dense brakes, along with scattered over wood of the hardier species of the dry deciduous forest, indicating the dominance of bamboo to be only secondary (Chaubey et al. 2015). The soil in this forest type is dry most of the year and was shallow and stony. The species is continuous throughout the hills of Eastern and Western Ghats and of central and south India, ascending to about 1000 m, but no natural groves found in Sri Lanka. It occurs in deciduous forests all over India except north-west Bengal, Assam, Chittagong Hill Tracts and Sylhet forest and also moist regions of the west coast. In South India, it is common in Andhra Pradesh, Tamil Nadu, Karnataka and only in northern part of Kerala. The species constitutes about 45 % of total bamboo-growing stock in India (Bansal and Nath 2002).

It is also widely cultivated in the different parts of India, Nepal, Myanmar and Thailand and in few areas of Bangladesh. Often planted in botanical or experimental gardens of Sri Lanka, Indo-China, Indonesia, Malaysia, the Philippines, Kenya, Central America and United States.

The species occurs in dry deciduous to mixed deciduous forests on well-drained soil, usually up to an altitude of 1000 m. It prefers sandy loams, hilly ground with pH 5.5–7.5 and of better quality in cooler climates.

4.6.1.3 Climatic Conditions

The required optimum mean temperature is between 20 and 30 °C, but species can withstand extremes as low as −5 °C and as high as 45 °C. Mature plants are frost hardy, but young and tender culms are affected. Optimum rainfall is between 1000 and 3000 mm with 300 mm per month during the growing season. The species can also tolerate abnormal drought conditions.

4.6.1.4 Uses

Dendrocalamus strictus is widely used for construction works and agricultural implements in India, especially as stakes to support sugarcane in the field at North India and is commonly used as raw material for making furniture, mats, baskets and novelty items. It is one of the suitable materials for pulp and paper industry. In Myanmar, when large culms are obtainable, they are much in request for masts for native boats.

The bamboo is widely cultivated as shelterbelt in drier areas in India and for the consolidation and support of embankment. The rain fall analysis study in *Dendrocalamus strictus* plantation in degraded ravine lands under semi-arid climatic condition of Central Gujarat, India, revealed that the throughfall varied from 43 to 72 %, stemflow varied from 7 to 22 % and interception losses varied from 12 to 50 % of the rainfall. The funnelling ratio varied from 22 to 91. The high stemflow amount (7–22 %) and funneling ratio (22–91) of *D. strictus* clumps in comparison to deciduous and coniferous plants makes better rainfall absorption and hydrologically best-suited plantation in degraded ravine lands (Rao et al. 2012). Thus, the species has considerable importance in the water resource management in the drier area.

4.6.1.5 Ethnic Utilization

The 3–5 m tall culms have few uses, but sections can be used as truncheons (*lathi*), usually for herding livestock. It is used to make covers of carts, bows, arrows and cordage. The dry stems are also used for torches and the production of fire by friction. The leaves are much sought after as food for buffaloes and are fairly good fodder for horses. The stem is used to make the buttermilk churning stick. Skin of the stem mixed with lime and lichens is used to make a bandage to heal knife and axe cuts.

The young shoots of *D. strictus* are edible. At Jharkhand, the shoot weight was found to vary from 425.5 to 1172.0 g, with about 52.4–71.8 % of edible part (Das 2014).

The seed is used in times of famine as food grains, and while wheat is sold at 12 seers for the rupee, bamboo seed is sold from 40 to 50 seers (Gamble 1896). In Manipur, roots are used as cleaning brush and as brooms (Singh et al. 2003).

4.6.2 Plant Data

4.6.2.1 Vegetative

Plant Habit

Clump more or less densely caespitose with sympodial, pachymaorph rhizome system (Fig. 4.6a). This is a polymorphic species with a wide distribution.

Culm Height

Broadly arched to erect, 7–20 m tall, much curved above half of its height.

Culm Diameter

A 3–8 cm in diameter.

Culm Thickness

The culms are very strong with narrow lumen or sometimes solid (Fig. 4.6b). The populations growing in the drier western part of India have comparatively solid culms, and in other parts, the culms are hollow with thick walls. In the valleys of Myanmar and South India, it reaches a large size with hollow culms, longer leaves and culm sheaths, but in the dry Deccan hills and Siwaliks, it is small and has nearly solid culms, small leaves and sheaths.

Culm Colour

Culm glabrous and with white powdery mass on the surface, glaucous green when young; dull green or yellowish when old.

Internode Length

A 30–50 cm, nodes somewhat swollen; without nodal roots, basal nodes often rooting. *D. strictus* being a tall and big-sized bamboo species, has series of more or less equal length internodes in the mid-culm zones similar to other tall and

---

Fig. 4.6 (**a**) Clump more or less densely caespitose with sympodial, pachymorph rhizome system. (**b**) The culms are very strong with narrow lumen or sometimes solid. (**c**) Culm sheath without auricles, sheath base firmly attached; *culm-sheath blade* narrowly triangular or lanceolate, hairy on both sides, usually reflexed, *ligule* narrow *leaf sheath* membranous, slightly hairy, no *auricle*, no bristles. (**d**) *Dendrocalamus strictus* has a large paniculate inflorescence with dense globular heads 3–5 cm apart, rachis rounded, smooth. (**e**) Caryopsis, like wheat grains, small ovoid to sub-globose and beaked, slightly brown shining, pericarp coriaceous. (**f**) As the age of the seedling increases, here it was 5 month old, subsequent rhizome structures became bigger and swollen, and also the neck of each successive rhizome moves deeper into the ground for producing big woody culms. At seedling stage, the papery sheaths carrying the rhizomes are usually pinkish in colour. (**g**(**1**), **g**(**2**)) Twelve-to-eighteen-month-old seedlings are best planting materials and carried manually to the nearby field or by truck to long-distance field for planting during rainy season. (**h**) The formation of congested clumps, in which the culms are packed tightly together and are often much bent and twisted, where clumps suffer injury, and in places where the soil is poor, dry or hardened

a
b
2 cm
1. 3 cm
d
e
Glumed
Deglumed
Seeds of DENDROCALAMUS STRICTUS
c
f

Fig. 4.6 (continued)

large size of *Bambusa* species (*B. balcooa* and *B. vulgaris*). Bamboo species having culms with such nodal architecture suggests that they prefer to grow directly under open sky and does not grow satisfactorily under the tall trees (Banik 2000, 2015a).

Branching

Branches several at each culm node, unequal, the central one is strongly dominant, of variable length and habit. Buds and branches present even at the base, in some cases lacking in the lower half to two-third of the culm. Upper branches curved, drooping.

Leaves

Linear lanceolate, leaf blades of variable size, small 2.5–5.0 cm long and 0.5 cm broad in dry localities, while big 25.5 cm long, 3.0 cm broad and persistent in moist localities; rounded suddenly at the base into short petiole, sharply acuminate, twisted point; pale green above, paler below, rough hairy above and softly hairy beneath, frequent 'pellucid dots' between the veinlets. The leaves fall in February to March, and the young new ones appear in April.

Culm Sheath

Variable, lower ones shorter, 8–30 cm long with golden brown stiff hairs, sometime glabrous in dry localities, ciliate at the edges, without auricles or very slightly auriculated, sheath base firmly attached; *culm-sheath blade* narrowly triangular or lanceolate, hairy on both sides, usually reflexed, may be constricted at base, *ligule* narrow. *Leaf sheath* membranous, slightly hairy, ending in a prominent callus, short ligule, no *auricle*, no bristles (Fig. 4.6c).

Vegetative Growth

The young culms generally emerge during June or even in July some times after the rains begin and emergence continue up to early part of November, with highest intensity in July to August. New culms of *Dendrocalumus strictus* were however noticed during December 1985 in Lucknow after late rains during October in that year (Chaturvedi 1986). But in south India, as may be seen on the eastern slopes of the Nilgiris (e.g. in the Coonoor valley), the new culms appear in September or October, probably with the first of the northeast monsoon rains (Gamble 1896). If rains are delayed, new culms keep coming up up to end of October. Generally, a clump may produce 3–6 culms in a year. However, on better sites, a full-grown well-developed clump may even produce 15–28 culms in a year.

Physical and Mechanical Properties of *Dendrocalamus strictus*

The average fibre length in *D. strictus* is 2.4 mm (Liese 1980).

Table 4.12 Some strength properties of *Dendrocalamus strictus*

	Culm source from four localities of India			
Strength properties parameter	Gorakpur UP	DehraDun	Bilashpur	Bhadra chalam
Moisture content (%) green	92.1	78.3	–	28.9
Moisture content (%) air-dry	9.9	12.4	9.8	8.5
Average specific gravity (green)	0.538	0.614	–	0.710
Average specific gravity (air-dry)	0.626	0.730	0.655	0.797
Fibre stress at elastic limit (kg/cm^2) green	425	520	–	361
Fibre stress at elastic limit (kg/cm^2) air-dry	665	868	658	933
Modulus of rupture (MOR); green (kg/cm^2)	753	958	–	649
Modulus of rupture (MOR); air-dry (kg/cm^2)	941	1498	1118	1278
Modulus of elasticity (MOE); green (1000 kg/cm^2)	64.6	107.5	–	122.1
Modulus of elasticity (MOE); air-dry (1000 kg/cm^2)	87.6	213.8	145.3	170.2
[a]C P to grain. Max Crsh Str; (green) (kg/cm^2)	344	384	–	343
[a]C P to grain. Max Crsh Str; (air-dry) (kg/cm^2)	629	684	570	749

Source: Tewari (1992)
Note:[a]C P to grain. *Max Crsh Str* compression parallel to grain. Maximum crushing stress

Limaye (1952) found that older culms of the species became 40–50 % stronger and stiffer than young ones. Maximum values were found in 3–6-year-old culms. The bamboo fibres were frequently used as reinforcement materials in composites. Jindal (1984) reported that bamboo fibre reinforced plastic composites at nearly one-eighth the density of mild steel and possess strength more or less equal to that of mild steel. Thermogravimetric analysis of *Dendrocalamus strictus* bamboo fibres indicated the fibre is suitable as reinforcement for making composites even with thermoplastic matrix materials whose processing temperature is less than 300 °C. The culm wood strength properties are given below (Table 4.12).

4.6.2.2 Reproductive

Flowering Nature

Dendrocalamus strictus usually exhibits gregarious flowering nature after a specific period of time. Sporadic flowering in parts of a continuous population, and even in parts of individual clumps, is of frequent occurrence in this species. Often a few clumps flower at a time, almost every year, in any locality, and such clumps then die off. These types of sporadic flowering, however, do not produce as much good seeds as when the gregarious flowering takes place.

Clements (1956) reported that *D. strictus* introduced in Cuba probably from the 1912 seeding in Garhwal (Uttar Pradesh) flowered in 1956, indicating a seeding cycle of 44 years. A plantation raised in Taiwan in 1912 from seeds obtained from Bihar (India) flowered in 1969; thus, Wang and Chen (1971) estimated the cycle of 57 years. A plantation of *D. strictus* raised during 1937 in the Forest Research Institute (FRI), DehraDun, flowered during 1987 indicating a cycle of 50 years (Dwivedi 1988).

More than one flowering cycle have been estimated for *D. strictus* from the available flowering records. In South India, it is 24–28 years (Kadambi 1949); in North East and Central India 40–44 years (Kadambi 1949; Gupta 1952); and Western India 65 years (Mathauda 1952). In Bangladesh, the species was introduced from the Angul District of Orissa and flowered synchronously after 45 years during 1983–1986 (Banik 1986, 2000) (Table 4.13). Thus, it appears the different populations of this species exhibit diversities in the duration of interseeding period. Precocious flowering is also known at the seedling stage.

Inflorescence

Dendrocalamus strictus has a large paniculate inflorescence with dense globular heads 3–5 cm apart, rachis rounded, smooth (Fig. 4.6d). *Pseudospikelets* present. Spikelets spinescent, usually hairy, 7–12 × 2.5–5.0 mm, with two to three fertile florets intermixed with many small sterile florets; *palea* ovate or obovate, lower ones two keeled, uppermost not keeled, six to eight nerved. *Stamens* three to six, filaments usually free, long-exserted, *anthers* yellow. The style is long, *ovary* stalked, purple feathery one to three *stigma*. The reproductive structures mature and expose themselves to the pollinating agents at different times (*dichogamy*). The gynoecium matures 3–4 days before the androecium (*protogyny*), effectively preventing self-pollination. So seed setting is very poor in isolated flowering clumps and also in sporadic flowering as in the case of most of the bamboo species (Banik 1986, 1997b). Flower blooming takes place over a period of 2–3 h and dependent on air temperature (Banik 1986) and only occurred between 0600 and 0800 h in summer, but in winter between 1100 and 1300 h (Nadgauda et al. 1993). *Dendrocalamus strictus* is anemophilous. Though bees are often seen to visit the flower, they fed on pollen but cannot pollinate the gynoecium due to protogyny. Although profound protogyny has its disadvantages in times of sporadic flowering, it can be useful in breeding programs because it eliminates the need for emasculation (Nadgauda et al. 1993).

Seed

Caryopsis, like wheat grains, small ovoid to sub-globose and beaked, slightly brown shining, pericarp coriaceous (Fig. 4.6e). Seed weight varies from place to place. In Chittagong, Bangladesh, 10 g contains 515 seeds; in India Seoni and Balaghat forest of Madhya Pradesh (MP) 1.0 kg had 32,000 seeds, so 320 seeds/10 g (NMBA

Table 4.13 Estimation of flowering cycle in *Dendrocalamus strictus* from the available flowering records

Country/locality	Flowering date (year)	References of flowering dates	Estimated flowering cycle (year)
India (north) *Utter Pradesh*			
Garhwal, outer Himalayan tract	1872–1876,	Troup (1921)	36–40 (1872, 1876–1908, 1913)
	1879	Gamble (1896)	
Oudh	1880	Gamble 1896)	
Bundelkhand	1908, 1911–1913	Troup (1921)	
Saharanpur, Siwalike	1883–1886 1926–1927 1870	Troup (1921) Kadambi (1949) Gupta (1952)	40–44 (1883, 1886–1926, 1927)
Sharanpur	1909–1910		39–40, 43 (1870–1909, 1910, 1949–1952)
(Girga—block)	1949–1952	Gupta (1952)	
Mirzapur (*UP*)	2004	Author observed	
Uttarakhand FRI *Dehradun*	1937 1987	Dwivedi (1988)	50 (1937–1987)
Himachal *Bilaspur* Solan	2004–2005 2004	Author observed Author observed	
India (East)			
Bihar: Porahat Div.	1921–1922	Anon. (1922)	46 (1922–1968) 46 (1967–2013)
Locality not reported	1968	Shah (1968)	
West Champaran,	1967 (Greg.)	Sinha (1967)	
Forest Division-I	2013–2015	Author observed	
Orissa: Angul Kalahandi Div.	1937–1939 1967, 1969	Sagereiya (1941) Mohapatra (1969)	30 (1937–1967)
India (west)			
Punjab: *Hoshiarpur*	1845–1911	Mathauda (1952)	65 (1845–1911)
Maharashtra Nagpur	1941	Sagereiya (1941)	41 (1941–1982)
	1956	Qureshi (1956)	
Chandrapur district	1982–1983 (Greg)	Khedkar (2005)	
(1.25 lakh ha)	1993–1995 (Greg)		
Bombay (Thana, Mokhada hill) Central Thana Div.	1901	Ryan (1901)	60 (1901–1961) or 30×2, might flowered in 1931 remain unnoticed
Jarida Range, Melghat Div.	1961	Desai and Subramanian (1980)	
Amravati Circle	1978–1980 1983	Tatwawadi and Kali (1983)	
Pune: Poona Univ. Garden	1990–1992	Nadgauda et al. (1993)	30 (1961–1992)
Gujarat Rajpipla	1986	Negi (1986)	
India (north-east)			
Assam : Cachar hills	1879	Tickell (1881)	40–45 (1879–1922, –1966)
	1922	Gupta (1952)	41–42 (1966–2007, 2008)
	1966	Gupta (1972)	
Tripura : all cultivated 4 clumps near Paratia	2007–2008	Author observed	

Table 4.13 (continued)

Country/locality	Flowering date (year)	References of flowering dates	Estimated flowering cycle (year)
Central India			
Madhya Pradesh Paniali *Khandwa* (North, South) *Raipur* *Jabalpur* Jabalpur forest circle— Barela and Katni Ranges	1865 1909, 1948 1910, 1954 1924, 1960 1930, 1965 1965 and 1974	Gamble (1896) Gupta (1952) Dwivedi (1988) Dwivedi (1988) Dwivedi (1988) Chaubey et al. (2013)	39 (1909–1948) 36 (1924–1960) 35 (1930–1965)
Bandhavgarh National Park Chhindwara *Hoshangabad*	1985–1986 1984–1987 1984–1987	Gopal (1989) Dwivedi (1988) Dwivedi (1988)	
Balaghat (North, South) *Balaghat, Seoni*	1916, 1963 2004–2006	Dwivedi (1988) NMBA (2009)	47 (1916–1963) 43 (1963–2006)
Bastar (South) Betul	1948, 1981 1940, 1968, 1984–1987	Dwivedi (1988)	33 (1948–1981) 28 (1940–1968)
Bilaspur Bilaspur North Harda	1895, 1960, 1942, 1976, 1984–1987	Dwivedi (1988)	34 (1942–1976)
Chanda	1940–1942	Kirpekar (1956)	
Mandla (South)	1900, 1921		21 (1900–1921)
Mandla (North)	1930, 1967	Dwivedi (1988)	
Mandla forest division (north)	1971–1973 (Greg)	Chaubey et al. (2013)	37 (1930–1967)
Seoni	1921		
Seoni (South)	1922	Dwivedi (1988)	42 (1922–1964)
Rukhad range, south Seoni forest division	2005–2006	Chaubey et al. (2013)	40 (1964–2005, 2006)
Shahdol (Umaria) *Shahdol*	1909, 1984 1984–1987	Dwivedi (1988) Dwivedi (1988)	75 (1909–1984) or 38 × 2, might flowered in 1870 remain unnoticed
India (south)			
Karnataka: Bhadravaati	1905–1908 1932–1933	 Kadambi (1949)	24–28 (1905, 1908–1932, 1933)
Andra Pradesh Nallamalai Hills (Kurnool dist)	1887 1987–1988	Gamble (1896), Sitaram (1988) Rao (1988)	
Gujarat Rajpipla	1993	Negi (1986)	
Hyderabad Golkonda hills	1890 1978	Gamble (1896) Uppin (1978)	
Tamil Nadu North Arcot	1891	Gamble (1896)	
Bangladesh *Chittagong* Hazarikhil (Introduced in 1939, seed source may be Angul) Dhalghat	 1983–1986 1993 (one clump)	 Banik (1986) Banik (2000)	 44–45 (1939–1983, 1986)

(continued)

Table 4.13 (continued)

Country/locality	Flowering date (year)	References of flowering dates	Estimated flowering cycle (year)
Myanmar			
Tharrawaddy	1865, 1888, 1906	Troup (1921)	
Tharrawaddy (Zigon, Minhla)	1895		
Tharrawaddy (Zigon)	1912–1913		
Thayetmyo *Henzada* *Thayetmyo* (Rangoon, *Henzada,* Yaw forest, *Mandalay,* Chindwin)	1906 1908, 1909 1914	Troup (1921)	
Mu forest	1911–1913, 1914	Troup (1921)	
Bago Yoma west of Taungoo and Oktwin District	1981–1982	Htun (1999)	30
Taiwan (Introduced, 1922 seed source may be Bihar, India)	1969	Wang and Chen (1971)	47 (1922–1969)
Cuba (Introduced, 1912 seed from Gharwal, India) Atkins Institution	1956	Clements (1956)	44 (1912–1956)

2009); in other parts of Madhya Pradesh 265 seeds/10 g (Bilaspur of MP), 258 seeds/10 g (Chhindwara M P), 244 seeds/10 g (Hoshangabad of MP) and 223 seeds/10 g (Betul MP) (Banik 2000).

Seed Germination

Seeds germinate within 3–7 days of sowing. At normal condition, seed viability remains up to 30–35 days after collection. However, seed life can be extended even up to 34 months when stored over silica gel or anhydrous calcium chloride in a desiccator or at 3–5 °C ambient temperature after reduction of its moisture content to 8 % (Varmah and Bahadur 1980). Seed lots with 67 % germination capacity were stored under these three conditions and exhibited 51, 54 and 59 % germination, respectively, after 34 months. The soaking-drying treatment with low concentration of disodium hydrogen phosphate (10^{-4} M) proved to be better than water in maintaining the vigour and viability of seeds of *D. strictus* (Sur et al. 1988).

Seedling Character

In *D. strictus,* pointed conical plumule emerges with sheathing scale-like leaves. It rapidly develops into a thin, wiry stem bearing single foliage leaves arising alternately at the nodes, the bases of the leaves sheathing the stem. Meanwhile, fibrous roots develop. More than one shoot, usually three to four, start developing from the base of the primary shoot forming a strongly geotropic rhizome system within 30–40 days of germination. As the age of the seedling increases, subsequent rhizome structures became bigger and swollen and also the neck of each successive rhizome moves deeper into the ground for producing big woody culms (Fig. 4.6f). At seedling stage, the papery sheaths carrying the rhizomes are usually pinkish in colour (Fig. 4.6f).

During gregarious flowering, huge numbers of seedlings are produced on the forest floor below the flowered mother clumps and start natural regeneration. Seedlings are to be protected from grazing and fire to obtain success in natural regeneration. Thinned-out wild seedlings also may be used as planting materials as has been practiced in other bamboo species (Banik 2000).

Time-to-time weeding, adding soils and FYM (30–45 g) in the poly bags may need application of fertilizer NPK [$N_{100}P_{50}$ K_{10}] through water both foliar and ground. After 5–6 months of age, seedlings need to be shifted from one bed to another so that developing rhizome system and roots cannot penetrate the ground below and neighbouring polythene bags, 12–18-month-old seedlings are best planting materials and carried to the field for planting during rainy season (Fig. 4.6g(1), g(2)).

Vegetative Propagation Methods

The guidelines for producing culm and branch cutting, collection and bagging of wild bamboo seedling and multiplication through macroproliferation and their nursery management are provided in Appendix I, and for details, reader may consult the INBAR Technical Report No. 6 (Banik 1995).

SL. no.	Multiplication method of *D. strictus*	Success rate (%) and comment
1	*Offset planting*: 1–2-year-old culms along with underground rhizomes are to be collected during summer (March to May). Retain basal 4–5 nodes of the culm portion and nursed at sand bed transit nursery, plant during July to August in the field	55–75, costly, large-scale plantation not possible for limited availability
2	*Culm cutting*: commonly 2-node culm segments are collected from 1 to 2-year-old culms during summer to rainy season (mid-February to September, best time March to May) and placed in propagation beds containing sand-rooting medium, under intermittent misting	55–65, comparatively costly than branch cutting as one full culm has to be purchased for making segmented cuttings
3	*Branch cutting*: Branches are usually collected after pre-monsoon shower (any time after May to June) and placed in propagation beds containing sand-rooting medium. Rooting hormone IBA or NAA (250 ppm) in talc formulation (available in market) should be used for better rooting	50–55, profuse rooting within 55–65 days. Cheap and efficient method
4	*Seedling macroproliferation*: Flowers after long interval so seedling availability is rare. However may be used when seeds/seedlings available	Effective technique, may be followed because seeds and seedlings are frequently available

Tissue Culture

D. strictus (explant)	Medium and environments	Results	References
(Embryo)	B5 + 2,4-D(2 and 6) B5 + IBA(1) + NAA(0.2)	Callus/embryoids Germination	Rao et al. (1985)
(Embryo)	MS + 2,4-D(1 and 10) + CW(10 %)	Cellus/embryoids	Dekkers and Rao (1989)
(Node)	MS + CW(10 %) + BAP(0.5) + Kn(0.2) MS + IBA(1;4 days), S(1/2) + AC(0.25 %)	Proliferated and rooted	Nadgir et al. (1984)
(Seedling)	MS + CW(5 %) + BAP(0.2) MS + IBA(0.1;2 days) MS(1/2) + AC(0.25 %)	Multiple shoots Root induction Rooting	Nirula and Rajbhandary (1987)
Single-node segments	MS + NAA) + IBA(1) + 2,4-D(0.5) + phloroglucinol	Multiple shoots Root induction Rooting	Chaturvedi and Sharma (1988)
Nodal explants from mature plants	(i) MS basal medium with 2 mg/l BAP. (ii) MS basal medium with 4 mg/l BAP and 15 mg/l (iii) MS+ IBA 5 mg/l	(i) Buds, which started growing, were transferred (ii) 3-fold multiplication (iii) Rooting only 20 %	Pandey and Singh (2012)
Nodal explants	MS+ BAP (2.0–5.0 mg l-l) MS+ BAP (2.0–5.0 mg l-l)	(i) Shoot bud initiation (ii) Shoot multiplication	Arya et al. (2002b)

Note: *2, 4-D* 2-4-dichlorophenoxyacetic acid, *CW* coconut water

Diversities and Conservation

Depending on the environmental and geographical source, three growth forms of clumps of *D. strictus* are recognized in India (Deogun 1937).

(A) *Common type*—(i) grows everywhere, ordinary form producing medium-sized culms with thick wall. (ii) Culms hollow with relatively thin walls, generally found in depressions, on cooler aspects and where conditions are more favourable. (iii) Clump does not attain a big size, culms solid or nearly so, usually found on ridges and on hotter aspects.

(B) *Large type*—grows within the forests of Uttar Pradesh (UP), Bihar and Orissa (India), has practically no side branches to a great height and seldom shows sign of congestion. The culms are big with long, straight and smooth internodes. Such types may be seen in certain localities, as, for example, Naukri, Lansdowne division in UP. In north Bihar Samastipur, Pusa, such type is also found and locally called as 'Dominee' bansh. Clump somewhat similar look to *Bambusa balcooa*.

(C) *Dwarf type*—this is of a small size and only rarely forms clumps typically found in Balaghat division of Madhya Pradesh (India) where it is known as *Karka* and a limited extent elsewhere. This is the poorest form of *D. strictus*. In the Nepalese Tarai, a small-stature form of *D. strictus* with solid culms and strong branching is quite common.

Plants in seedling populations vary, apparently both phenotypically and genetically. Seedling populations are seen to vary widely in stature, and habit of culms, thickness of culm wall, texture and pubescences of the culm sheath, branching habit, size of foliage leaves and disposition towards congestion of culms in the clump. Some variations are also not uncommon in seedlings, like (i) dwarf grass-like bushy with many thin shoots and (ii) usually dwarf wiry shoots with variegated leaves (Banik 2000). Both these types can be grown in earthen pots and placed in the house lawn and balconies of buildings for ornamentation. So it seems *Dendrocalamus strictus* is a polymorphic species (McClure 1966).

The study of genetic diversity of *D. strictus* for increasing the productivity of bamboo plantation in Vindhyan Region of Uttar Pradesh was confined to diameter and volume of culm, which are the major traits for better quality of bamboo. The rhizomes of 1-year-old culms were collected from the different Candidate Plus Clumps (CPCs) as selected in Renukoot, Mirzapur and Obra Forest Divisions. The collected rhizomes were raised at Research Nursery, Kotwa, Mirzapur for *ex situ* conservation. Significant variations were observed in diameter of the culm and volume along with culm height among all the 12 CPCs. It provides a clear-cut indication for good selection of CPCs for cultivation at Mirzapur and adjoining places for improvement and multiplication of good-quality bamboo vis-a-vis enhancing the yield (Kumari et al. 2001).

'It has been found in an interesting, nearly glabrous-flowered, variety, in the Great Cocos Islands by Dr D. Prain (Voyage of the 'Investigator')' (Gamble 1896).

No planned *conservation programme* of different genetic sources and variations is yet observed. The local people have been conserving the species to some extent in their homestead and settled forest areas. The species is domesticated in the villages as well as grows wild in the forest and varieties need to be properly identified and conserved.

4.6.3 Cultivation and Management

4.6.3.1 Site Selection and Preparation

The species can grow throughout deciduous forests in different countries of the region, except moist to heavy rainfall region of northeast India, CHTs and other similar localities.

4.6.3.2 Planting Space and Pit Making

Close spacing 2.5 m×2.5 m and 3 m×3 m are used for quick stabilizing the soil on hill slopes, the river and canal bank. In the closer spacing treatment (1 m×1 m), 9.9 culms per clump were obtained as compared to 6.5 and 5.2 in 2 m×2 m and 3 m×3 m spacing, respectively, at 572 days after plantation (Patil and Patil 1988). For obtaining good-sized culms, 4 m×4 m to 5 m×5 m are used in normal forestry plantation. The pit size is usually 45–60 m^3 depending on the size of planting material. Twelve- to eighteen-month-old seedlings are best planting materials (Fig. 4.6g(1), g(2)); however, both culm and branch cuttings are also good when seeds/seedlings are not available. The offsets are conventionally used for small scale including homestead plantations.

Among different bamboo species, *Dendrocalamus strictus* has been found quite promising bamboo plant in the ravine lands (Pande et al. 2012). Along three major Indian rivers, namely, Mahi ravines at Vasad (Gujarat), Chambal ravines at Kota (Rajasthan) and Yamuna ravines at Agra (UP) encompassing about 15 ha gullied land at each location to understand the utility of bamboo plantations on extremely degraded ravenous lands for controlling soil erosion on beds and banks of gullies and streams. Seedlings of *D. strictus* were planted at 4 m×4 m in gully beds of watershed and also at close spacing 2 m×2 m in two rows on Bori bunds. It has been observed that bamboo plantation-based interventions absorbs more than 80% of rainfall (Mishra and Krishna Rao 2012). Lowest runoff, soil loss and nutrient loss occurred in bamboo plantation with small earthen check dams followed by bamboo with trenches. Bamboo-based interventions reduced the nutrient losses which vary from 50 to 67%. Economic analysis has been carried out using data from three major ravine systems, viz. Mahi, Chambal and Yamuna, to examine economic viability of plantation under different soil conditions. The analysis has suggested a cash outflow ranging from INR 30,550/ha to INR 48,000/ha from the 7th year onwards to individual stakeholders in the region, in addition to the benefits accrued to society in terms of value of nutrient (INR 2125–5555/ha) saved through soil conservation and incremental soil carbon build-up (INR 41,000/ha) with the recommended harvest practice of harvesting one-third old culms per clump over the life of plantation (Pande et al. 2012). The high stemflow amount and funneling ratio of bamboo plants in comparison to deciduous and coniferous plants make better rainfall absorption and hydrologically best-suited plantation in degraded ravine lands. Maximum silt deposition was recorded in gully treating with bamboo plantation with earthen gully plugs.

4.6.3.3 Fertilization and Soil Preparation

Then soil preparation is to be completed during December to January. When digging a pit (planting hole), the underneath and surface soil should be separated and put at the two sides of the hole. A more effective fertility strategy is to select a moist site with a rich organic soil. The pits are to be filled with cow dung/FYM (farmyard manure) (10 kg), urea (20 g), triple superphosphate (20 g) and muriate of potash

(10.0 g) and soil. The dug earth must be used to fill 2/3 the height of the pit, and during planting, the surface soil must be added near the root zone of plants. The earth in the hole must be thoroughly mixed with farmyard.

All operations like weeding, intercropping, loosening of soil, watering, mounding and mulching, etc. may be followed as mentioned under *B. balcooa, B. bambos, B. cacharensis.* and *D. longispathus* considering the local climatic and site condition

4.6.3.4 Thinning and Pruning

Dendrocalamus strictus clumps are usually found to be congested due to crowded growth of culms. It is important to thin the culms yearly, especially for congested clump, because this helps to increase the number and quality of culm production for the next 2 or more years (Kadambi 1949). Selective thinning of older culms (4 years and above) from the central part of the clump has to be done during January to April, simultaneously followed by soil working and fertilizer application.

4.6.3.5 Pest and Diseases and Their Control

Diseases like damping off, wilt, seedling stem infection, leaf blight, leaf rust are seen to attack this species. Application of Bavistin in combination with Fytolan is effective in controlling disease (Mishra and Krishna Rao 2012). Defoliators cause some damages, but they are considered minor pests.

4.6.3.6 Coppicing and Culm Congestion

The green culms produce coppice shoots after cutting. These are thinner than the culms and are sometimes referred to as switches. These are covered with culm sheaths and are sometimes difficult to distinguish from the actual culms. Coppice shoots are also produced by injury to young culms, and green culms of all ages may produce coppice shoots. If dry culms are not removed from a clump, the coppice shoots produced in such a clump may cause congestion. The formation of congested clumps, in which the culms are packed tightly together and are often much bent and twisted, is frequently seen under certain conditions, especially in village lands and on the outskirts of the forest where clumps suffer injury, and in places where the soil is poor, dry or hardened (Fig. 4.6h). Congestion in bamboos is one of the most serious problems. In this congested condition, the clumps become reduced especially in the circumference due to ill management practice, such as uncontrolled cutting round the periphery, no protection from browsing of the young culms round the edge of the clump by cattle, the ground becoming hardened, continual removal of the young shoots for food and digging up culms with rhizomes attached for making walking sticks. Such constant injury to the periphery of the culm causes the death of the rhizomes and, in time, a dense mass of dead rhizomes prevents the living rhizomes from spreading outwards. The latter accordingly develop within the clump, where also the new culms are produced year after year, with the result that congestion takes place and the new culms may bend in all directions in their efforts to penetrate the dense mass of older culms. Congested clump shoots are whippy and thinner than the normal culms. In extreme cases, if mature bamboo culms are not harvested in time, they prevent new shoot growth and as a result clumps become congested with degenerated culms. If the congested clump is clear-felled in one stroke, good-quality culms can be obtained in greater numbers after 4 years (Banik 2015c).

4.6.4 Specific Spot Characters for Dendrocalamus strictus Field Identification

1. Clumps big to medium in size, congested to slightly open, culm glabrous and with white powdery mass on the surface, glaucous green when young; dull green or yellowish when old.
2. Culms are very thick walled to solid (Fig. 4.6b).
3. No auricles on the culm sheath.
4. Sheath base is firmly attached to the culm node.
5. Absence of thorn-like branchlets and absence of hairs on the leaf sheath of larger variety can distinguish from *B. balcooa*.
6. Unattended clumps are often congested (Fig. 4.6h).
7. Young emerging shoot is brown with thick, dark brown hairs; apex short.

Chapter 5
Gigantochloa Kurz ex Munro

The Greek word *gigantos*, a giant, and *chloa*, grass. *Gigantochloa* is a genus of giant arborescent or scandent bamboos. Culm sheaths are generally stiff, auricled, usually dark and hairy, often with loose hairs and long blade.

Distribution is mainly in Southeast Asia (Myanmar, Thailand, Laos, Vietnam, mainly South China, Malaysia, Singapore, Brunei, Indonesia, Philippines) and also in Northeast India and Bangladesh. The number of 37 species is known under the genus *Gigantochloa* (Orhnberger 1999). Out of these, only one bamboo species is described in this chapter as the species growing naturally on the hills of the north-eastern and eastern part of South Asian region including its popularity to the local indigenous people, and ecological importance in the scrub vegetation also has high demand in utilization.

5.1 *Gigantochloa andamanica* (Kurz) Kurz

[*Gigantochloa andamanica* (Kurz) Kurz, For. Fl. Burma 2:556. 1877; Naithani in Indian For. 125:1130. 1999; Alam MK in Bamboos of Bangladesh: field identification manual. BFRI Chittag 2001. *Bambusa andamanica* Kurz in *J As Soc Beng*. 39(2):88. 1870. *G. auriculata* (Kurz) Kurz, For. Fl. Burm 2:557. 1877. *Oxytenanthera nigrociliata* (Buse) Munro in *Trans Linn Soc Lond*. 26:128. 1868. Nomenclature of Indian species of *Oxytenanthera* Naithani in *J Bomb Nat His Soc* 87:439-440. 1990; *G. maxima* var. *minor* Holttum; Holttum (1956) restricted the genus *Oxytenanthera* to its only African member (the type of the genus), *Oxytenanthera abyssinica*, having a hollow style (ovary appendage) which is unlike that found in Asiatic species of that genus, all of which were reassigned to *Dendrocalamus* or *Gigantochloa*.]

R.L. Banik, *Silviculture of South Asian Priority Bamboos*, Tropical Forestry,
DOI 10.1007/978-981-10-0569-5_5

5.1.1 General Information

5.1.1.1 Vernacular and Local Names

Kalibans or kailyai (Tripura and Assam, India), *wah-Thaibok or wasut* (Garo Meghalaya, India), *kalichari or kali* (Chittagong, CHT, and Sylhet, Bangladesh) and *paribas or belangi* (Orissa, India).

5.1.1.2 Natural Distribution and General Habitat

Gigantochloa andamanica grows naturally in Tripura, Assam, Meghalaya, Bihar, Chhattisgarh, Madhya Pradesh, Maharashtra and Karnataka of India and in the forests of Sylhet and Chittagong. This is an evergreen and semi-deciduous medium-sized tufted bamboo with comparatively smaller diameter having prominent nodes. The species is found to grow gregariously on the flat ground and low elevations along the streams in the moist semi-deciduous forest of Chittagong Hill Tracts (CHT) as observed in Kassalong valley and Sitapahar of Kaptai and the evergreen forests of Myanmar. Characteristically the species is plentiful and found as a congested clump in the fringes and on both sides of Chittagong Cox's Bazar Road near the fringes of Chunati and Harbang forests (Banik 1994a). The species has the capability to tolerate the repeated cutting, while most of the other bamboo species fail to survive. *Gigantochloa andamanica* is also found to grow in isolated clumps on the disturbed scrubby and drier hills of Chittagong and Keochia, Bandarbans, Hyanko, Ramgarh, Kaptai and Rangamati of CHT. The species could also be seen gregariously in the forests of scrubby, exposed, deforested areas of Tripura and Cox's Bazar. Some isolated clumps and patches are found growing on the disturbed and drier hills between Tepania Eco Park and Bishramganj of South Tripura (Banik 2004b).

The species has the capability to tolerate the repeated cutting and occasional ground fire, while most of the other local bamboo species (*B. tulda*, *D. hamiltonii*, *D. longispathus*, *S. dullooa*, etc.) fail to survive; in that context the species has ecological protective role on the scrubby hilly dry forest ground. The green crown and compact caespitose clump with underground rhizome system act as barrier against strong rains of monsoon from hitting the dry ground and soil erosion. This bamboo is not common in Meghalaya, but it can be seen in two places in West Garo Hills, i.e. Danakgri in Nehru Park near Tura and Phulbari area (Naithani 2007).

5.1.1.3 Climatic Conditions

The species grows well in moist and moderately high rainfall (4000–5500 mm) areas with temperature range from 9 to 40 °C and seen up to 300 m altitude.

Uses

It is used for basket making and also extensively used in housing and construction works due to hardy and less branching nature of the culms.

5.1.2 Plant Data

5.1.2.1 Vegetative

Plant Habit

Gigantochloa andamanica is an evergreen and semi-deciduous medium-sized tufted bamboo. The clump is upright and straight with persistent culm sheath, tufted with pachymorph rhizome system (Fig. 5.1a). *Young shoots* are green, and *sheaths* are covered with glossy black pubescence with a medium glabrous patch; *blades* are erect imbricating with naked auricle.

Culm Height

6–15 m tall and erect.

Culm Diameter

2.5–5.3 cm.

Culm Thickness

At the basal internode, culm wall thickness is 0.85–1.49 cm, at mid-culm zone 0.44–0.67 cm and at the top 0.18–0.23 cm.

Culm Colour

Glossy green to dark green and sometimes marked with longitudinal yellow stripes (Fig. 5.1b).

Internode Length

Nodes are prominent and greater in diameter than internodes; nodal line is single and horizontal; there is a circular band of brown pubescence below the node; internode is 20–40 cm long.

Branching

Branch bud is oval, about 1.5 × 1.5 cm; the tip is rounded and slightly keeled; keels are naked; branches are from lower mid-culms to upwards.

Leaves

The leaves are 15–30 cm long, 2.5–5 cm broad, lanceolate, rounded or tapering at base into a short petiole, ending in a subulate twisted point, somewhat scabrous above and little hirsute at first but afterwards glabrous; margins are scabrous; main vein is prominent, and secondary veins are 8–12 and inconspicuous; *sheath* is stiff and hairy at first but afterwards smooth, terminating in two thick glossy auricles; *ligule* is narrow and inconspicuous.

Fig. 5.1 (**a**) The clump of *Gigantochloa andamanica* is erect and straight with persistent culm sheath, tufted with pachymorph rhizome system. (**b**) Culms of *G. andamanica* are glossy green to dark green, sometimes marked with longitudinal yellow stripes. (**c**) Inflorescence is a panicle of spicate verticils; rachis is pubescent especially above and 2.5–7.5 cm between verticils; spikelets are usually few in the verticils. (**d**) In the flowering clump of *G. andamanica*, the spikelets are narrow, cylindrical and conspicuously black or purple fringed at the edges of the glumes. (**e**) Fresh glumed and deglumed caryopsis (seeds) of *G. andamanica*. Deglumed caryopsis is blackish brown, narrow and cylindrical with slightly wider base and 30-day-old seedlings. (**f**) A 5-month-old seedling of *G. andamanica* ready for multiplication through rhizome separation technique known as macroproliferation. (**g**) Regeneration of wild seedlings in the forest floor after flowering of *G. andamanica* clumps during 2004 at Tepania Eco Park Tripura

Fig. 5.1 (continued)

The amount of leaf fall from *Gigantochloa andamanica* is 5.6 ton per hectare.

Culm Sheath

Culm sheaths are green when young and yellow when mature, persistent, oblong-cylindrical, 15 cm long, striate, ciliate at the margins, glabrous or covered on the back with appressed stiff brown hairs, slightly narrowed upwards and truncate at the apex; *sheath proper* is cylindrical-elongated and 10–14 cm long; top is truncate to round; adaxial surface is glabrous, while abaxial surface is covered with patches of blackish-brown pubescence with median vacant patch; when young, sheaths ciliated the patches; *auricles* are two, stiff and naked; ligule is narrow and intact; *imperfect blade* is elongated, conical and 3–5 cm long; adaxial surface is covered with black pubescence. *Leaf sheaths* are thick and slightly keeled on the back, with silvery pubescent; margin is ciliated; auricles are naked; *blades* are lanceolate and 6–30 × 1.4–4.0 cm.

Vegetative Growth

Only 10 % culms emerge in the month of May, and a maximum number emerge during the month of July to September and least in October and November; no emergence from December to April.

Table 5.1 Some physical and mechanical properties of *Gigantochloa andamanica*

	Culm		
Parameter	Top	Middle	Bottom
Moisture content (%)	104	118	129
Specific gravity (based on green weight)	0.51	0.49	0.48
Specific gravity (based on oven-dry weight)	0.74	0.69	0.66
Shrinkage in wall thickness (%)	5.9	8.1	9.6
Shrinkage in diameter (%)	2.4	3.0	4.8
Compressive strength, green (kg/cm^2)	301	287	257
Compressive strength, air dry (kg/cm^2)	417	387	346
Modulus of elasticity (MOE); green (1000/cm^2)	169	131	119
Modulus of elasticity (MOE); air dry (1000/cm^2)	224	150	131
Modulus of rupture (MOR); green (kg/cm^2)	415	459	514
Modulus of rupture (MOR); air dry (kg/cm^2)	530	580	721

Sattar et al. (1992)

Physical and Mechanical Properties of Culm

The average fibre length of *G. andamanica* (syn. *Oxytenanthera nigrociliata*) is 3.6 mm (Liese 1980). On the basis of study made in BFRI Chittagong, it appears that this bamboo has high value of compressive strength and modulus of rupture (Table 5.1).

5.1.2.2 Reproductive

Flowering Nature

It appears from the past records the species flowers after about 30±5 years in Myanmar and Chittagong Hill Tracts (Table 5.2). Amarpur and Udaipur localities of Tripura are border areas to CHT of Bangladesh where the species also flowered at the same time in 2008–2010; thus, it is likely that the species in Tripura also flowered after same interval of about 30 years (Table 5.2).

Inflorescence

It is a large compressed, sometimes leafy, panicle of spicate verticils; rachis is pubescent especially above and 2.5–7.5 cm between verticils; spikelets are usually few in the verticils (Fig. 5.1c). Spikelets are narrow, cylindrical, 2.5 cm long by 25–5 mm in diameter and conspicuously black or purple fringed at the edges of the glumes (Fig. 5.1d). Empty glumes are two to three, fertile flowers are two to three and a terminal imperfect flower; *empty glumes* are ovate-acute and *flowering glume*

Table 5.2 Estimation of flowering cycle in *G. andamanica* from the available flowering records

Country/locality	Flowering date (calendar year)	References of the flowering dates	Estimated flowering cycle (years)
Myanmar			
Tenasserim	1849	Gamble (1896)	30
Pagoda hill, Moulmein	1879	Gamble (1896)	(1849–1879)
Bangladesh			
CHT	1978	Hasan (1979)	32
Pablakhali	2008–2010	Author observed	(1978–2010)
India			
Andaman	1875	Kurz (1877)	24–30, 32
Tripura: Charilam and Tepania	2004–2007	Banik (2004a)	(1978–2004, 2008,
Amarpur and *Udaipur*	2008–2010	Author observed	2010)

Note: *CHT Chittagong Hill Tracts*

is lanceolate and acuminate; *palea* is shorter than, or as long as, the flowering glume, narrow, obtuse, three keeled and ciliate. Stamens are exserted; tube are at first thick and membranous; *anthers* are 6–8 mm long and purple; *stigma* is one to three, plumose and short.

Seeds

Caryopsis is linear to oblong, 14–17 mm long, 2.5 mm broad and glabrous; a 10 g contains on an average 265 glumed seeds.

Seed Germination

Fresh seeds germinate within 5–7 days of sowing, with 80–85 % germination in the germination bed (Fig. 5.1e).

Seedling Character

It is like any other seedlings of *Bambusa* and *Dendrocalamus* species.

Vegetative Propagation Methods

The guidelines for producing culm and branch cutting, collection and bagging of wild bamboo seedling and multiplication through macroproliferation and their nursery management are provided in Appendix I, and for details, consult the INBAR Technical Report No. 6 (Banik 1995).

Sl. no	Multiplication method of *G. andamanica*	Success rate (%) and comment
1	*Offset planting*: collect 1–2-year-old material during March to May, and nurse at sand bed nursery; plant in July to August	65–75; costly; large-scale plantation is not possible for limited availability of planting materials
2	*Culm cutting*: rooting improves with the treatment of hormone IBA (250 ppm) or NAA (100 ppm)	40–50
3	*Branch cutting*: rooting improves with hormone treatment	35–45
4	*Seedling macroproliferation* (Fig. 5.1f): both wild seedlings (Fig. 5.1g) and seed-raised seedlings need to be collected and maintained in nursery into polythene bags containing mixture of soil, sand and FYM at the ratio 2:1:1, and after 5–6 months of age multiplied through rhizome separation (macroproliferation technique)	70–85; cheap and very efficient method
5	*Macroproliferation of cuttings*: may be practised when a cutting has 4–5 shoots	Possible but rarely practised due to poor success (20–30 %)

Tissue Culture

Not known.

Cytology

Not known.

Diversity and Conservation

It appears the species exhibits some differences morphologically between the localities of CHT and Tripura. The CHT and Sylhet type has more blackish hairs than those of Tripura. A detail study of these types is needed including the stock growing in Myanmar. The indigenous hill people in these areas are maintaining and conserving the species, and thus it is obvious that the species is important to the hill lives. The detailed ethnobotanical, socio-economic and ecological importance of the species is not yet known properly.

5.1.3 Cultivation and Management

5.1.3.1 Site Selection and Preparation

Due to its hardy nature, *G. andamanica* can be planted from hilltop, slope and valleys. However, at young stage, the planting materials like moist condition for better growth and survival. Hilly land with pH 4.5–5.5 is a suitable site for the species.

5.1.3.2 Planting Space and Pit Making

The spacing 3 m × 3 m has been found as suitable spacing for growth, and as a result more clumps can be grown per ha of land. Pit size (45 × 45 × 45 cm or 60 × 60 × 60 cm) would depend on the size of planting materials. All planting activities should be done in rainy season (June to August). All the maintenance works like weeding and vine cutting, loosening of soil, mounding and mulching are to be practised regularly. Other silvicultural practices may be followed as prescribed under *B. balcooa*, *B. cacharensis* and *Dendrocalamus longispathus*.

5.1.4 Specific Spot Characters for Identification of Gigantochloa andamanica in the Field

1. The species has close clump with mostly persistent culm sheath (Fig. 3.1a).
2. Culm sheath with naked auricle and patches of black hairs on the back.
3. Occasionally a few yellow stripes are visible on the basal culm nodes (Fig. 3.1b).
4. Young shoot has shining black hairs; its apex is long, pointed and conical.

Chapter 6
Melocalamus Benth. & Hook. f.

Melocalamus word roots from the Greek word *melon*, an apple, and Sanskrit *kalam*, a reed for writing, referring to a reed with globular fruits.

Melocalamus Bentham in Bentham & J D Hooker, *Gen Pl*, 3.2. 1883:1095, 1212. It is naturally found in the areas between Eastern Himalayas and Vietnam including South China, Yunnan and Guangxi; Bangladesh to northwestern Thailand and Myanmar; Northeast India—Tripura, Mizoram, Manipur and Meghalaya. Nine species are known to the genus *Melocalamus* (Orhnberger 1999). Type species: *Melocalamus compactiflorus* (Kurz) Bentham is found in the moist hills of South Asia region.

6.1 *Melocalamus compactiflorus* (Kurz) Benth. & Hook. f.

[*Synonym(s)*: *Dinochloa compactiflora* (Kurz) McClure in Kew Bulletin 1936, p. 253, *Melocalamus compactiflorus* (Kurz) Benth. & Hook. f., Gen. Pl.3:1212. 1883: Gamble, *Ann Roy Bot Gard Cal*. 7:94, 1896 and *Pseudostachyum compactiflorum* Kurz in *J As Soc Bengal*. 42: 252, 1873]

Note The genus *Melocalamus* was monotypic till 1985 when Majumder described one more species, *M. indicus*, distributed in Cachar, Manipur and other adjoining parts of Assam. In the field *M. indicus* can be distinguished from *M. compactiflorus* by the deciduous culm sheaths (Majumder 1985). The genus *Melocalamus* was merged with *Dinochloa* by McClure in 1937. Holttum (1958) and others did not accept the merger and believed that it was erroneously transferred to the *Dinochloa* by McClure in 1937. A population of this bamboo at FRI, Dehradun, flowered in the late 1980s; after examining the authentic material, this mistake has been rectified; some misidentifications have also been corrected (Bahadur and Naithani 1983).

R.L. Banik, *Silviculture of South Asian Priority Bamboos*, Tropical Forestry,
DOI 10.1007/978-981-10-0569-5_6

6.1.1 General Information

6.1.1.1 Vernacular and Local Names

Lota bansh and *lata bansh* (Chittagong Hill Tracts, Bangladesh); *daral* (Sylhet, Bangladesh); *saril* (Tripura, India); *tumoh* (Jaintia Meghalaya, India); *kale o* (Karen, Burmese). n*achin wa* and *usawi* (Kachin, Burmese), *wa nwe*, *wa nwe kok* (Myanmar); *ha ge* ('ha' means climbing and 'ge' means wild bamboo; Hani community of Xishuangbanna, Yunnan, South China); *phai khruea wan* and *phai chang chang* (Thailand).

6.1.1.2 Natural Distribution and General Habitat

Melocalamus compactiflorus is commonly distributed in moist deciduous and sub-evergreen forests of Assam, Meghalaya, Manipur, Mizoram and towards further south up to Cox's Bazar of Bangladesh, Myanmar and also in South China (Yunnan) and Northern Thailand. The species also confined to less-disturbed area of Jampui Hills of Tripura bordering to Mizoram (Banik 2004b). The moist and shady sites inside the less-disturbed forests are the natural habitat for this species. As a requirement this bamboo likes partial shade and needs tall trees to climb on it. The tropical, subtropical to medium and low hilly ranges and valleys having moisture and organic matter are natural habitats of *Melocalamus compactiflorus*.

6.1.1.3 Climatic Conditions

It grows in hills, frequently between 1200 and 1800 m altitude; however, it is also found to occur in small hills of Cox's Bazar below 500 m. The plant prefers fringes of perennial streams for both moisture and partial sunlight in forests of heavy rainfall areas as observed on the lower hill slope by the side of the road leading towards Cherapunjee in Meghalaya.

6.1.1.4 Uses

It is much used as rope, especially in making bridges. Shoes/sandals and baskets are made of the outer green fibrous layers of the culms of this climbing bamboo.

6.1.2 Plant Data

6.1.2.1 Vegetative

Plant Habit

M. compactiflorus is one of the handsome climbing bamboos growing naturally on the trees in liana-like fashion in the forests of northeastern part of India and Bangladesh. This is a bamboo with arching culms growing on the supporting trees.

Fig. 6.1 (**a**) *Melocalamus compactiflorus* is a bamboo with arching culms growing on the supporting trees. Culms are brownish in colour, while the young turns green as it becomes older; white band appears below the nodes; there are no branches at the three to six nodes; then up to the mid-culm zone, branches are two in number; and at the upper part of each node, it possesses three branches. (**b**) Branches are thin arising from the culm node firstly perpendicularly up to one to two branch node and then making 45–60° angle. (**c**) Culm sheath is persistent, cylindrical, smooth or covered with white appressed hairs; blade is more or less equal to the length of the sheath, recurved, long-acuminate and often rounded at the base. (**d**) Seeds are comparatively large, subglobular, 2.5–3.8 cm broad (diameter), fleshy, chestnut-like and glossy green to brown, with the base supported by the persistent glumes and pericarp that are fleshy and thick

Culms are brownish in colour, while the young turns green as it becomes older; white band is present below the nodes (Fig. 6.1a).

Culm Height

Tall 10–33 m.

Culm Diameter

1.2–2.5 cm.

Culm Thickness

Somewhat solid.

Culm Colour

Grey green, rough.

Internode Length

The internode is long 30–62 cm; nodes are somewhat swollen, and whitish pubescent band is present at the base of the five to ten basal nodes (Fig. 6.1a, c). Internodes up to 10–18 are more or less straight than gradually becoming scorpoid. This scorpoid habit is very prominent in the upper portion of the culm.

Branching

Its branches are thin and two to three or more. There are no branches at the three to six nodes; then up to the mid-culm zone, branches are two in number; and at the upper part of each node, it possesses three branches. Branches are arising from the culm node firstly perpendicularly up to one to two branch node and then making 45–60^0 angle (Fig. 6.1b).

Leaves

It is oblong-lanceolate, 15–25 cm long × 2.5–5 cm, subulate towards apex and ending in a twisted point; base is rounded; petiole is 3–5 mm long and hairy; leaf sheath is striate, covered with many appressed white hairs when young and glabrous when old. The leaves are comparatively big and provide shelter for birds and some small mammals.

Culm Sheath

It is 15–20 cm long, 5–8 cm broad, persistent, hard, brittle, cylindrical, smooth or covered with white appressed hairs; blade is more or less equal to the length of the sheath, recurved, long-acuminate (Fig. 6.1c) and often rounded at the base; auricle is narrow, dark, crescent shaped and fringed with stiff bristle; ligule is narrow and intact. Both blade and sheath are present up to one-third of the culm, and then only sheaths on the remaining upper nodes.

Vegetative Growth

Culm emerges during May to August. The average number of culms per clump is 20 or sometimes may have many culms.

6.1.2.2 Reproductive

Flowering Nature

It is a gregariously to isolated flowering species. It appears that the species exhibits short flowering cycle (7–16 years) in Myanmar. During 1977 seven seedlings of *Melocalamus compactiflorus* were collected from Cox's Bazar forest and planted at the BFRI, Chittagong Bambusetum (line no. 41). The species flowered and seeded in Cox's Bazar during 1974–1975; thus, the collected seedlings were 2–3 years old in the year 1977. These seedlings were 23–25 years old in the year 2000 and did not flower (Banik 2000); even till June 2015, it also did not flower. So, it proves that the species exhibits long flowering cycle, maybe (as estimated) 47–50 years in Bangladesh and 53 years in Assam, Manipur, Mizoram (India) and Myanmar, say, 50±5 years (Table 6.1; Banik 2000); in fact all the places are neighbour to each other.

Inflorescence

A large compound interrupted the panicle of small subglobose heads, often leaf-bearing heads with several fertile and many sterile spikelets. The rachis is very pubescent, flexuose, flattened on one side, 1.5–3.7 cm in length and 0.7–1.5 cm in diameter. Spikelets are very small, 0.3–0.5 cm long, 0.3 cm broad and glabrous, with two fertile florets and one terminal sterile abortive floret. There are three, large and ovate lodicules. *Stamen* is free, and filament is short; *anthers* are yellowish; the *ovary* is ovoid-globose, surmounted by short thick style, two *stigmas* or three plumose.

Table 6.1 Estimation of flowering cycle in *Melocalamus compactiflorus* from the available flowering records

Country/locality	Flowering date (calendar year)	References of the flowering dates	Estimated flowering cycle (year)
Myanmar Martaban, Karen hill Arakan, Yoma Ruby Mine District Bhamo, Byingyi Thingyan Division Hlaing Yoma hill	1871 1878 1894, 1902 1917, 1920 1926 1933	Gamble (1896) Gamble (1896) Gamble (1896) Bahadur and Naithani (1983)	55 (1878–1933)
Bangladesh *Cox's Bazar* (western border area of Arakan) Cox's Bazar (Paner chara block)	1928 1974–1975	Trevor (1928) Banik (2000)	50 (1878 of Arakan–1928) 47 (1928–1975)
India locality not stated *Assam* Cachar *Assam*	1902 1921 2014	Blackwell (1902) Bahadur and Naithani (1983) Das et al. (2014)	53 (1921–1974) 50 (1974–2014)
Dehradun FRI *Manipur* *Manipur*, Tamenglong hill *Mizoram* (northeastern border of Cox's Bazar, Bangladesh)	1980 1974 2003 1981	Bahadur and Naithani (1983) Malick (1974) Author observed and collected seeds in 2003 Ram and Gopal (1981)	53 (1921–1974) 53 (1928–1981)

Seed

It is large, subglobular, 2.5–3.8 cm broad (diameter), weight varied from 3–12 g, fleshy, like chestnut, and glossy green to brown, with the base supported by the persistent glumes and pericarp that are fleshy and thick (Fig. 6.1d).

Seed Germination

The seed germinates promptly, 30–50 %, and is short lived but some may remain viable upto 6 months when stored inside dry sand as reported in *Melocanna baccifera* (Banik 1998b, 2000); raised seedlings in 1974–1975 at BFRI (Bangladesh Forest Research Institute), Chittagong. Seed germination often takes place before falling from the plant (Bahadur and Naithani 1983).

Seedling Character

Comparatively thick plumule, usually 1, may be 2–4, and somewhat succulent stem, with bigger leaves.

Vegetative Propagation Methods

The guidelines for producing culm cutting, collection and bagging of wild bamboo seedling and their nursery management are provided in Appendix I, and for details consult the INBAR Technical Report No. 6 (Banik 1995).

It was observed in *M. compactiflorus* species culm segments with two to three nodal buds are planted in the soils mixed with sand (1:3) which produce rhizomes, roots and culms within 6–9 months. However, when two to three node segments (having fresh buds) after treating with rooting hormone (IBA 200 ppm, NAA150 ppm) were placed in sand propagation bed, they produce 55–70 % rooted planting materials within 6–7 weeks under intermittent misting.

Diversities and Conservation

No systematic exploration on the ecological diversities and genetic diversities of *Melocalamus compactiflorus* have been undertaken in its natural habitat. Due to clear felling of high forests and tall trees, the habitat of this climbing bamboo has been mostly destroyed, and as a result the species could be classified as rare and thus potentially threatened (Banik 1994a). Immediate action should be taken to protect the natural belt. The supporting trees should not be felled in the natural belt of *Melocalamus compactiflorus* as the species exhibits clambering habit.

As, for example, in Bangladesh, the distribution of the species is mostly confined to the area of Teknaf and Panerchara forests of Cox's Bazar. Even 50 years before (late 1960s), *M. compactiflorus* was a common bamboo species to be found in these forests. The species flowered and died gregariously during 1974–1975 in these forests of Cox's Bazar. The natural regeneration of this bamboo was seriously hampered during 1978–1980 because large areas of these forests were clear felled and burned for making plantation of oil palm and fast-growing exotic tree species such as *Eucalyptus* sp., *Acacia auriculiformis* and *A. mangium*. As a result, this bamboo species has been eliminated from most of these natural belts (Banik 2000).

For threatened bamboo species, protection of their natural populations in situ is apparently the only feasible conservation technique at present, as ex-situ conservation techniques are currently not practical for bamboos (Stapleton and Rao 1996).

6.1.3 Cultivation and Management

Planting sites of *M. compactiflorus* should be selected along the fringes of perennial streams near the semi-deciduous trees so that the plant can have sufficient moisture and partial sunlight. The site should be under heavy rainfall areas in protected forests.

The spacing, as such, cannot be fixed for making the pits as the plants need support from trees nearby. However, 2×2 to 3×3 m may be maintained as spacing.

6.1.4 Specific Spot Characters of *Melocalamus compactiflorus* for Field Identification

1. Climbing bamboo with solid culm (Fig. 6.1a).
2. Culm nodes somewhat swollen with whitish pubescent band at the base.
3. The basal internodes are straight and then of scorpoid habit.
4. Perpendicular and angular branching (Fig. 6.1b).
5. Seeds are big and chestnut-like (Fig. 6.1d).

Chapter 7
Melocanna Trin.

Melocanna roots from the Greek word *melon*, an apple, and Greek word *kanna*, a reed, which means a reed-like bamboo with apple-like fruits. The pear-like fruit of *Melocanna* has attracted the attention of many botanists. However, it is suggested that the fleshy fruit may have been derived independently in several sub-tribes including *Melocanninae* (Soderstrom and Ellis 1987). In the field the species can be recognized by its open clump and erect culms with persistent culm sheaths that have long narrow imperfect blades. The genus is a native to the northeast region of India, Chittagong Hill Tract and Arakan Range and also introduced in pilot plantations in other parts of subcontinent and in botanical gardens and eco-parks in Southeast Asia, Royal Botanic Garden at Kew, London, Europe and North America.

7.1 *Melocanna baccifera* (Roxb.) Kurz

(Synonym: *Melocanna bambusoides* Trin.)

[*Melocanna baccifera* (Roxb.) Kurz, Prelim. Report For. Fl. Pegu, Append. B. 1875; Skeels in US Dept Agri Bur Pl Industr. Bull 223:50. 1911.; *Beesha baccifera* (Roxb.) Kunth in Jour. de Phys. Chim. Hist. Nat. 95:151. 1822. *M. bambusoides* Trin. In Spreng. Neue Entdeck. 2:43. 1821.]

Roxburgh (1814, 1819) presented the original description and illustration of the species, as *Bambusa baccifera* Roxb. Trinius (1821) first described the genus *Melocanna* and named the species as *Melocanna bambusoides* Trin. On the basis of *Bambusa baccifera* Roxb., the correct binomial was published by Kurz (1875) as *Melocanna baccifera* (Roxb.) Kurz

R.L. Banik, *Silviculture of South Asian Priority Bamboos*, Tropical Forestry,
DOI 10.1007/978-981-10-0569-5_7

7.1.1 General Information

7.1.1.1 Vernacular and Local Names

Berry bamboo (English); *muli* (Tripura, West Bengal, India); *wathwi* (Kokborok, Tripura); *watrai* (Garo, Meghalaya, India); *usylli* (Khasi Meghalaya); *tyrlaw* (Jaintia Meghalaya, India); *tarai*, *muli* and *wati* (Assam, India); *saneibi* and *moubi* (Manipuri, India); *rieng* [Rongmai Naga,Tamenglong (Manipur)]; *turiah* (Nagaland, India); *mautak mau* (Mizo, India); *tador dort* (Arunachal, India); *philinge bans* (Eastern Nepal), *lahure bans* (Western Nepal; no common name known from Central Nepal); *lahure bans* (Bhutan); *muli*, *nali*, *tengra muli*, *paiya*, *bazali* and *ekuzha* [Bangali, Chakma, Chittagong Hill Tracts (CHT), Sylhet and Mymensingh forests, Bangladesh]. Various tribes inhabiting in CHT have different names for *Melocanna baccifera*. The *Bawm* tribe calls the species as *mautak maau*; Chakma tribe, *ekuzha banz* and *egojya bansh*; Lusai, *mau hrau*; Marma, *kaiang waah*; Murong, *kawoo thum*; Tanchangya, *paba baith*; and Tipara, *Hruthui yeaha*. The Magh tribe of Myanmar calls the species as *kayoung wa* (Banik 1998a). Thus it appears local ethnic people have been identifying bamboos by their own 'binomial'—one 'generic' and the other is 'species' name. However, the name *muli* is common and well known to most of the hill-tribe people and plain land Bengalis in Northeast India, West Bengal and Bangladesh (CHT, Sylhet, north Mymensingh).

7.1.1.2 Natural Distribution and General Habitat

Melocanna baccifera is naturally growing throughout the hill forests of Bangladesh, Myanmar and northeast (Assam, Arunachal, Meghalaya and eastern part of Tripura, Nagaland, Manipur, Mizoram) of India. In fact, no natural belt of the species is found in Arunachal Pradesh (AP); the migrated Chakma tribe from CHT have been seen to cultivate the species in Manabhum RF under Lohit district in Diyun Circle, southern part of AP. In East Khasi Hills, Jaintia Hills and Ri-Bhoi districts, it is rarely planted by the local people. There is also gregarious distribution of this species from East Sikkim and northern part of West Bengal. The 'natural home' of *M. baccifera* is believed to be CHT where this species grows gregariously covering large tracts of land (McClure 1966). The species occurs as an undergrowth to many tree species and also forms a pure stand by aggressive nature of its underground rhizome in areas after burning. The plant thrives satisfactorily on moist sandy, clay loam alluvial soils, well-drained residual soils, sandy rough slopes and the top of the hills. It indicates the wide adaptability and hardy nature of the species. However, the culms are very healthy and tall when grown in the lower slopes and valleys with rich moist soil as observed in Mamit and Lunglei of Mizoram.

It is also cultivated in southeastern Tarai part of Nepal and southern border of Bhutan, probably introduced from Bangladesh; occasionally cultivated or introduced and planted in many botanical and private gardens all over the world, including Hong Kong, Indonesia, Taiwan and South America. Author assisted in introducing this species into the Botanic Gardens of Singapore and Makiling Botanic Gardens of Los Baños, Philippines, through the seeds collected from CHT during 1994 and 1997, respectively (Banik 2010).

7.1.1.3 Climatic Condition

Melocanna baccifera thrives satisfactorily on moist sandy, clay loam alluvial soils, well-drained residual soils, sandy rough slopes and the apices of hills. The species can also grow from highly weathered deep clay soil to shallow and to very deep loamy soil, having pH 4.5–6.0. The species naturally thrive well and grows luxuriantly in the area (Northeast India, CHT and Arakan, Pegu) with annual rainfall of 2500–6000 mm and an average temperature maximum of 37.0 °C and a minimum of 5.0 °C with a long dry season each year from November to March. The species grow quite well on the hills of Tamenglong, Manipur, even up to 1600 m in height. Moreover, a number of introductory field trial conducted during 2003–2010 through seeds of *muli* have shown very satisfactory growth in areas far away from its natural habitat Northeast India to other localities in India, such as Uravu, Wyandu (Kerala), Sindhudurg (Goa, Konkan) and Dehradun (Banik 2010).

The species has rarely been cultivated in the homesteads of plain tract of Assam, Tripura and Bengal due to its intolerance to flood, as the species cannot survive flooding for more than 2 weeks (Banik 2000). Besides the farmers generally avoid growing *muli* bamboo in their homestead gardens as the species outnumbers other crops and trees due to its aggressive (running) nature of culm production and clump expansion behaviour. Except in a few hilly villages at CHT and Northeast India, *M. baccifera* has rarely been tried for domestication in the homesteads of plains, and as such most of the populations of the species are in wilderness.

7.1.1.4 Uses

Although thin walled, the culms of *M. baccifera* (common name muli bansh) are strong and naturally durable, and it has the great advantage of being straight and having only very slight knots. Therefore this bamboo is much used for roofing, thatching and matting in house construction in Northeast India, Bangladesh and Myanmar. It has been estimated that 80 % of rural houses existing in this vast tract of land are made of bamboo, and about 90 % of a house is constructed by muli culms. In this context Gamble (1896) described this species as 'one of the most universally used for building purposes'. He further reports 'Major Lewis says that white ants [termites] do not touch it'. The species is widely used for fencing mats, granary walls, house wall, poultry and piggery cages, cattle sheds and roof ceiling. This bamboo has great demand in cottage industries, cloth hanging rod, pulp and paper and rayon mills in India, Bangladesh and Myanmar. About 15 to 50 million culms are being used annually for the production of pulp and rayon, respectively, in Bangladesh. About 80 years ago, Raitt (1929) documented the possibility of making pulp from *M. baccifera* (muli bamboo) and also enclosed a sample piece as an exhibit in the Indian forest records. The fibre characteristics and the pulp and paper making properties of muli bamboo are reported as fibre length 2.78 mm, fibre diameter 15.60 μm, lumen diameter 3.55 μm, alkali used 25 %, screened pulp yield 43.8 %, breaking length 5480 m, burst factor 40.0 and tear factor 210.7 (Varmah and Bahadur 1980).

Oye and Mizuno (1967) applied a prehydrolysis sulphate process to *muli bamboo wood* and produced quality rayon-grade pulp with α-cellulose 95.0 %, β-cellulose 4.0 %, pentosan 3.0 %, extracts 0.03 %, ash 0.068 %, CaO+MgO 0.24 % and brightness 91.

With the tests and analyses conducted by Netpro at the Indian Institute of Science (IISc), Bangalore have shown that 1.2 kg of *muli* bamboo biomass can generate 1 kwh of electricity.

Young shoots of the species are edible and tasty. Moreover, smoothness or fewer amounts of hairs on the sheaths of the shoots in this species is an advantage, since the shoots with hairs in other local species are difficult to handle. The hairs often tend to stick in the fingers as the sheaths are removed from the shoots. The shoots are yellowish green to yellowish brown; sheath margin and top pinkish; ligule horseshoe shaped; blades flagellate and glabrous; and the shoots are somewhat slightly bitter to sweet in taste and pleasant flavour at raw state (Banik 1997a). The weights of sheath cover and edible portions of *muli* shoot (Fig. 7.1n) along with other edible characters are described in Table 7.1.

In Mizoram, Tripura and CHT, local market *muli* bamboo supplies about 55 %, 45 % and 75 % edible shoots, respectively, out of the total available shoots of different bamboo species.

7.1.1.5 Ethnic Utilization

The tender shoots are very much liked by tribal people in CHT, Mizoram, Manipur and Tripura. The local preparation is as follows: chilli (dry or green), fish (dry or fresh), bamboo shoot or any vegetables and salt are put inside the internode of the bamboo and then mixed thoroughly by beating with a bamboo stick. The paste is served as sauce/pickle in the dishes. *Tripuri* people of CHT called it as *moso pring chenai* (Banik 1998a). To get cure from the bite of any animal, paste is prepared from the tip of a very young culm of this bamboo and placed on the bite spot. When there is a deep cut/injury, the green *Melocanna* bamboo skin is lightly scuffed and put on the injured spot, and then it is covered by a bandage. Patient should drink water sufficiently. It helps in stopping the bleeding and joining the cut without any stitch. The *Tipra* people locally called it *yeaha thia uchua*. The creamy white-coloured embryo of *Melocanna* bamboo seeds is taken as medicine for strength.

The Dimasas, also known as Dimasa Kacharies, constitute the largest ethnic group in North Cachar Hills and linguistically belongs to the Tibeto-Burman group. This ethnic people call *M. baccifera* as *wa-thi* and uses culm for water purification rituals especially during *Misengba* ceremony. The offering of rice beer (*judima*) in cups made from the culms to the deities in various religious ceremonies is a common practice (Medhi and Borthakur 2013).

Fig. 7.1 (**a**) Mature, well-developed clumps of *M. baccifera* are diagnostically open and diffuse in nature. (**b**) The rhizome is pachymorph and diffused with widely elongated rhizome necks. (**c**) Thin wing-like growth is present on both left and right margins of the bud. (**d**) Dense tuft of slender and thin subequal branches 25–40 develop from each node of the culm in *M. baccifera*. (**e**) In culm sheath ligule is horseshoe shaped and blades are flagellate. (**f**) Purple-coloured stigmas, the major portion of which is exposed. (**g**) Flowering and fruiting take place simultaneously. (**h**) Fruit is big, green, smooth and not solid but possesses a seed cavity filled with endosperm towards the stalk end. (**i**) In *M. baccifera* juvenile leaves are bigger one and half times more in size than those of adult clumps. (**j**) In seedlings of *M. baccifera*, rhizome becomes well developed at 6 months of age. (**k**) The shoot types possessing a yellowish culm sheath are usually preferred as edible shoots, while in others the sheaths are comparatively deep brown and have bitter and astringent taste. (**l**) Seedlings are raised at the nursery in April to May by sowing seeds and maintained and outplanted during July to August. (**m**) The harvested bamboos of *M. baccifera* and other bamboos are transported through rafting in the hilly streams and rivers in Chittagong Hill Tracts and also in Northeast India. (**n**) The longitudinal section of young shoot shows the internal edible creamy-coloured meat portion. On the basis of fresh weight, the outer sheath cover amounts to 44.2 %, and the remaining 55.8 % contains edible inner meat portion of a shoot

Fig. 7.1 (continued)

Fig. 7.1 (continued)

Table 7.1 Quality of edible shoots of *Melocanna baccifera*

Quality of edible shoots of *Melocanna baccifera*						
	Fresh weight %		Edible portion (meat)			
Basal diameter (cm)	Sheath	Edible part	Taste	Flavour	Colour	Texture
4.7	44.2	55.8	S. Bt – Sw	Sw – As	Cw	Cr – Tn

Note: taste and flavour: *S* slight, *Bt* bitter, *Sw* sweet, *As* astringency
Colour: *Cw* creamy white; texture: *Tn* tender, *Cr* crisp

7.1.2 Plant Data

7.1.2.1 Vegetative

Plant Habit

Melocanna baccifera culms are erect, arising singly in a variety of distances from a common creeping rhizome and young clumps (up to 3–4 years of age) usually densely caespitose at first, becoming more open on the periphery as they develop. Mature, well-developed clumps diagnostically open and diffuse in nature (Fig. 7.1a). The rhizome is pachymorph and diffused, with widely elongated, slender and obterete necks (1.0–2.0 m) (Fig. 7.1b; Banik 2000) which have the ability to spread and quickly cover the vacant space of the hills by producing culms provided the plant is not heavily disturbed. Therefore, the rhizomes of many different clumps of *M. baccifera* intermingle with each other forming an underground network, and in the forest it is impossible to demarcate the boundary of a clump. This condition usually makes impossible for other plant species to invade the area. Such underground rhizome network helps protect soil erosion in the hills of Northeast India, some parts of lower Tarai, CHT and Myanmar and from heavy monsoon rain (Banik 1997c, 2010). Thus, from ecologic, social and economic viewpoints, this bamboo is an important multipurpose resource for this region.

Culm Height

It is upright and straight 8–22 m. The young culms (less than 1 year old) are completely unbranched; apical thin part is drooping with four to six leaves, and persistent straw-coloured culm sheaths on the nodes are clearly visible from a distance during October to April.

Culm Diameter

This commonly varies from 3 to 7 cm. In Manipur tall (18–22 m) and large-sized culms (7–11 cm diameter at breast height) are observed on hills in both sides of the river Barak while going to Aziuram from Tamenglong (Banik 2010).

Culm Thickness

The culms are hollow and thin walled; the average thickness at mid-culm zone is 0.3 cm, very thin (0.16 cm) at the top part, but thick walled (2.3 cm) in the two to four basal nodes.

Culm Colour

It is grassy green and turns into straw colour with age.

Internode Length

The internode is 10–60 cm with a maximum elongation at mid-culm zone, and nodes are not raised. The bud on the culm node is flat and appressed with the culm wall. Thin wing-like growth is present on both left and right margins of the bud (Fig. 7.1c).

Branching

Dense tuft of slender, thin subequal 25–40 branches develop from each node (Fig. 7.1d) after 6–9 months of emergence. The degree of precocious branching is pronounced and confined from mid to upper part of the culms.

Leaves

It is oblong-lanceolate, 15–30 cm long, 2.5–5.0 cm broad and apex acuminate; base is oblique or somewhat truncate; adaxial surface is glabrous, and abaxial surface is glabrous or sometimes sparsely pubescent; margins are finely ciliate. Leaf sheath is thick with ash colour; it has prominent very long erect wavy bristles and 8–25 mm long auricles; the ligule is short, and petioles are narrow. The amount of leaf falls from *M. baccifera* is 6.0 ton per hectare.

Culm Sheath

It is fairly large 10–15 cm long, yellowish green when young and yellowish brown when old, brittle, felted, symmetric and narrow; auricle is prominent, and blade is arising from a concave depression. Culm sheaths are covered with white hairs at first, and the ridge is present on the outside of the sheath where the blade is attached; blade is narrow, usually 15–30 cm long, 2–3 cm broad, sword shaped and longer than the sheath. *Ligule* is serrate and 1–2 mm tall. In young shoot ligule is horseshoe shaped and blades are flagellate (Fig. 7.1e).

Vegetative Growth

A clump of *M. baccifera* may start producing culms any time between May to mid-November. The culm emergence is maximum in July to mid-September. The culm completes elongation within 45–60 days of emergence; rate of elongation may be as much as 40 cm/day (Banik 1993a). In 5–7 years of age, a clump usually produces 7–15 culms per year; even some clumps have been found to produce 62–78 culms annually. Unlike thick-walled bamboo species, the natural mortality of emerging culms in *M. baccifera* was found to be very low, 8–10 % only (Banik 1983a). Soon after emergence, culms elongate very slowly up to 1.0–2.0 m within 20 days and then gradually gain speed until they attain the optimum size, and thereafter the rate of elongation gets slowed down quickly. The culm elongation mostly takes place in the night. In Chittagong the maximum elongation is observed to be at 44 cm in a day (Banik 2000). It was further observed that every day only the middle internode elongated maximum and moved gradually in successive order from the base to the tip. The lower internodes to the middle one rarely elongated (Banik 1999b). Total culm elongation period was 55–60 days. The diameter with which culm emerges remains unchanged throughout its life. The size of the diameter of a culm is determined by the size and vigour of the bud present in the mother rhizome from where it originates.

In *M. baccifera* culm sheaths start dislodging basipetally and bud break progresses up to one-third of the culm. Sheaths persist on the lower two-third of culm and sheath blades are loosely fitted. In the next year, culm sheaths further dislodge, and bud break and branching continue up to two-third portion of the culm. Most of the time, nodal buds remain dormant on lower one-third portion (three to seven basal nodes) of culm with persistent culm sheaths. The natural life of culms in a well-developed clump may vary from 5 to 11 years. Elongation of rhizome neck of *M. baccifera* goes on below the ground either slowly or actively throughout the year irrespective of seasonal variations (Banik 1999b).

Physical and Mechanical Properties of Culm

The values shown in Table 7.2 are of 3-year-old culms as at that age *M. baccifera* showed best values in all the parameters. However, maximum compressive strength was found in a 4-year-old culm, so the values of this parameter are presented in Table 7.2 only for that age of culm.

The highest MOE (modulus of elasticity) and MOR (modulus of rupture) were observed in *M. baccifera* than those of other three forest-grown natural associate bamboo *B. tulda* (local name *mitinga* or *jati*), *D. longispathus* (local name *ora* or *rupai*) and *Gigantochloa andamanica* (local name *kali*, *kallayai* or *kalichari*); so the species is the most utilized bamboo species in matting, thatching, roofing, walling, etc., in the region.

Bamboo (*M. baccifera*) mat plywood with the length, width and thickness of 2.4 m × 1.2 m × 7 mm was made using liquid urea formaldehyde resin. The physical and mechanical properties of such plywood were compared with the existing market plywood in Bangladesh. The average density of bamboo mat plywood was much higher compared to the existing market plywood (Ashaduzzaman et al. 2011). The MOR and MOE of *Melocanna* bamboo mat plywood were three and six times higher compared to commercial timber (*Bombax ceiba*) plywood. Interestingly, the specific MOR and MOE were significantly higher than those of market plywood. The high strength values might be due to the long fibre length of the bamboo.

Table 7.2 Some physical and mechanical properties of culms of *Melocanna baccifera*

Parameters	Culm position		
	Top	Middle	Bottom
Moisture content (%)	71	88	102
Specific gravity (based on green weight)	0.64	0.60	0.55
Specific gravity (based on oven dry weight)	0.75	0.71	0.70
Shrinkage in wall thickness (%)	4.8	5.6	9.5
Shrinkage in diameter (%)	2.9	4.1	5.4
Compressive strength, parallel to grain, green (kg/cm^2)	460	419	391
Compressive strength, parallel to grain, green air-dry (kg/cm^2)	647	611	575
Modulus of elasticity (MOE); green ($1{,}000/cm^2$)	239	191	178
Modulus of elasticity (MOE); air-dry ($1{,}000/cm^2$)	281	228	188
Modulus of rupture (MOR); green (kg/cm^2)	622	647	728
Modulus of rupture (MOR); air-dry (kg/cm^2)	687	700	782

Source: Sattar et al. (1991)

7.1.2.2 Reproductive

Flowering Nature

The species flowers gregariously in nature; flowering pattern proceeds like a 'wave' from one area to another covering large tracts of land. Flowering wave may continue for 10–15 years; isolated clumps may also rarely flower. The species has more than one flowering genotypes in the population; some are exhibiting flowering cycle of 30 ± 5 years, others 45–48 years and a few 60–65 years (Table 7.3).

Inflorescence

The inflorescence is a large compound panicle, usually remains drooping in nature. Thin leafy branches produce leafless elongated floral shoots at the apical region. The floral buds are borne at nodes along one side of the axis of a floral shoot, and thus the pseudospikelets are produced on one side of the branches. The culms in a clump do not always flower synchronously; some clumps may have culms in two or more different phenological states. The *spikelet* is cylindrical and about 40–60 mm long. *Palea* is convolute, mucronate and seven veined. *Lodicules* are two and fimbriate. The *ovaries* in florets are diagnostically prominent, are visible and have elongated styles. The ovary is ovoid and narrowed upwards into an elongated style which is divided into two to four hairy, recurved and purple-coloured *stigmas*, the major portion of which is exposed (Fig. 7.1f). *Stamens* are five to six, and filaments are short, free and 7–9 mm long. *Anther* is yellow, obtuse and 2–4 mm long, not much exposed or emerged out. Pollen grains are globose, with a wide variation in diameter size, from 41 to 86 μm.

Table 7.3 Estimation of flowering cycle in *Melocanna baccifera* from the available flowering records

Country *states/provinces*	Flowering date (calendar year)	Reference of flowering dates	Estimated flowering cycle (years)
Myanmar: *Arakan*	1863–1866 (Greg.), 1900, 1902, 1904–1905 (Spor.)	Kurz (1876), Brandis (1906), Troup (1921)	36–39, 35–37 (1863, 1866–1900, 1902)
Arakan, Yoma Prome, Henzada	1910–1913 (Greg.)	Troup (1921)	44–47 (1863, 1866–1910, 1913)
Arakan	1915–1916 (Greg.)	Troup (1921)	44–52 (1863–1909, 1916)
	1909–1916	Raitt (1929)	45–51 (1864–1909, 1916)
Arakan (some part)	Up to 1928 not flowered since 1866	Raitt (1929)	About 60 (1866–1928)
Rakhine	1933	Htun (1999)	
Chin Hills	1915–16, 1960	Htun (1999)	44–45 (1915, 1916–1960)
Place not reported	1900, 1957–1959 (Greg.)	McClure (1966)	About 60 (1900–1959)
All over country	Up to 1998 not flowered after 1957	Htun (1999)	More than 41 (1957, 1960–1998)
Bangladesh: *Chittagong (south)*	1864 (Greg.) 1901–1905 (Spor.), 1908–1912 (Greg.)	Troup (1921)	44–48 (1864–1908, 1912)
	1952 (Spor.), 1958–1959 (Spor.)	Hossain (1962)	51–54 (1901, 1905–1952, 1959)
	1957–1959 (Greg.), 1960–1961 (Greg.) 2002–2007 (Spor./Greg.)	McClure (1966), Hasan (1973) Author observed	48–52 (1908, 1912–1960, 1961) 46 (1961–2007)
Cox's Bazar	1908–1912 (Greg.), 1959–1960 (Spor.), 1960–1961 (Greg.) 2001–05 (Spor./Greg.)	Troup (1921), Hossain (1962), Hossain (1962) Author observed	48–53 (1908, 1912–1961) 45 (1960, 1961–2004, 2006)
Chittagong (north)	1863–1866 (Greg.), 1901–1905 (Spor.)	Kurz (1876), Gamble (1896), Brandis (1906)	35–38 (1863, 1866–1901, 1905)
Hathazari, Nazirhat Fatehabad	1988–1997 (Spor.)	Banik (1998b)	28–32 (1960, 1961–1988, 1997)
Chittagong Hill Tracts (CHT)	1863–1866 (Greg.), 1927 (Spor.)	Kurz (1876), Brandis (1899), Hossain (1962)	61 or (30 × 2) = (1866–1927)
Rangamati	1930 (Greg.), 1935 (Spor.)	Nath (1930), Hossain (1962)	64 or (32 × 2) = (1866–1930)
Kassalong	1958–1959 (Spor.), 1959–1960 (Greg.)	Hossain (1962), Hasan (1973)	31 (1927–1958) 30 (1930–1960)

Bangladesh: *Chittagong Hill Tracts* (*CHT*) *north*	1957–1958 (Spor.), 1958–1961 (Greg.) 1998 (Spor./Greg.) 2005–2007 (Greg.)	Hossain (1962) Banik (2000, 2010)	36 (1961–1988)
Kamalchari, Hyanko Haludia, Fatickchari Naranhat, Shisak	1986–1997 (Spor.)	Banik (1989, 1991a, 1998b)	29–39 (1957, 1958–1986, 1997)
India: *Tripura* (*south*) Bagafa, Sabrum (border CHT north, Hyanko), Silachari Killa, Matabari Sabrum	1958–1959 (Greg.) 1995–2003 (Initial Spor.) 2002–2007 (Greg.) 2008 (Greg.) 2016 (Final sporadic)	Banik (2004a) Banik (2004a) Banik (2010) Banik (2010) Author observed	35 (1959–1995) 43 (1959–2003)
Mid-Tripura (*west*) *Agartala*, Shekarkot Bangla khal, Adnagar Indranagar, *Narsingarh*, Tulakona, Anada Nagar, Duke, Jampuijala, Kamalghat, *Mohanpur*	2000–2005 (Spor./Greg.) 2002–2003 (Spor.) 2004–2007 (Greg.) 2005–2007 (Greg.) 2006–2008 (Greg.)	Banik (2010) Banik (2004a) Banik (2010)	40–47 (1959–2000, 2003, 2005, 2007, 2008)
Khowai, Kalyanpur, Teliamura, Nuna Chara	2003; 2006–2007 (Greg.)	Banik (2010)	44 (1959–2003)
North Tripura: Ambassa, Ratacherra Kumarghat, Kanchanbari, *Pecharthal* Sunaimuri, *Panisagar*, Attaramura, Longtrai	2004–2006 (Spor./Greg.) 2004–2006 (Spor./Greg.) 2006–2008 (Greg.)	Banik (2010)	45 (1959–2004) 47 (1959–2006)
Assam *Jorhat*	1863–1866 (Greg.) 1892–1893 (Spor.) 2000–2001 (Spor.)	Gamble (1896) Brandis (1899), Tripathi (2002), Pathak (2002)	27–29 (1863, 1866–1892, 1893)
Lushai hills, Assam Garo and Khasi Hills	1864 (Greg.) 1889 (Greg.) 1900, 1902 (Spor.)	Brandis (1906) Troup (1921) Parry (1931)	25 (1864–1889)

(continued)

Table 7.3 (continued)

Country *states/provinces*	Flowering date (calendar year)	Reference of flowering dates	Estimated flowering cycle (years)
Garo Hills, *Cachar Hill* Lushai hills, *Assam*	1910–1912 (Greg.), 1917–1919 (Spor.)	Hossain (1962) Hossain (1962)	19–21 (1889–1910, 1912) 26 (1892, 1893–1917, 1919)
Meghalaya Garo Hills (south)	1917–1919 (Spor.) 2004–2008 (Spor./Greg.)	Author observed	88 or (44 years × 2) (1919–2007)
Cachar Hills, Halflong or *Haflong*	1952–1956 (Greg.), 1967 (Spor.), 1988 (Greg.)	Gupta (1988)	40–44 (1910, 1912–1952, 1956) 32–36 (1952, 1956–1988)
Barak Valley	1912, 1958 (Greg.) 1997, 2002 (Spor.)	Yadava (2002) Banik (2004a)	46 (1912–1958) 39, 44 (1958–1997, 2002)
Surma Valley	1958–1961 (Greg.)	Hadfield (1958), Nath (1959, 1960, 1962)	48–49 (1910, 1912–1958, 1961)
Lushai hills (presently *Mizoram*)	1911–1912 (Greg.) 1958 (Greg.)	Troup (1921) Yadava (2002)	46–47 (1864–1911, 1912) 46–47 (1911, 1912–1958)
Mizo hills, *Cachar Hills Assam* (famine due to gregarious flowering of *muli* is called *mautam*, and *tam* means famine) *Mizoram* (*mautam*)	1815 1863–1866 (Greg.) 1892–1893 (Spor.) 1900–1902 (Spor.) 1910–1912 (Greg.) 1933 (Spor.) 1959–1960 (Greg.)	Lalnuntluanga et al. (2003) Chatterjee (1960) Troup (1921) Chatterjee (1960)	42–46 (1815–1863, 1866) 26–30 (1863, 1866–1892, 1893) 47 (1863–1910, 1912) 31–33 (1900, 1902–1933) 48–50 (1910, 1912–1960)
Kolasib Sabual, Mamit	1959–1960 (Greg.) 2001–2004 (Spor.) 2005–2008 (Greg.)	Chatterjee (1960) Anon (2007) Banik (2010)	45–49 (1959, 1960–2004, 2008)
Manipur Tamenglong (near Barak Valley)	1967 (Spor./Greg.) 2003–07 (Spor./Greg.)	Nath (1968) Banik (2004a, 2010)	48–50 (1917–1967) 40–44 (1958, 1967–2003, 2007)
Dehradun: New Forest (seeds introduced in 1892; reported by Gamble 1896)	1892 1958–1960 1991–1992	Gamble (1896) Vaid (1972) Sharma (1992)	66 or 33 years × 2 = 1892–1958 so might flowered 1925–1930 was unnoticed 31–34 (1958, 1960–1991, 1992)

West Bengal: Siliguri Sebak, Duars	1960 2006–2008	Chatterjee (1960) Banik (2010)	46 (1960–2006)
Calcutta Botanic Garden (introduced date not known)	1863–1866	Gamble (1896)	
Bangladesh: *Sylhet*	1910–1912 (Greg.), 1957–1958 (Spor.), 1958–1959 (Greg.)	Troup (1921), Hossain (1962), Hasan (1973)	46–47 (1910, 1912–1957, 1959)
Sylhet: Dholaichara, Madhav, Chara, Putijuri (border of Ambassa, *Tripura*)	1996 (Spor.), 1997 (Spor.) 2003–2008 (Spor./Greg.)	Banik (1998b, 2000) Banik (2004a, 2010)	40–41 (1957–1996, 1997)
Mymensingh: North Forest, range of Rasulpur (border to south *Garo Hill*, *Meghalaya*)	1974 (Spor.) 1975 (Spor.)	Hasan (1973) Banik (1998b)	63–64 (32 × 2) (1910, 1912 of Garo Hills, 1974, 1975)
Nepal – in eastern and central districts	2009 (first time report from Nepal)	Shrestha (2009)	
Sri Lanka Royal Botanic Gardens, Peradeniya (introduced from Assam in 1910)	1957–1958	Ramanayake and Weerawardene (2003)	47–48

Note: *Spor.* sporadic flowering, *Greg.* gregarious flowering

Soon after blooming in the apical floral shoot, all the leaves below it on the branches turn yellow and gradually wither. The buds on the axils of the withered leaves then start producing short panicles; both flowering and fruiting take place simultaneously (Fig. 7.1g). Accordingly within a few weeks, the main and secondary branches become leafless and form a large compound panicle. Thus finally all branches on the culms become leafless and produce flowers. All the leafless flowering culms in a clump look like a giant inflorescence. The clumps of *M. baccifera* take almost 1 year to complete flowering and then die.

Complete flowering clumps of *M. baccifera* do not produce any culms in the current year of flowering. All the flowering clumps under study (1990–1991) at Chittagong Hill Tracts did not produce any new culms or produce a very few numbers of culms. However, in the previous years of flowering year, i.e. during 1989, 1988 and 1987 (1, 2 and 3 proceeding years of flowering), the average culm production per clump was 8.0, 17.0 and 12.0, respectively (Banik 1991a). Thus it appears that flowering clumps usually do not produce any new culm, but in the immediate preceding year culm production decreases significantly. Similar situation was observed in Mamit area of Mizoram where flowering started sporadically in 2006 and gregariously during 2007–2008. Before flowering in 2005, the culm number of more than 3 years old, 2 years old and 1 year old in the current year per hectare was found as 11100 (in 2001), 2850 (in 2002), 5660 (in 2003), 4500 (in 2004) and 4310 (in 2005), respectively. In 2006 the number of new culms decreased to 3000 only per hectare, and all of them were very thin, 1.0–2.0 cm in diameter (Shibata et al. 2008). The culms of various age groups including 9–12-month-old young culms also flowered at a time in the same clumps.

The florets of *M. baccifera* are mostly open in the morning up to 10 a.m. Generally upper floret opens first and the lower floret opens after 2–5 days (Banik 1998b). Anthers come out in the morning and burst in the afternoon at 3 p.m. to 5 p.m.

Fruit (Seed)

Kurz (1876) described the fruit of *Melocanna baccifera* as 'a large pear-shaped and fleshy' and 'berry like'. Later the fruits of this species have been termed as '*bacciform caryopsis*' or '*bacoid caryopsis*' (Tsvelev 1976). In botanical Latin '*bacca*' means a berry, '*bacciformis*' means shaped like a berry and '*bacciferous*' means bearing berries. Fruits are green with smooth surface, large, obliquely ovoid, onion shaped, thick and fleshy at stalk end, with the apex terminating in a curved beak not covered with any glumes. Variations in size and weight exist among the fruits of a population. In a sample of 1000 fruits collected from Hyanko forest area (CHT), the length, diameter and weight varied from 35 to 110 mm, 22 to 60 mm and 7 to 151 g, respectively. In one case individual fruits weighed 300 g. However, the seeds are generally within the weight range of 40–75 g. One medium-sized full-grown clump produces 5–7 kg of 'seeds' in one flush and 25–40 kg within the whole flowering period before dying. The fruit has thick

(7–13 mm) white cream colour fleshy fused pericarp and mesocarp filled with starch. The fruit is not solid but possesses a seed cavity towards the stalk end (Fig. 7.1h). The endosperm of much reduced embryo is with relatively large scutellum containing starch grains.

Seed Germination

Mature seeds of *M. baccifera* germinate well (60–80 %) under partial shade usually within 3–7 days. In addition to normal germination, species also exhibit 'vivipary' type of germination especially at the end of fruiting period and later part of rainy season (Banik 1991a); the seeds produced in May to June were comparatively bigger (length 6.9 + 0.3 cm, diameter 4.1 + 0.2 cm) and heavier (weight 55.3 + 5.46 g) and also germinated with higher rate (79.8 + 6.5 %) than those produced in September. Seeds produced in September were small (length 5.2 + 0.5 cm, diameter 3.3 + 0.4 cm), were light in weight (17.8 + 2.6 g) and had poor germination capability (46.6 + 11.4 %). The fresh seeds when stored in an air-conditioned room retained viability up to 45 days, while it was only 35 days at normal room condition and prolonged further up to 60–70 days when stored with dry sand in jute bags (Banik 1994b, 2010). The seeds can be carried with sand in the jute bags during long-distance transportation to minimize the damage and to retain their viability. The seeds should be properly sown by placing below (1.0 cm) the soil surface preferably horizontally or may be vertically keeping swollen portion up. Seedling survival is in a maximum (70–75 %) when raised from the seeds heavier than 50 g, but it drops (50 %) when raised from light-weight (7–16 g) seeds. Different types of abnormalities such as, rootless plumules, stunted radicles and radicles growing upwards, etc., are not uncommon in the seedlings produced from light-weight seed (Banik 1991a).

Seedling Character

The thick (4–6 mm) plumule with pointed tip is usually one and sometimes two to four in number, and roots are being produced from the thick end of the fruit. Within 4 weeks, the plumule elongates rapidly (80–100 cm) into a stem bearing two to three leaves alternating at the nodes. Seedlings up to 9–10 months of age have tender unbranched stems. Juvenile leaves are bigger one and half times more in size than those of adult clumps (Fig. 7.1i). New shoots are bigger and taller than the older ones. The germinating seeds remain attached to the seedlings for 80–100 days to support the initial vigorous growth. Rhizome development starts after 30–40 days of seedling age and becomes well developed at 6 months to support the growth of new culms (Fig. 7.1j). While seedlings are kept at nursery for 10–12 months, there is a need to be shifted from one bed to another after 3–4 months of age frequently to minimize the intermingling of rhizomes and roots to other seedlings in the nursery beds. As a result, the transportation of seedlings to the field will be easy with lower or no damage in the root and rhizome system. Seedling height should be controlled by cutting the stem tips above five nodes at 3 months of age for easy transportation with less stem damage (Banik 2000, 2010).

Vegetative Propagation Methods

Normal offset planting gives only 20–25 % survival. There is no or poor success in culm or branch cutting, ground layering and marcotting methods. *Part clump or rhizome assembles* with two to three offsets connected with each other and has to be collected as a unit for successful planting (45–55 %) in the field during monsoon (Banik 1995, 2000, 2010).

Tissue Culture

It was observed that in *M. baccifera* shoot developed at faster rate from explants, were 7–9 cm long and required taller glass bottles or test tubes for good culture growth.

Melocanna baccifera (explant)	Medium and environments	Results	References
Immature embryo from developing fruit	MS (4 % sucrose and vitamins of B_5 medium) + 2,4-D (1.0–5.0 mg/l) + BAP (0.1–3.0 mg/l)	After 2–3 weeks shoot and root produced	Banik (1991b)
Branch node with bud (adult clump) treatment of 0.1 % benomyl solution and then soaking in 0.075 % $HgCl_2$ for 10 min	1. MS + 2,4-D (10 mg/l) + BAP (0.5 mg/l) + CM (10 %); tissue survived and grew in higher concentration of growth regulator (2,4-D 10 mg/l and BAP 5.0 mg/i) after successive transfer 2. MS + BAP (5.0 mg/l) + NAA (1.0 mg/l); rooted plantlet developed in 6–7 months from the first inoculation	1. Produced shoots 70–80 % within 1 week Plantlets produced after subsequent sixth transfer culture with 25–30 days of interval 2. Shoots rooted, hardened and outplanted	Banik (1991b)
Single node segments (18-year-old clumps) pretreated with the mix of bavistin and streptomycin and then by surface disinfection with 0.1 % $HgCl_2$ for 10–12 min	(1a) Liquid MS + 20 μM BAP (1b) Further multiplied on liquid MS +15 μM BAP and 3 μM Kn at a rate of 2.99 folds, every 4 weeks (2) MS ½ strength + 25 μM IBA in vitro raised plantlets hardened and acclimatized	(1a) Differentiated shoots (1b) Bud break with multiple shoots (2) in vitro shoots rooted (60 %)	Kant et al. (2009)

Note: *MS* Murashige and Skoog medium, *2,4-D* 2,4-dichlorophenoxyacetic acid
BAP 6-benzylaminopurine, *Kn* 6-furfurylaminopurine, *IBA* indole-3-butyric acid, *CM* coconut milk, *AC* activated charcoal and *NAA* naphthalene acetic acid

Cytology

2n = 72.

Diversity and Conservation

In the forests sometimes small-sized clumps usually 5–8 m tall with many numbers of small diameter (1–3 cm at mid-culm zone) culms of *M. baccifera* have been found to occur naturally. Local forest dwellers in Bangladesh call these small-sized clumps of *M. baccifera* as *tengra muli or nali or bazali*. This type regenerates successfully even in heavily burnt and grazed areas (Banik 1994a, 2010).

It appears that in *M. baccifera* two types of clumps exist in nature and are judged on the basis of emerging shoot characters. In one type shoots possess a yellowish culm sheath and are usually preferred as edible shoots. In the other, the sheaths are comparatively deep brown and are not usually favoured as food due to their bitter and astringent taste (Fig. 7.1k; Banik 1994b).

Enormous genetic variabilities are expected within this vast natural habitat (in the Northeast India, from Sylhet, CHT and Cox's bazar to Arakan range of Myanmar) due to such diversities in flowering cycle, which may be regarded as different cohorts or flowering populations. With *Melocanna baccifera* being a cross-pollinating bamboo species covering a vast geographical region, there must exist variations in population(s) or cohorts with diverse flowering intervals (cycle) with enormous genetic variability within vast natural habitat of the species (Banik 1998b). When flowering occurs the exact area coverage should be marked and identified, and a map using Geographic Information System (GIS) must be prepared. Cohort mapping would help in estimating the interseeding periods and predicting future flowering dates in various cohorts located in different areas of the region. This information would be found very essential in the systematic management of gregarious flowering vegetation of *M. baccifera* covering vast region of the subcontinent. It is also important to collect seeds and seedlings from each of the identified cohorts located in different parts of the region, and their year-wise centralization at one protected place through raising plantations (minimum of 100 plants) in blocks as in the next flowering time these cohorts are likely to flower and seed block by block. Such centralized plots may act as 'seed orchard' or 'seed stands' of *M. baccifera* (Banik 1997b, 2008). Plantations raised from different cohort source would not flower and die at a time in future. Such diverse flowering populations (cohorts) also offer opportunities for selection and improvement of the species. The existence of diverse duration of interseeding populations would also offer possibilities of *frequent availability of seeds* in *muli* bamboo in the next flowering time. Keen observations on seeding along with localities and their documentation are important in this regard. Further, raised *muli* plantations having diverse populations would maintain a *wide range of genetic base* (Banik 1997b, 2010). There are important factors to identify the suitable age for harvesting of culms (Table 7.4).

Table 7.4 *Morphological characters for culm age determination of M. baccifera*: The change of morphologic characters in relation to the age of culm is described below

Melocanna baccifera	
1st year	*Culm sheath*: all nodes of the culm are covered with culm sheath except the thin tip
	Bud break and branches: no bud break on the culm node; two to four tip buds on the culm produce drooping large leaves directly on the nodes; no branches
	Culm: green branchless straight culm with drooping tips having two to four leaves
2nd year	*Culm sheath*: sheath persists on the lower two-third culm; blade of the sheath is loosely fitted
	Bud break and branches: buds on the one-third upper portion of the culm nodes break and produce many thin branches in assembly
	Culm: green
3rd year	*Culm sheath*: sheath may persists loosely only on basal one-third portion of culm (three to five internodes); mostly black colour; blade of the culm sheath is shed off
	Bud break and branches: further buds break up to upper two-third of the culm, and many thin branches are produced in assembly with comparatively less number of leaves
	Culm: dull dark green
4th year	*Culm sheath*: absent
	Bud break and branches: buds are mostly dead on the basal culm nodes. Some branches in the branch assembly die mostly on the upper portion of the culm. Leafy branches are less, and dead leaf scars are present on the branches
	Culm: light green to yellowish straw colour

7.1.3 Plantation Raising

Compact *block or line plantations* of *Melocanna baccifera* (*muli* bamboo) bamboo in close spacing (2×2 m, 3×3 m) help in protecting soil from erosion on the hill slopes to *reduce siltation* or *prevent soil cutting from riverbanks*, which are perhaps the major cause in reduced carrying capacity of rivers and flooding during peak rainy season. Therefore, creation of continuous green cover of *muli* bamboo vegetation to a large extent is synonymous with *environmental protection* on the hills.

Site Selection and Field Preparation Plantations of *muli* bamboo need to be raised for augmenting the bamboo resources in the region. For quick regreening of the hills and river catchment areas, the following sites can be selected:

- Denuded hills and degraded areas
- Logged over forest

This bamboo does not survive under deep shade. Direct sowing of seeds may be done in the well-thinned or widely spaced forest plantation. Lower slopes of the hills are good planting sites; upper slopes have to be avoided. It requires gentle terrain sites, with deep, loose and fertile sandy loam. Dry and barren site and rocky or too sticky soil are not suitable to be selected as planting land for this species.

However, this bamboo being a hardy species can grow fairly on the hilltops and steep slopes, but does not survive in saline habitats or waterlogged conditions.

Planting site should be well drained, moist and preferably rich in organic matter. Bamboos' new canal banks can also be selected. The jungle should be cut and cleaned in the planting sites.

(i) *Direct sowing of germinating seeds for raising muli* (*Melocanna baccifera*) *plantation*: In the sites having no problem of predations, *direct seed sowing* in the pits would reduce the plantation costs. The technique of direct sowing of seeds of *muli* in the planting pits ensures better stocking in a plantation.

Seed Collection Seed quality is known to have a great impact on the quality of planting stock. Collection of *muli* seeds should be started from later part of May and may be continued up to August. The weather conditions during seed collection also influence seed health. Storage conditions also affect seed viability as seeds stored at a low temperature (at 5–7 °C) have higher germination ability than those stored at a normal room temperature. Usually 25–40 kg seeds are produced per clump before it dies. The seeds need to be collected by plucking directly from the plant or by shaking the clumps, and due to this the mature seeds fall on the ground. As an extra advantage, this also reduces the availability of seeds to the rats, porcupine and wild bore on the ground. Thereby it assists in controlling the population increase of such animals.

To ensure better stocking, more than one seed may be sown in one pit. However, planting of seedlings will overcome the problem of seed predation. Seed sowing and/or seedling planting should be done in the rainy days, June to August.

Pit Making and Seed Sowing In the barren hills and scrubby forests, close planting is suggested, and the planting pits are to be dug at 3×3 m spacing, so 1111 pits per ha. The dug out soil is kept in two parts, one part containing the 20 cm top layer soil and the other part with inner rest 10 cm layer soil. The planting pit size may be 30×30×30 cm. To take care of any causality due to possible damage in handling of seeds, two seeds may be sown in each pit on the contour lines in hills. Thus for raising *Melocanna* plantation in one hectare of land, 2222 germinated seeds are required, and for this (considering germination rate of 80 %) about 2780 seeds have to be sown in the beds.

Sowing and Refilling of Pits Sowing should be done by placing the seeds horizontally at least 0.5–1.0 cm below the soil surface and be covered with dug out top layer soil (fertile humus soil) for better nutrition. The earth in pits should be refilled making small mound to avoid water stagnation during heavy rain as this will rot the seeds. Seeds should be sown horizontally 1.0 cm below the soil surface for better germination. The seeds can be treated with neem formulations before sowing to protect against the predation from animals and insects.

Direct sowing of seeds should be restricted only to the areas having no or less population of such animals. Therefore, to avoid predation problem from wild animal, sometimes seedlings can be planted instead of direct sowing of seeds.

(ii) *Planting of seedlings for raising of muli plantation*: The *muli* seedlings are raised at the nursery in April to May by sowing seeds and maintained, later outplanted during July to August (Fig. 7.11).

7.1.3.1 Aftercare and Management of Plantation for Culm Production

Immediately after raising the plantation of *M. baccifera* either by direct sowing of seeds or by planting seedlings, proper nursing and management of the plants are essential to maintain desired good stocking (survival) and their growth (Banik 2010).

Weeding Cutting and cleaning of vines and weeds around the planting pits are essential for maintaining the competition-free condition for germinating seedlings. Weeding is to be carried out in July and August, when seedlings are 2–3 months old. After planting in June to August, three times weeding and vine cuttings are done in the first year and two in the second year. Afterwards only one weeding has to be done every year around the planted clump. In the second year, two ring (0.5 m radius) weedings are to be done once in April to May and the other in July to August.

Mulching It is important to make sure that the bamboo does not dry out during the first summer after planting. The signs of drying out are apparent when the bamboo leaves roll up like cigarette papers and become very narrow looking. To reduce the loss of moisture due to evaporation from the planting pits, fallen tree leaves, barks, bamboo leaves, rice or wheat stalk and chaff or hay, cut grass, etc., alone or mixed, are placed as 7.0–10.0 cm thick layer and spread at 1.0–1.5 m radius on the ground around the plant, thus covering the exposed ground (*mulching*) of the pit. Mulching conserves moisture in the pit and underground temperature and also checks the weed growth around the bamboo plant. After each ring weeding at the end of rainy season, proper soil work has to be done around each of the planted seedling/cuttings. Immediately after that mulching should be done around the base of the seedling; this is needed for the completion of annual drought period (December–March). Fertilizer NPK may be applied in the form of urea, superphosphate and muriate of potash at the rate of 60, 40 and 40 g per plant in three split doses at intervals of 3 months.

Protection against fire is ensured by providing *fire line* including its maintenance throughout the dry season. *Block fencing* may be done to protect the seedlings from grazing, and the extent of fencing would depend on the intensity of grazing and amount of area to be protected.

Intercropping The common plain land conventional agricultural crops have never been selected as intercrops as *Melocanna* plantations are, generally, raised in the forest hilly lands. The seedlings of this species exhibit somewhat caespitose clump-forming nature up to 3 years of planting, and after that clumps start expanding in all direction and within a few years cover the space between the plants. So weeding may be avoided by cultivating the legume crops [*Cajanus cajan*, lentil (*Lens culinaris*), *Sesbania* sp., etc.] in the early ages of plantation. At the early stage before developing a thick canopy, possible intercropping can be given to tapioca, *Dioscorea* and other vegetable crops like ginger, chilly, etc. The intercropping of soyabean (*Glycine max*) was found to improve the soil condition and influenced the bamboo growth.

As *M. baccifera* is a fast-growing and quick-harvesting crop, about ten culms are produced per clump in the first year of harvest (5 years after plantation). Within the first 5 years of plantation period on an average, one clump can produce a total of 20–25 culms. A 3–4-year selection felling cycle has been followed for harvesting *Melocanna* bamboo from the forest. The harvested bamboos are transported through rafting in the hilly streams and rivers to the destination of market depot (Banik 2000, 2010, 2015c; Fig. 7.1m).

7.1.4 Specific Spot Characters of *Melocanna baccifera* in the Field for Identification

1. Clump is open and diffuse, and culms are erect, arising singly in well-separated distances (0.5–2 m apart) from the ramification of an underground rhizome system (Fig. 7.1a).
2. Rhizome necks are greatly elongated even up to 2.5 m (Fig. 7.1b).
3. The bud on the culm node is flat and appressed with the culm wall. Thin wing-like growth is present on both left and right margins of the bud (Fig. 7.1c).
4. Culm sheaths have ridge on the outside of the sheath where the narrow sword-shaped blade (longer than the sheath proper) is attached (Fig. 7.1e).
5. Branching is usually confined from mid to upper part of culm. Tufts of subequal thin branches 25–40 are developed from each node (Fig. 7.1d).
6. Young shoot is yellowish brown to brown; sheath margin and top is pinkish; ligule is horseshoe shaped; blades are flagellate and glabrous. Longitudinal section of a newly emerged edible shoot is showing creamy white colour inner shoot meat (Fig. 7.1n).
7. The florets are prominent and have elongated purple-coloured stigma clearly visible from apart (Fig. 7.1f).
8. Fruits are big in size, onion shaped with a beak, fleshy and green and weigh 7–150 g per fruit (Fig. 7.1h).
9. Leaf is oblong-lanceolate, has no cross-veins and has larger leaf blade in seedlings than adult plants (Fig. 7.1i).

Chapter 8
Ochlandra Thwaites

Members of the genus *Ochlandra*, referred to as reed bamboo, are thin-walled and thickly clumped endemic bamboo of the South Western Ghats of India (10 species) and Sri Lanka (1 species). Among the different species occurring under the genus *Ochlandra* in India, *O. travancorica* and *O. scriptoria* are found to grow in abundance. *Ochlandra scriptoria* (Dennst.) CEC Fisher is an endemic reed bamboo of Western Ghats (Koshy and Harikumar 2001) and found along the stream banks in the lower elevations. This is a small and thin reed widely distributed in the Southern Kerala and in smaller portions in the northern part (Basha 1991). All the species have relatively large fruits with thick pericarp, such as those of *Melocanna*, but smaller, and also a large number of floral parts, especially the stamens (as many as 120 in one floret). Little is known about the genus, and more intensive investigation is required (Dransfield 1980). However, two species, *Ochlandra travancorica* and *O. stridula*, are described below due to their socio-economic and ecological importance in southern India and Sri Lanka.

8.1 *Ochlandra travancorica* Benth. & Hook. f.

8.1.1 General Information

8.1.1.1 Vernacular and Local Names

Etta, kar-etta, vei, irul (Tamil India), reed bamboo, elephant grass of Travancore (English)

R.L. Banik, *Silviculture of South Asian Priority Bamboos*, Tropical Forestry,
DOI 10.1007/978-981-10-0569-5_8

8.1.1.2 Natural Distribution and General Habitat

Ochlandra travancorica grows naturally in the wet evergreen and semievergreen forests of Mysore (India) and on the mountains of southern India (Kerala and Tamil Nadu) in Tinnevelly and Travancore at 1000–1065 m elevation. It is also planted in Tamil Nadu and at Peradeniya in Sri Lanka. Pure patches grow also as impenetrable thickets along the sides of rivers and streams, covering many kilometres of the mountains where other tree species are not allowed to come up; even elephants attempt hard to get through it. The species is a big size reed occurring widely as an undergrowth in the low-level semievergreen and evergreen forests.

Ochlandra travancorica and *O. wightii* are important members of the group thriving only in the wet hill slopes and stream sides of Western Ghats. The existence of these two endemics is dependent on the sustenance of their native habitats. Habitat destruction, anthropogenic activities, invasion of alien weeds and animal predation of seeds and seedlings reduce their regeneration The attack of *M. micrantha* and wild pigs (*Sus scrofa*) gave the most deleterious impacts besides human disturbances on natural regeneration of reed bamboo (Gopakumar and Motwani 2013). The reed bamboo (*O. travancorica*) vegetation plays a great role on overall development and fertility of soils through its protective umbrella of canopy, litter and root mat in the soil of Western Ghats of India (Sujatha et al. 2008).

It is the most important associate of the tropical evergreen forests and attains maximum growth in the very wet type of evergreen forests. They are adapted to moist hill slopes and banks of watercourses. In the evergreen types, the most important associations of *Ochlandra* include *Hopea parviflora*, *Cullenia exarillata*, *Canarium strictum*, *Dipterocarpus indicus*, etc. In the semievergreen type, the associations include both evergreen and deciduous species in the top canopy such as *Terminalia* spp., *Xylia* sp., *Sterculia* sp., etc. *Ochlandra*, being shade tolerant, grows well even under the closed canopy of evergreen forests. In the tropical wet evergreen forests, the reed *Calophyllum* association, a localized edaphic climax type is observed. Here the reeds are found occurring in considerable stretches in marshy areas. The canopy of trees chiefly belongs to *Calophyllum elatum, Hopea glabra, Bischofia javanica* and *Eugenia* spp. (Kumar 1988).

8.1.1.3 Climatic Conditions

The species prefers diffused sunlight and running water and therefore is found in the fringes of perennial streams in subevergreen and evergreen forests of heavy rainfall areas. It is an efficient soil binder.

8.1.1.4 Uses

It is much used for making mat, basket and other articles of local use. More than 25,000 artisans are involved in bamboo mat weaving and craft using basic raw material of *B. bambos* and *O. travancorica* which are available in plenty and supplied by Bamboo Corporation (Rao et al. 2008). In the traditional sector, *Ochlandra* bamboo supports the livelihood of over 300,000 workers belonging to the socially and economically weaker sections of the society (Seethalakshmi and Gnanaharan 1998). Leaves are used as fodder by both domestic and wild animal. Koshy et al. (2010) attributed satiation of large predators such as elephants as one of the reasons for larger leaf sizes in *Ochlandra* plants.

8.1.2 Plant Data

8.1.2.1 Vegetative

Plant Habit

An erect, shrubby or arborescent, reed-like gregarious bamboo (Fig. 8.1a).

Culm Height

Generally 2–6 m tall, maximum height recorded is over 15.0 m. Reeds in the catchments are usually bigger in size and length.

Culm Diameter

Mid-culm 2.5–5 cm. Concentration of reed is substantially in 9.0–18.0 cm girth class with maximum distribution in 5.0–10.0 m height class (Noushad 2008).

Culm Thickness

Very thin walled, 2.0–3.0 mm; maximum weight is 4.0 kg with an average of 2 kg per reed.

Culm Colour

Grey green, rough.

Internode Length

Long 50–80 cm, sometimes evens up to 1.0 m; nodes are somewhat swollen and marked with base of fallen sheaths.

a
c1
c2
Ochlandra travancorica
b1
d
b2

Branching

Thin, subequal and many.

Leaves

Oblong-lanceolate. 15–45 cm long and 5–10 cm broad, rounded often unequal at the base, petiole 0.7–0.9 cm, long setaceous acuminate twisted tip.

Culm Sheath

Fifteen to 20 cm long, thin, longitudinally wrinkled and striate, when young covered with many appraised golden or blackish hairs, glabrous when old, blade narrow 3–7 cm long; ligule narrow and entire. In *O. travancorica* the blade is abruptly bend backwards, awl shaped without hairs. The auricle is short and unconspicuous, with numerous stiff bristles. However, in *O. scriptoria* the blade is erect, glabrous and narrow, Auricles are small falcate with stiff bristles.

Vegetative Growth

Culm emerges during May to August. Average number of culms per clump is 20 or sometimes may have many culms.

8.1.2.2 Reproductive

Flowering Nature

Ochlandra travancorica is monocarpic bamboo species and flowers gregariously at an interval of 7, 10 and 15 years cycle (Table 1). The gregarious flowering of *Ochlandra* spp. in southern India (Blatter 1929–1930) was almost phased with peak draught years: 1868 (1868), 1875–1876 (1876–1877), 1882 (1884), missing year of flowering (1891), 1896 (1899) and 1905 (1905). It appears the frequency of draught

Fig. 8.1 (**a**) The clump of *Ochlandra travancorica* is an erect, shrubby, reed-like gregarious bamboo (**a**). (**b**(**1**, **2**)) Inflorescence of *Ochlandra travancorica* is a sub-verticillate, spicate panicle with a few large fertile ovate or oblong-ovate; broad, glabrous spikelets and a few much smaller sterile ones. Stamens are very many (55–130), monadelphous at first, afterwards long exserted (**b**). (**c**(**1**, **2**)) The seeds (*fruits*) of *Ochlandra* are large, 5 cm long, 2–3 cm broad, brown at maturity, oval-oblong and wrinkled, with 4–5 cm long stiff beak, pericarp fleshy enclosing seed, the whole surrounded by the persistent glumes and palea (**c**). (**d**) Two-year-old seedlings of *Ochlandra* sp. in polythene bags (**d**) (**a**, **c**(1) and **d**: Courtesy of Dr Muralidharan KFRI, India)

Table 1 Estimation of flowering cycle in *O. travancorica* from the available flowering records

Country/locality	Flowering date (calendar year)	References of the flowering dates	Estimated flowering cycle (year)
India	1868, 1905	Blatter (1929–1930)	7
South India	1875,1882	B., T. F. (1887)	7, 7×3=23 (1882–1905)
Travancore	1982	Venkatesh (1984)	7 (1875–1882)
Pooyan kutty	1985, 1986 1998–2001 2002, 2004–2006	Basha (1991) *Seethalakshmi (*2006*)*	80=or 20×4 (1905–1985, 1986)

years decreased in that region, and there have been no further reports of gregarious flowering until 1982 (Venkatesh 1984). The pattern would accord with the general classification of *Ochlandra* as sporadic or continual flowering. *Ochlandra* is considered by Holttum (1956) to have a relative primitive morphology among Asian bamboos and its flowering pattern as primitive as well.

Inflorescence

A sub-verticillate, spicate panicle with a few large fertile spikelets and a few much smaller sterile ones. *Spikelets* are ovate or oblong-ovate, 5.0 × 10 long, 2.5 cm broad and glabrous. *Palea* is shorter and narrower, acute, faintly 2-keeled. *Lodicules* 3, unequal. *Stamens* are very many (55–130), monadelphous at first, afterwards long exserted (Fig. 8.1b(1, 2)). *Anthers* are 2.54 cm long, narrow, hairy apiculate. Spirally twisted together 5–6 plumose *stigmas*. The *ovary* is narrow and smooth. The observation (Venkatesh 1984) on breeding system of *O. travancorica* showed dichogamy (maturing of male and female reproductive organs in different periods) with protogyny (female reproductive organ matured first). Time gap between ripening of stigma and anthers is 2 days. There is no self-incompatibility and pollen grains showed 90 % viability in this wind-pollinated species. Monoporate pollen was observed. Cross-pollination was observed. *Ochlandra travancorica* produced better quality and quantity of pollen, viability was nearly 99 %, longevity was more than 30 min and the seed setting was very high (Beena 2011). Seeds were viviparous with no dormancy, and good natural regeneration was found in the species.

Seed

Very large, 5 cm long, 2–3 cm broad, brown, oval-oblong, wrinkled, with 4–5 cm long stiff beak, and pericarp fleshy enclosing seed; the whole is surrounded by the persistent glumes and palea (Fig. 8.1c(1, 2)). *O. travancorica* var. *hirsuta* produces fruits; the weight of a fruit varies from 24 to 32 g.

About 40 fruits per kg in *O. travancorica*, seed germination up to 75 % and longevity up to 120 days. Fruits are small in *O. scriptoria,* 640 fruits/kg (Seethalakshmi and Kumar 1998).

Seed Germination

Germinates promptly. Seeds are viviparous with no dormancy, and good natural regeneration is found (Beena 2011).

Seedling Character and Natural Regeneration

Seedlings are somewhat of similar look to those of other bamboo species (Fig. 8.1d). The natural regeneration of *Ochlandra travancorica* and *O. wightii* is restrained by many factors. These factors include invasive growth of *Mikania* weed, fruit predation by boars, rodents and larvae of *Achroia grisella*, seedling predation by elephants and other herbivores, human activities such as clear-felling, spreading fires, cleaning the bamboo grounds after flowering, construction of river embankments, establishing plantations and forest encroachments. The fruiting of endemic *O. wightii* supported the population of an endemic rodent *Platacanthomys lasyrus* (Gopakumar and Motwani 2013).

Vegetative Propagation Methods

Among all the known methods, culm-cutting technique has been found most suitable for different species of *Ochlandra* (Surendran and Seethalakshmi 1985, Seethalakshmi et al. 1988).

Culm cutting The 2-node culm segments are made (from 2- to 3-year-old culm) leaving 5–7 cm on each side beyond node. The branches from the remaining portion of the culm are removed carefully without damaging the axillary buds on the nodes. About 100 ml solution of rooting hormone is poured into the culm cavity (through an opening at the centre of internode), and the opening is sealed by wrapping tightly with a polythene strip. The culm segment treated either with 10 ppm coumarin (March to May) or 1000 ppm Naphthalene acetic acid (NAA) (April) sprouted from the nodes within 2 weeks and gave 40–50 % rooting after 1 month of sowing in the rooting beds. The cuttings without hormone also produce roots (50 %) in June. However, the growth of cuttings treated with NAA 100 ppm was better as more number of taller culms were produced per clump (Seethalakshmi et al. 1988). The rooted cuttings are maintained in the nursery for 1 year for profuse rooting and rhizome formation with numerous new shoots before field planting.

The performance of cuttings is better in the field than the seedlings.

Tissue Culture

Ochlandra wightii (explant)	Medium and environments	Results	References
Isolated embryos from mature seeds	(a) MS + BAP (0.5 mg l^{-1})	(a) Isolated embryos germinated (85 %) 1–2 shoots in 30 days	Bejoy et al. (2012)
	(b) ½ MS + BAP (0.5 mg l^{-1}) and TDZ (0.5 mgl^{-1})	(b) Shoot multiplied to an average of 9.8 shoots in 60 days, also roots developed	
	(c) Cultures then treated with KN (1 mg l^{-1}) and sucrose (4 %)	(c) Rhizome induction in 35 days from 65 % of seedling Around 880 plantlets generated in 9 months from ten embryos isolated from mature seeds	
Nodal segments from in vitro seedlings	(a) MS+ BAP (2 mg l^{-1}) and KN (0.5 mg l^{-1})	(a) Average 3.6 shoots produced in 40 days; 20 % culture developed 1–2 roots	Bejoy et al. (2012)
	(b) ½ MS + IBA (0.5 mg l^{-1})	(b) Rest 80 % rooted. Plantlets hardened and transplanted to the clay pots. 80 % survival after 6 months. Plantlets with in vitro rhizomes showed 100 % survival	

Note: *BAP* 6-benzyl amino purine, *TDZ* thidiazuron, *KN* kinetin, *IBA* indole-3-butyric acid

Cytology

2*n* = 72.

Diversities and Conservation

The different populations of reed (*Ochlandra* spp.) bamboos need to be studied to identify the diversities and superior types of each species. Reed bamboo prevents top soil from eroding and acts as reinforcements to banks of swift-flowing and annual-flooding waterways. These bamboo species have to overcome many biotic constraints in survival. They protect many forest faunas by providing shelter and food. These bamboos are integral to the ecological stability of the wet land region of South Western Ghats including other similar sites of South India and Sri Lanka.

Therefore, serious efforts are needed to conserve these reed bamboos for ecological and socio-economic benefit for the region.

8.1.3 Cultivation and Management

The management of reed bamboo forests (*Ochlandra* spp.), in general, involves a selective felling system with a felling cycle of 4 years as immature culms are unsuitable for handicraft and cottage industries (basket making and weaving). However, with the emergence of pulp and paper industries as major consumer of reeds, the selective felling system often ignored and payment based on weight favour clear-felling system (Kumar 1988). As a result immature culms are also harvested and the clumps become weak and less productive. The same area is exploited year after year by various agencies. Besides reed area has to be protected from fire and other biotic pressure. Reed bamboos (*Ochlandra travancorica*) are worked on a 3-year felling cycle in Kerala. The felling rules prescribe (Manoharan and Trivedi 2008) that (i) bamboos and reeds adjacent to the stream banks and located on slopes above 300 gradients are not worked; (ii) culms of age more than 2 years are only felled; (iii) the felling of culms is done on a horseshoe pattern; (iv) no felling is done during the regeneration period of June, July and August and (v) all dead, malformed culms are removed irrespective of age.

8.1.4 Specific Characters for Identification of Ochlandra travancorica in the Field

1. An erect, shrubby, reed-like bamboo (Fig. 8.1a).
2. A bamboo with very long culm internode.
3. Leaves broadly oblong-lanceolate with long setaceous acuminate tips.
4. Large inflorescence, many more (monadelphous) stamens, may be up to 130 (Fig. 8.1b(1, 2)).
5. Spirally twisted stigmas and big caryopsis (Fig. 8.1c).

8.2 *Ochlandra stridula* Moon ex Thw.

[The reed bamboo of Sri Lanka *Ochlandra stridula* is taxonomically conspecific to *Ochlandra scriptoria* of Western Ghats, and therefore, according to the rules of priority, the latter species needs to be synonymized under the former (Kumar 2011).]

8.2.1 General Information

8.2.1.1 Vernacular and Local Names

Bata, Bata li (Sri Lanka).

8.2.1.2 Natural Distribution and General Habitat

The species is indigenous to Sri Lanka and covers a vast area of the country in the south and west. It grows extensively covering hundreds of square miles in the rainforests of the wet lowlands and lower monotone areas below 1500 m. The species is also found in a small area in the southwestern of the country extending from the south of Mahaoya, down to Galle and Matara and eastwards to the foothills of Adam's Peak. Within the rainforest it grows in a wide variety of habitats ranging from forest gaps and rocky ridgetops to stream edges (Zoysa 1994). Often in a major component of secondary successional communities following shifting cultivation, the species can occur in dense thickets over large area. *Ochlandra stridula* is found extensively in the wet lowlands of the southwestern region. The other species under *Ochlandra* are found in the high-altitudinal montane areas of the central hill country of Sri Lanka.

8.2.1.3 Climatic Conditions

Rainfall of 1500–5000 mm and a temperature of 17 °C.

8.2.1.4 Uses

Ochlandra stridula is mainly used for the production of basket ware, storage boxes for paddy, linen and carrier baskets for food, winnowing fans, food covers, milk strainers, flutes, blinds, tea plucker's baskets and other materials. The culm is simply cleaned and split into long strips and then used for making lantern frames during the festival of 'Vesak' in the month of May (Zoysa 1994). It has been reported that 79 % of the people who harvest bamboo from state forests use the material for themselves in the manufacture of crafts with household importance like baskets, winnowing fans, etc. (Kariyawasam 1999). The main raw material collected for the purpose is *O. stridula.* The availability of this bamboo for the handicraft industry is fast decreasing. The leaves make excellent thatch, and that is called *Batta gass* roofing and housing. The leaves have been used for thatching, particularly among the low-income groups. The raw material has to be transported from far distances at an increasingly high cost.

Bamboo flutes are also made from the internodes of this species and are of cultural importance in folk music. In areas where *Ochlandra stridula* grows freely, whole culms have been traditionally used as wattle and strips for tats, blinds and inner partitions.

8.2.2 Plant Data

8.2.2.1 Vegetative (Mostly Adapted from Soderstrom and Ellis 1988)

Plant Habit

Ochlandra stridula is a close-growing gregarious shrubby bamboo. Culms crowded erect below and arched above with tips bending over.

Culm Height

Upright and straight 2–6 m high.

Culm Diameter

Six to 20 mm in diameter.

Culm Thickness

Thin walled

Culm Colour

Young culms are light-brown maroon and sparsely covered with hairs with a whitish appearance. Mature culms are green, with a tough and gritty surface.

Internode Length

Scabrous nodes that are 30–50 cm long are not prominent but with a circular band of whitish bloom, geniculate.

Branching

Mid-culm upwards.

Leaves

Broad, oblong-lanceolate tapering to an acuminate apex, 20–33 cm long and 3.5–7.0 cm broad and rounded at the base into a short broad petiole; ending above in a long setaceous, scabrous point; smooth above and slightly rough beneath; main vein is thin, secondary vein is 10–12 pairs and transverse veinlets are none; *leaf twigs* are 50–80 cm long, bearing 8–14 leaves, *leaf-sheaths* striate and smooth when old.

Culm Sheath

Cylindric, glabrous, purple when young, rounded at the top and on either side with small falcate auricles; 7–20 cm long and 3–10 cm wide. *Imperfect blades* are subulate, recurved and, in small new shoots, very long; *ligule* is short.

Vegetative Growth

Clumps can grow up to 2–3 m tall with 2 years after planting of seedlings.

8.2.3 Reproductive

8.2.3.1 Flowering Nature

The species appears to have annual flowering cycle, but individual clumps do not flower annually. Flowers are loosely arranged on the inflorescence.

8.2.3.2 Inflorescence

A spicate terminal panicle, usually a leaf-bearing branchlet; the spikelets are verticillate; few are fertile and many sterile; rachis is long, glaucous above the joints; spikelet is 2.5 cm long, 0.3 cm broad, cylindro-conical with one fertile flower; *empty glumes* 2–3, convolute, mucronate, ciliate on the edges; *flowering glume* is similar but larger; *palea* is membranaceous with lodicules 6–12 or more. *Stamens* are many, often 30, yellow and brown with green tip when mature; *filament* is free, flat and wavy in the upper half; *anther* is finally exserted, bifid, at the apex and mucronate. The *ovary* is narrow and smooth; the beak of the perigynium is enclosing the *style,* which is surmounted by 4–5 short plumose *stigmas*.

8.2.3.3 Seeds

Ochlandra stridula fruits (caryopsis) are ovoid, about 2.5 cm long, surmounted by a long 2.5 cm beak, smooth or wrinkled, and supported by the persistent glumes. The pericarp is hard and thick. The fruits are sensitive to drying.

The most important native species that supports the cottage industry in Sri Lanka, *O. stridula,* is effectively propagated by seed. The species shows annual flowering and fruiting cycle, but individual clumps do not flower annually. The fruits mature in small quantities and are moisture sensitive unlike other bamboo seeds where seeds are gregarious and very dry. *O. stridula* seeds should be kept moist from harvest and sown immediately or potted directly, watered frequently and kept in partially shaded area. Germination takes 3 weeks. Seedlings are ready for field planting in 2–3 months.

8.2.3.4 Seed Germination

The seeds must be kept moist and sown immediately after collection. It germinates within 3 weeks, better under partial shade, and needs proper watering.

8.2.3.5 Seedling Character

Seedlings start producing new shoots within a few weeks and ready for planting within 2–3 months.

8.2.3.6 Vegetative Propagation Methods

Not much work has been done as seeds are available almost every year.

8.2.3.7 Tissue Culture

Not much work has been done in developing the micropropagation technique for the species.

8.2.3.8 Diversities and Conservation

Not seen any documentation.

8.2.4 Cultivation and Management

The species is relatively tolerant of nutrient poor soil and ideal for introducing into small holdings and home gardens. It can be grown as fences along the boundaries, on eroding stream edges or steep and rock slopes. *Ochlandra stridula* is effectively propagated by seed.

The current methods of harvesting are also wasteful. Only the young pliable culms are used, while the mature culms are discarded, even though these can be used for other purposes. Although on a country-wise basis *0. stridula* is becoming scarce, it can still be found abundantly in a few localized areas. In the tree plantation areas, for instance, it is often considered as a weed and its eradication is promoted. Consequently in these areas, there is often a misconception that this resource is plentiful. The conversion of existing natural stands into managed clumps is one of the major research needs. Detail investigations are needed to determine both silvicultural and ecological requirements of this species and management strategies for augmenting poor areas.

8.2.5 Specific Spot Characters for Identification of Ochlandra stridula

1. A close-growing gregarious shrubby bamboo
2. Culms crowded, erect below and arched above with tips bending over

Chapter 9
Schizostachyum Nees

[The genus *Schizostachyum* name comes from the Greek word *schistos*, cleft or divided, and *stachys*, ear or corn, referring to the spaced grouping of the spikelets. Culm is slender, thin walled and usually drooping at tips. Young shoots and especially the culm sheaths are covered with appressed golden brown, light or pale brown or white hairs.

The genus *Schizostachyum* was described by Nees in 1829, with one species *S. blumii* Nees from Java. McClure Blumea 2: 86–94, 1936. Since then many species have been described. The genus is found from South China, throughout Malaysia to the Pacific Islands. It is also found in the natural forests of Northeast India and Chittagong Hill Tracts and Sylhet of Bangladesh].

9.1 *Schizostachyum dullooa* (Gamble) Majumdar

Schizostachyum dullooa (Gamble) Majumdar in Karthikeyan et al., Fl. Ind. Enumeratio: Monocotyledonae. 281. Bot. Survey of India., 1989.

[Synonyms: *Teinostachyum dullooa* Gamble in Ann. Roy. Bot. Gard. Calcutta 7:101. Pl. 89. 1896. *Neohouzeaua dullooa* (Gamble) A. Camus in Bull. Mus. Hist. Nat. Paris 28: 101. 1922.]

9.1.1 General Information

9.1.1.1 Vernacular and Local Names

Dullu (Assam, Tripura, India); *wadlok* (Kokbarok, Tripura); *pogslo* and *puksalu* (Lepcha, India); *shlu* and *siej la* (Khasi, Meghalaya, India); *tarang* (Jaintia, Meghalaya, India); *wadrow* (Garo, Meghalaya, India); *unap* and *unal*, (Manipuri, Manipur, India);

R.L. Banik, *Silviculture of South Asian Priority Bamboos*, Tropical Forestry,
DOI 10.1007/978-981-10-0569-5_9

guh [Rongmai Naga, Tamenglong (Manipur)]; *rawthla* (Mizoram, India); *dalu bans* (CHT, Chittagong, Sylhet); *tokhre bans* (Nepal, Bhutan); *wa-byauk*, *gya-wa* and *thai-kwaba* (Myanmar); *lakhra* (Kachin). The species name "*dullooa*" might have taken from the Assamese word "dulla" meaning hollow, dula also name of the big basket usually made of this bamboo for storing food grains. As the species has been extensively growing along the bank of hilly stream '*dalu khal*' flowing in southern Sylhet and Patharia Forest Reserve, it might have been named as *S.dullooa* (Banik 2000).

9.1.1.2 Natural Distribution and General Habitat

The species is distributed in the moist semievergreen forests of northeastern region of India (Assam, Cachar, Manipur, Meghalaya, Mizoram, Sikkim, Tripura) to Sylhet, Chittagong and Chittagong Hill Tracts (CHT) of Bangladesh. It grows on small hills (altitude of 1200 m) and in high altitude up to 2000 m of Jaintia Hills and southern part of Nepal and Bhutan. In Bhutan the species has been found up to 1700 m altitude. The species has also been found at 2000 m altitude on Taungme, Ruby Mines district of Myanmar. The species is also found in Northern Vietnam near Quang Yen. It prefers shade and occurs luxuriantly on the moist, well-drained gullies in between the hills and fertile valleys also as understory in the deciduous forests of Bagafa (Hrishyamukh) in South Tripura and Gandachara of northwest Tripura, Cachar, Meghalaya, Manipur, Sylhet, Chittagong and Chittagong Hill Tracts (CHT) (Banik 2000). Moist and shady valleys are preferred sites for the species. The lower halves of clumps remain in the moist and shade conditions inside the gullies and produce new shoots in such partly dark sites where animals like voles, water rats, field mice, rodents, etc., are attracted to feed on the tender shoots; and such moist habitats are usually ideal places for pythons both for food and shelter during rainy seasons. The vegetation of *S. dullooa* thus provides an important ecological support role for the python population on the hills of Northeast India, CHT and Sylhet.

The species is not available on drier, open and scrubby hills. However, *S. dullooa* is also capable of growing on the coarse-textured soil provided there is moisture and may be seen on soils originating from sandstones (Rao and Ramakrishnan 1987).

The specimen of *Schizostachyum dullooa* in Kew herbarium was collected by Hooker around the 1880s from Sitakunda Hill of Chittagong (Gamble 1896). But at present, in 1999, one can seldom find a clump of *dalu* in Sitakunda Hills. Clear felling, repeated cutting and fire have destroyed the trees in the forests and exposed the ground to direct sunshine. As a result the land has become gradually drier, overhead shade has also been destroyed and habitat has become unfavourable for this bamboo (Banik 1999a). Similar situations are also been observed in the different forests of Myanmar, CHT, Mizoram, Meghalaya, Tripura and other areas of Northeast India.

It prefers shade and generally occurs on the moist, well-drained and fertile valleys as understory in the deciduous forests of Bagafa (Hrishyamukh) in south Tripura and Gandachara of northwest Tripura and different parts of CHT. Recent gregarious flowering and simultaneous large-scale death of clumps created alarming situation for the species.

9.1.1.3 Climatic Conditions

Prefers annual rainfall 3000–6000 mm, average temperature range 5–35 °C and well-drained soil with low pH (4.5–6.5).

9.1.1.4 Uses

Culms are used as umbrella sticks and for making mats, baskets and novelty items, small boxes to carry pan leaves (*Piper betle*) and animal cage for carrying pigs (Fig. 9.1h). Such products are shiny and light and have great demand in the local and international market. In Tamenglong (Manipur) near Mokru bridge point in Nungkao area, *S dullooa* poles 15–21 m long having 0.85–1.2 m long internodes with diameter 5–7 cm are usually stacked on the bank of the Mokru river near the bridge area. These bamboos are harvested from Rangku Long about 12 km from Makru bridge point and transported there through river. Each bundle contains 30, and raft may contain 500–700 culms. The culm internode is cut into 18, 55, 60, 65, 70, 75 and 80 cm long segments (Fig. 9.1c). Each of the segments is split into four to six strips. Thus, seven to nine internodes were split, and all the pieces were tied in a bundle, and each bundle was carried by truck to nearby lower Assam Silchar market and from there to wholesale market at Calcutta (Kolkata) by air. Every year about Indian rupee 20 crore trades have been going on involving 0.3 million people from harvest, processing and transportation. These strips are mainly exported to the Middle East, Dubai and other countries and are mainly used for making kite structures. Sticks are also used locally in incense stick industry. The mats made of this bamboo have special demand in outside market due to its shiny and bright creamy colour.

9.1.1.5 Ethnic Utilization

Tribal peoples use the long hollow internode for carrying the water in the hilly terrain because it is light in weight and no powdery substances are present inside the lumen of the internode. The culm wall skin is very sharp and is used as a bamboo knife to cut off the umbilical cord of a newborn baby. Young culm internodes are used for making cake from rice powder. The internode filled in fresh rice powder is roasted over fire and the inner soft cylindrical cake is collected by splitting the internode. The used rice are sticky in nature. The cake is chopped in to pieces and cooked with milk and molasses and locally called as *Chunga pitha* in CHT and Sylhet (Banik 1998a, 2000). Long internode cylinder is also made as flute.

The shoots of *S. dullooa* are also bitter in taste and very rarely used as food. The percentage of (on fresh weight basis) sheath cover and edible portion of a freshly collected shoot is 47.8 and 52.2, respectively. The shoot meat possesses creamy colour, and the taste is bitter to slight bitter with crisp to tender texture (Banik 1997a, 2000).

Fig. 9.1 (**a**) *Schizostachyum dullooa* is a congested medium-sized clump, more or less densely caespitose; culms with especially elongated internodes and the young culm tips are long and thin and may hang down to the ground (scrambling or pendulous) (**a**). (**b**) In *S. dullooa* the culm is glossy green to dark green covered with silvery white pubescence; whitish band (1.0–1.5 cm) below the nodes is distinctly visible in the culm (**b**). (**c**) *Schizostachyum dullooa* is easily distinguishable by its very long (0.5–1.5 m) thin-walled, glossy green to bottle green internodes; an internode is more than half of the height of a man (**c**). (**d**) Branch bud on the culm node is flat appressed on the internode wall and has two thinned out growths; internode wall is covered with scattered shiny white-appressed hairs and white band just below the node; a dense tuft of slender, thin subequal 3–35 branches develop in a complement from each node (**d**). (**e**) The flowering *S. dullooa* is mostly gregarious (**e**) and occasionally sporadic. (**f**) The ripe deglumed seeds are blackish brown and small, elongated grain-like, somewhat broader and flat base with cylindrical top terminating in a long beak formed by the persistent base of the style, covered with glabrous glumes. The length of glumed seed (grain part) is 1.0–1.9 cm, and that of persistent style is 1.1–1.7 cm (**f**). (**g**(**1**, **2**)) In addition to the raising of seedlings from seeds, the thinned out bare-rooted wild seedlings of *S. dullooa* were also collected from the forest floor and brought to the nursery and then transplanted into the polythene bags and maintained in the temporary nursery under partial shade near the planting site; after few months in rainy season, these are going to be used for raising large-scale plantation (**g**). (**h**) The animal cages are made with the slivers of *S. dullooa* and sold in the local hill market for carrying pigs (**h**)

Fig. 9.1 (continued)

9.1.2 Plant Data

9.1.2.1 Vegetative

Plant Habit

The species is congested medium-sized clump and more or less densely caespitose with pachymorph rhizome without or with very small necks; especially the young culm tips are long and thin elongated internodes, may hang down to the ground (scrambling or pendulous) with four to six big leaves (Fig. 9.1a) and peep above the gullies to trap the sunlight to support growth. Architectural design of the plant was found to be adopted to capitalize on the high-light regime of the early successional environment, and based on the leaf characteristics the species, *S. dullooa* is described as periodic growth-evergreen type (Rao and Ramakrishnan 1988).

Culm Height

Tall 8–22 m., culms erect, or leaning, or scandent.

Culm Diameter

3–7 cm.

Culm Thickness

Very thin walled and 2–5 mm from tip to base of culm.

Culm Colour

Dark green covered with silvery white pubescence; whitish band (1.0–1.5 cm) below the nodes distinctly visible in young culm, glossy when dry (Fig. 9.1b). The culm towards the sun attains slightly reddish colour while in the deep valleys and shady sites becomes deep green.

Internode Length

Schizostachyum dullooa is easily distinguishable by its very long (0.5–1.5 m) (Fig. 9.1c) thin-walled internodes covered with scattered shiny white or appressed hairs, and the wall skin is very sharp when cut. Culm nodes are slightly prominent and smooth. The length of internodes is increasing successively from base to the mid-culm zone followed by a sharp decrease in length towards the culm tip which is represented by a sharp peak in the middle of the internode curve. It indicates that the species is more or less shade tolerant and, in fact, grows along the moist banks of streams associating with tree shade inside the forest (Banik 2000, 2015b). This species is usually not found on the open hilltops, and under exposed conditions the internodes are comparatively short, and plant growth is stunted. Rao and Ramakrishnan (1988) have also reported that *S. dullooa* is relatively shade-tolerant bamboo species.

Branching

Usually from mid-culm to top. Dense tuft of slender, thin subequal 3–35 branches in a complement develop from each node (Fig. 9.1d).

Leaves

Leaf is big, usually with variable size; oblong-lanceolate, acuminate, rounded base into a 5–10 mm and pale yellowish petiole. The young leaves measure 25.0–30.0 × 3.5–4.6 cm, and old leaves measure 20.5–25.5 × 2.4–3.5 cm. Midrib is pale; nerves are six to ten pairs. Leaves are not easily separable from the branch. Clumps exhibit semi-deciduous to deciduous habit when grown on the hilltops or at dry-site condition.

Culm Sheath

Broad, leathery, thick and striate, with scattered white-appressed hairs prominent above, rounded at the top and then somewhat concavely truncated and loosely fringed with comparatively tall copper-coloured bristle. The sheath size varies with the culm size, from 15 cm long and 10 cm broad up to 30 cm long and 25 cm broad. *Imperfect blade* is narrow and linear-lanceolate (8–17 cm long and 0.8–1.8 cm broad at the base) with dense brown hairs, subulate and reflexed. Culm sheath is not auricled. *Ligule* is hairy, 3–5 mm long, fimbriate and light brownish. Leaf sheaths are striately veined; outer margin is hairy.

Vegetative Growth

In *Schizostachyum dullooa* culm emergence usually takes place during June to October, with maximum number of emergence in July to September. The breaking of bud dormancy on the node of a culm was found to be in basipetal order, gradually moving towards the base and being completed in 3-4 years. A clump may produce 3–6 culms per year. The culm life in a clump is 4–7 years (Banik 2000).

Physical Properties of Culm

The average fibre length of *Teinostachyum* sp. (syn. *Schizostachyum* sp.) is 3.6 mm (Liese 1980). Data on different strength parameters of the culm wood are not available.

9.1.2.2 Reproductive

Flowering Nature

Flowering mostly gregarious (Fig. 9.1e) and occasionally sporadic. The gregarious flowering in Chittagong (in the year 1999) moved towards South Tripura, Bagafa (in the year 2002–2003) (Banik 2004a), and then flowering waves moved towards north Tripura's Gandachara, bank of Juri River in 2004. Hasan (1973) reported that *Schizostachyum dullooa* flowers gregariously after 47 years of interval in Chittagong Hill Tracts. It appears from the reported flowering dates mentioned in the table below that the species flowers gregariously after 45 ± 5 and 35 ± 2 years, sporadically after 15 ± 2 years, showing two to three flowering populations may exist in the region (Table 9.1).

Inflorescence

A panicle of spiciform branches bearing verticils of few spikelets. *Spikelets* are 1–2.5 cm long and softly pubescent; glumes hairy eight to ten reeved. *Stamens* 6, monadelphous; *anthers* are 7–9 mm long; tip is smooth; *stigma* 3, short and red; the *ovary* is elongate and glabrous.

Seed

Small, elongated grain-like, somewhat broader and flat base with cylindrical top terminating in a long beak formed by the persistent base of the style and covered with glabrous glumes (Fig. 9.1f). The length of glumed seed (grain part) is 1.0–1.9 cm, and that of persistent style is 1.1–1.7 cm. The breadth and width of basal part are 0.14–0.19 and 0.18–0.26 cm, respectively. The ripe deglumed seeds are blackish brown (Fig. 9.1f). The weight of a hundred glumed seeds and deglumed seeds is 3.03 gm and 2.51 gm, respectively. Generally, 10 gm contains 393–410 seeds.

Table 9.1 Estimation of flowering cycle in *Schizostachyum dullooa* from available flowering records

Country/locality	Flowering date (calendar year)	References of flowering dates	Estimated flowering cycle (year)
Myanmar Monastery at Hawyaw in Katha District, Upper Burma	1892	Gamble (1896)	
India *Assam* Cachar, Mizo hill, Cachar forest reserve	1951–1953 1962 (Greg.) 1967–1968 2009 (Greg.) 2010–2011 (Greg) 2012 (final Spor.)	Gupta (1972) Nath (1962) Gupta (1972) Nath and Das (2010) Author observed	14–17 (1951–1968) 47 (1962–2009) 49 (1962–2011)
Mizoram	1962 (Greg.) 2010–2013	Nath (1962) Author observed	48 (1962–2010)
Tripura (South) Bagafa: Takkatulsi RF, Hrishyamukh, Sreenagar, Belonia, Amlighat *Tripura* (towards north) Ganganagar, Juri riverbank [not far from Mizo hills, Cachar)	2002–2003 (Spor.) 2004 (Greg.) 2009	Banik (2004) Banik (2004) Banik and Sharma (2009)	42 (1962–2004)
Meghalaya Khasi Hills	1968	Rao and Ramakrishnan (1988)	
Arunachal	2004–2005	Author observed	
Nagaland	2004–2005	Author observed	
Bangladesh *Chittagong Hill Tracts* Rangamati	1880 (Greg.) 1927 (Greg.)	Gamble (1896) Trevor (1927)	44 (1880–1927)
Kassalong Reserve Shishak	1974 (Greg.) 1990 (Spor.)	Hasan (1973) Banik (2000)	47 (1927–1974)
Chittagong Hazarikhil (Rangapani tea garden area), nearer to border South Tripura, India	1999 (Greg.)	Banik (2000)	37 (1962–1999)

Seed Germination

Do not dry the seeds under sunlight; rather saw immediately in seed germination beds. The fresh seeds (within 30 h of collection from plants) germinate within 5–7 days of sowing at 60–75 %; sowing after 7 days of collection gives about 50 % germination. It loses viability very quickly; seeds show only 0–5 % germination after 20 days of storage.

Seedling Character

Grass-like, thin stem with long internodes. In addition to raising seedlings in nursery, wild seedlings of four- to six-leaf stage can be collected from the naturally regenerated areas of the species after completion of flowering. A plentiful pure vegetation of *S. dullooa* growing naturally on the bank of Madhuchara stream flowing in the gully of Kuicha Tilla of Tekkatulshi hill in South Tripura was seen in large-scale death due to gregarious flowering in the year 2002–2003. Many seedlings were coming up from huge amount of seeds fallen on the ground below the flowering mother clumps. The place was revisited in February 2010, and found a small area scatteredly covered with a few regenerated medium-sized young clumps of the species. Clearing of forests and raising of rubber plantation destroyed many of the regenerating seedlings.

During gregarious seeding time, huge *wild seedlings* are produced compactly on the ground below the flowering mother clumps. These seedlings need to be thinned out for healthy growth by minimizing the competition among themselves. The thinned out bare-rooted seedlings (Fig. 9.1g(1)) should be brought to the nursery for transplantation in to the polythene bags (Fig. 9.1g(2)), and after few months these are used for large-scale plantation raising programmes.

Vegetative Propagation Methods

The guidelines for collection and bagging of wild bamboo seedling and multiplication through macroproliferation and their nursery management are provided in *Appendix I*, and for details, consult the INBAR Technical Report No. 6 (Banik 1995).

Tissue Culture Techniques

Not known.

Diversities and Conservation

Due to overexploitation and destruction of habitat, the availability of *Schizostachyum dullooa* bamboo becomes rare. Since 2002 the species has been flowering and died at Hrishamukh (South Tripura), then in Ganganagar near Gandachara and recently (2008–09) in Kanchanpur. Before that, in 1998, it flowered gregariously and exhibited mass scale death in Chittagong Hill Tracts. At present *S. dullooa* is considered as a threatened bamboo species of Northeast India and nearby Bangladesh forests (Banik 1994a, 2000). Therefore, it is urgently needed to protect and maintain some patches of vegetation of this species for economic and social reasons. With similar reasons some tribal families at Ambassa, Tripura, have been maintaining and conserving some patches of this species on the nearby hills.

9.1.3 Cultivation and Management

9.1.3.1 Site Selection and Preparation

Plantation can be raised along the bank of hilly streams in the gully sites. Moist lower hill slopes with shady condition can be also selected. Hilltops and dry places are very unsuitable site for survival and growth of the species.

9.1.3.2 Planting Space and Pit Making

Wide spacing (5×5 m, or more) is not necessary; in fact it adversely affects the clump growth. Compact plantation with 3×3 m spacing is good for erect culm growth. Pits should be dug at 3×3 m spacing in two to three lines above the flush flood level of rainwater in the streams; otherwise, during heavy rain planted seedlings may get washed away.

9.1.3.3 Fertilization and Soil Preparation

Fertilization and soil preparation are not required; sites having fertile humus ground should be selected.

9.1.3.4 Weeding and Vine Cutting

During the first 3 years of plantation, proper weeding has to be practised.

9.1.3.5 Aftercare and Tending Operations

Planted stocks should be protected from fire and grazing.

Remaining all other silvicultural practices like loosening of soil, mounding and mulching may be done at drier sites.

9.1.4 Specific Spot Characters of Schizostachyum dullooa for Field Identification

1. The bamboo is easily distinguishable by its very long 1.0–1.5 m internodes with slightly prominent nodes (Fig. 9.1a).
2. The young pendulous culm tips each with four to six leaves at the apex are visible in more or less densely caespitose clump.

3. Culms are very thin walled, 2–5 mm and dark green with the presence of fine-appressed scattered shiny silvery white (or colourless) hairs on the young culms, glossy when dry.
4. Dark green covered with silvery white pubescence; whitish band (1.0–1.5 cm) below the nodes is distinctly visible in young culm, glossy when dry (Fig. 9.1b).
5. The long internode is used as container, especially for carrying drinking water in the hills as there is no powdery substance inside the internode.

Chapter 10
Thyrsostachys Gamble

During 1891 a handsome bamboo of 'graceful habit and modern size' flowered in Katha District, Upper Burma. The flowering specimens and seeds of species were collected by J. W. Oliver, Conservator of Forests, Burma, and sent to the J. S. Gamble, Conservator of Forests, School Circle, to the director of the Imperial Forest School, Dehradun, and also some to the Royal Botanic Garden, Calcutta. After studying the flowering specimens of 1891, Gamble (1894, 1896) placed that bamboo under a new taxonomic genus *Thyrsostachys* and named it as *Thyrsostachys oliveri*, in the honour of Mr. Oliver (Banik 2014). Thus, the taxonomic nomenclature of the species was published as *Thyrsostachys oliveri* Gamble. The genus *Thyrsostachys* is naturally distributed in Thailand and Myanmar. It is also popular in cultivation in many different parts of Indian sub-continent. Only two species, namely, *Thyrsostachys oliveri* and *Thyrsostachys siamensis*, are known. Both the species are described below due to their socio-economic and industrial demand in the South Asian region.

10.1 *Thyrsostachys oliveri* Gamble

[Gamble in Ind. For. 20:1., 1894 and in Ann. Roy. Bot. Gard. Calcutta 7:58. 1896 and in Hook. f., Fl. Brit. India 7:397, 1897; Brandis, Indian Trees 671, 1906; Camus, Les Bambusees 137. 1913; Bor in India For Rec. (n.s.) Bot. 2(2):222.1941; Varmah and Bahadur in Ind. For. Rec. (n.s.) Bot. 6(1):4. 1980]

10.1.1 General Information

10.1.1.1 Vernacular and Local Names

Kanakaich (all over Tripura, India) (Banik 2004b), *bushai* (Chebipara/Karuamura/Amtali, Tripura), *kalangsi* (Kokbarok, Tripura), *nala bauns* (Orissa, India), *thanawa* (Myanmar), *phai ruakdum* (Thailand)

R.L. Banik, *Silviculture of South Asian Priority Bamboos*, Tropical Forestry,
DOI 10.1007/978-981-10-0569-5_10

10.1.1.2 Natural Distribution and General Habitat

Thyrsostachys oliveri believed to be a native species to Upper Burma and Thailand. This bamboo covers extensive areas on well-drained ground in the mixed deciduous forests of Upper Burma, often associated with teak and other important trees (Troup 1921). It does not occur in very moist types of forest, but is a characteristic of intermediate-type forest of Myanmar and the deciduous forests of Northern Thailand.

Thyrsostachys oliveri has been grown in southern Manipur (Moreh, Imphal, Churachandpur, Chandel, Imphal-Kohima road of Manipur) along the border of Myanmar due to its multiple uses and ornamental plantation in sacred places. In unknown year of the past, probably, from there planting propagules of this useful bamboo were carried by human beings to other part of Northeast India and Bangladesh (Banik 2004b, 2014). This species has been extensively cultivated in the rural homesteads of greater Sylhet District and some parts of Comilla District.

This bamboo has been isolatedly cultivated in the east (West Bengal, Orissa, Bihar), north (Uttar Pradesh), northeastern (Tripura up to Arunachal Pradesh), and in Uttarakhand, Kerala, Andaman, and even in peninsular India. It is a lowland to medium-altitude species.

10.1.1.3 Uses

It is mainly used as fishing rods, javelin, pole vaults and material for making furniture. Also it has been used in construction, as props, for making tools and industrial baskets. This bamboo is also planted as ornamental plants in the gardens, parks and hotels and also in sacred places. In Thailand *T. oliveri* is an important bamboo for edible shoot production (Ramyarangsi 1987). The percentage of (on fresh weight basis) sheath cover and edible portion of a freshly collected shoot is 46.2 and 52.8, respectively. The shoot meat possesses creamy white colour, and the taste is slightly bitter and astringent with crisp to tender texture (Banik 1997a, 2000).

The upper narrow, strong portion including the tip of culms, about 3–3.5 m long, is cut from the culms, mainly used as fishing rods and javelin and also exported to other parts of India and Maldives. The remaining basal 4.5-, 2.8- and 1.5 m (15, 9, 5 ft) portions are sold as poles for different construction works. Normally the tip portion of the culm of this bamboo is not fully straight; to make it straight and attractive, each piece is rubbed with muster oil mixed with washing soda and then placed over the heat of coal burnt in local country-made furnace(Fig. 10.1f(1)). The tip part then becomes comparatively soft and needs to be continuously pressed at all the nodes by a piece of wood to make it straight (Fig. 10.1f(2)). One skilled labour per day can make such 100 straight poles.

At present (2014) annually on an average 600 trucks, each carries about 11,000–15,000 bamboo poles from Tripura to Kolkata, and from Kolkata these are sold to different parts of India, like Delhi, Punjab, UP, Andhra, Kerala, etc., and also exported abroad. The demand of this bamboo in the local furniture (Fig. 10.1g) factories is also getting high.

Fig. 10.1 (**a**) *Thyrsostachys oliveri* clump is a handsome, caespitose and erect bamboo with persistent culm sheaths that wrap the stems from the base to the tip. (**b**) Offsets of *T. oliveri* are collected in March–April (spring) nursed in a rural community transit nursery beds having sand medium with regular watering like misting; it is maintained till planting is done in June–August (rainy days). (**c**) Thinned out newly produced sprouts from planted offsets of *T. oliveri* are collected by farmer to utilize as fodder. (**d**) In wide spacing (5×5 m, 6×6 m) plantation of *T. oliveri*, clumps are intercropped up to 2 years successfully with ginger, turmeric, cowpea, peanut, maize and mustard.(**e**) Storm-affected uprooted clump of *T. oliveri* is rehabilitated by mounding of soil at the base, and in the next season all new culms emerged straight. (**f**(**1**, **2**)) Normally the tip portion of the culm of *T. oliveri* is not fully straight; to make it straight and attractive, each piece is rubbed with muster oil mixed with washing soda and then placed over coal-burnt fire in local country-made furnace. The tip part then becomes comparatively soft and needs to be continuously pressed at all the nodes by placing in the groove of a wooden stump to make it straight. (**g**) Different types of furniture, such as sofa sets, are made of *T. oliveri* culms, in the drawing room of author

Fig. 10.1 (continued)

10.1.2 Plant Data

10.1.2.1 Vegetative

Plant Habit

Thyrsostachys oliveri is a large, handsome, caespitose and erect bamboo (Fig. 9.1a).

Culm Character

Culms are 15–25 m tall, straight 2.5–5.0 cm in diameter, and bright green with whitish silky surface when young; dull green to grey green, rough or yellowish on maturity. *Internode* 7.0–22.0 cm long; in some cases, it was found to be 55.8 cm; nodes are hardly thickened. The culm wall is very thick at the base and gradually becomes thin walled towards the top (Table 10.1).

Table 10.1 Culm diameter and wall thickness in relation to intermodal position of culms in *Thyrsostachys oliveri*

	Culm node number (from base to top)											
Parameter (cm)	2	14	22	30	36	45	46	52	57	61	66	71
Internode length	7.0	15.0	19.5	21.0	22.0	14.7	14.0	13.5	13.0	10.0	7.5	4.5
Culm diameter	6.6	4.56	4.48	4.06	3.56	2.81	1.74	1.12	0.89	0.61	0.45	0.36
Wall thickness	2.61	1.01	0.98	0.75	0.61	0.55	0.49	0.36	0.32	0.23	0.16	0.15
Lumen diameter at culm centre	0.65	2.46	2.72	2.75	2.11	1.74	0.99	0.67	0.46	0.32	0.21	0.06

Branching

Branches fascicled at the nodes, lower ones ascending and upper ones horizontal. Branching is always at the one third to one fourth in the upper part of the culms, making the bamboo attractive.

Leaves

Linear-lanceolate, acuminate, 8–20 cm long, 1.2–1.5 cm broad, light green, base rounded into a short petiole, both surface rough, hairy beneath and scabrous on the margins. It sheds leaves during January to March.

Culm Sheath

Culms with persistent culm sheaths that wrap the stems from the base to the tip (Fig. 10.1a). The culm sheaths are fibrous, felted, symmetrical, imbricating at the base, green when young, turning orange and finally brown, persistent, clothed on the back with thick white stiff pubescence, rounded at the top, 17–22 cm long and wider (5–17 cm) at the base; margins are thin and ciliate; *ligule* is 4 mm high, dentate; no distinct *auricle*, very small; *blade* conical, narrowly triangular, 4–8 cm long, 1–3 cm broad, recurved, awl shaped and hairy on both surfaces. The young shoot may have yellow lines on the sheath; lower culm nodes are often bearded with thin roots.

Vegetative Growth

Culm emerges during May to August. A clump may have many 15–25 culms. It is less vigorous in the initial stage of clump development but shows a steady culm increment once established. The growth of weeds particularly vines is maximum due to its less-dense crown, and weeds need to be removed to avoid competition for moisture and nutrient.

Physical and Mechanical Properties of Culm

Some strength properties of *T. oliveri* bamboo are compared to two major bamboo furnitures of India (Table 10.2).

A higher-specific gravity, MOR (modulus of rupture) and crushing strength suggest that *T. oliveri* can be used for load-bearing structure and in crossbeams especially in building as compared to *D. stocksii* and *D. strictus* (Seethalakshmi and Kumar 1988).

10.1.2.2 Reproductive

Flowering Nature

The species flowered in Myanmar in 1891. Seeds were sent to Calcutta and Dehradun, approximately 1500 km apart, and the clumps, which were raised at both the places, flowered simultaneously in 1939–1940. Flowering starts in November, and seeds start ripening at the end of January. It appears from the past flowering records that the species flowers gregariously after 48–50 years (Table 10.3). The last flowering reports are from Dehradun in 1989–1991, but no flowering reports are available from other parts of India and Bangladesh. However, from the available flowering history, it appears that *T. oliveri* bamboo growing in Northeast India (especially Tripura and nearby areas) may flower soon in near future or may be of different genotype with longer interseeding period.

Inflorescence

It is a large compound curved thyrsoid panicle, bearing bracts at the nodes, with usually one long and two shorter flower-bearing spikelets and one or more sterile ones. *Spikelets* 1.80–2.54 cm; fertile flowers 2-3; empty glumes 2; flowering glumes long. *Lodicules* 2, lanceolate, acuminate, ciliate, very thin. *Stamens* long exserted, drooping filaments purple; *anther* yellow. *Ovary* yellow, stalked, surmounted by a long *style*, bearing 3 broad feathery *stigmas*.

Table 10.2 Strength properties of *Thyrsostachys oliveri* in comparison to other two bamboo furnitures of India

	Bamboo species		
Strength properties	*Thyrsostachys oliveri*	*Dendrocalamus stocksii*	*Dendrocalamus strictus*
Specific gravity	0.758	0.691	0.631
Modulus of rupture (kg/cm^2)	917.7	610	734
Max. crushing stress (kg/cm^2)	478	386	359

Source: Vishwanath and Arade (2014)

Table 10.3 Estimation of flowering cycle in *Thyrsostachys oliveri* from available flowering records

Country/locality	Flowering date (calendar year)	References of flowering dates	Estimated flowering cycle (year)
Myanmar Katha Dist. Upper Burma Mandalay, Ruby Mine dist.	1891 Greg. 1939, 1940 Greg. 1902–1904, 1911–1914 Spor.	Gamble (1896)), Bor (1941), Troup (1921)	48–49 (1891–1939, 1940)
India Royal Botanical Garden, *Calcutta* (seed of 1891 from *Myanmar*)	1938–1939	Bor (1941), McClure (1966)	47–48 (1891–1938, 1939)
Dehradun (seed of 1891 from *Myanmar*)	1940,1941 1989, 1991	Bor (1941), Tewari (1992)	48–50 (1940, 1941–1989, 1991)

Seed

Cylindrical with a somewhat broader tops ending in a long beak formed by the persistent base of the style, glabrous, 8–10 mm long and 1.5–2.0 mm in diameter, *germinates* promptly with 80–90 %. The weight of 40 seeds varies from 0.62 to 0.73 g.

Seedling

Grass-like seedlings as in most other bamboo species.

Vegetative Propagation Methods

Mainly by offset planting. Successful (40–45 %) culm cutting is produced from 2- to 3-node segments collected from 20 to 30 month-old culms. Rooting success is about 30–50 % in branch cutting and ground layering.

Planting of offset is the conventional method of cultivation as the materials are small 0.5–1.0 kg, cheap and available in thousands and can carry many at a time unlike offsets of most other bamboo species. The offsets are usually collected from 12 to 18 month-old culms during March–April; each offset should have a minimum of two culm nodes with intact basal rhizome part. Instead of direct outplanting, these offsets are to be maintained under shed-net roof (60–70 % light) in transit nursery beds (1.2 m wide, 21 cm deep and 6–10 m long), which contain sand media (Fig. 10.1b) for 2–3 months till the arrival of next rainy season (June–August). During this time, offsets in the beds are regularly watered mechanically or preferably through intermittent misting.

The guidelines for producing culm and branch cutting, collection and bagging of wild bamboo seedling and multiplication through macroproliferation and their nursery management are provided in *Appendix I*, and for details consult the INBAR Technical Report No. 6 (Banik 1995).

Tissue Culture

Species (explant)	Medium and environments	Results	References
T. oliveri nodal buds	1. MS + BAP (1.0 mg/l)	1. Shoot bud initiation and developed	Islam and Rahman (2005)
	2. Shoot further cultured on MS + BAP (1.0–5.0 mg/l) with number of subculture	2. Multiplication of shoot	
	3. Shoots grown on ½ MS + NAA (1.0–3.0 mg /l)+ IBA (1.0-5.0 mg/l)	3. Profuse rooting in shoots, TC plantlets hardened, acclimatized and then outplanted	

10.1.3 Cultivation and Management

Farmers in Northeast India, especially in Tripura state, have been earning annually Indian rupees 2.5 lakh (0.25 million) per ha from fourth year onwards by selling the bamboo poles, making it a profitable farming. During the 1980s about 35 years back, there were only one to two families cultivating, in total about 1–2 ha; at present this bamboo has been cultivated extensively as a farm crop covering several hundreds of hectares in Tripura and as it became a cash crop to the farmer. The requirement of poles has been increasing with the time to cope pace with higher demand in external markets and local furniture industries.

10.1.3.1 Site Selection and Land Preparation

The land can be prepared in three ways, such as overall land preparation, strip land preparation and land preparation for spot planting. In steep sloping fields, at desired spacing, the spot can be cleared of weeds and shrubs with 1.0 m radius around the planting spot, and then digging of pit is done without much disturbing the surrounding soil. In Mohanpur, Katlamara of Tripura farmers have been cultivating *Thyrsostachys oliveri*, in their flat farmland by thoroughly ploughing the land. The strip land preparation is generally adopted on hill slope planting. To prevent the soil erosion and water loss, strip land preparation can be made parallel with the contours.

10.1.3.2 Planting Space and Pit Making

Usually close spacing either 1.25 (for narrow culm production) or 1.5 m between plants and 2 m line to line are maintained to produce straight, less branchy and more number of culms per hectare of land. Such spacing also results in solid and comparatively narrower culms for making fishing rods. The planting pit is usually of 25.0×25.0×25.0 cm in size. Further clumps growing in moist and shady places will produce big and less solid culms than grown under the direct sunlight and upland dry sites. In Tripura farmers have been maintaining the closer spacing either 1.25 or 1.5 m in raising *T. oliveri* plantation in their farmland to produce straight, less branchy and more number of culms per hectare of land. About 2960 propagules are planted per hectare, and after 3 years of plantation age by selective felling, nearly 8000 culms could be harvested.

In many villages of eastern Bangladesh (greater Sylhet and Comilla Districts), clumps of this bamboo species are commonly planted in a line at closer spacing (0.8×0.8 m) along the boundary of homesteads as a live fence and also as a veil for the families especially for female members (Banik 2000).

10.1.3.3 Field Planting

The offsets are easily taken out from the sand media of transit beds without or least damage and planted in the field during rainy season (June–July) for higher (90–98 %) survival rate.

10.1.3.4 Aftercare and Tending Operations

(a) *Thinning of sprouts:* After 1–2 weeks of planting, a number of tender new coppice sprouts/shoots with young leaves are first produced from the nodes of offsets/rhizomes. Usually one to two healthy sprouts is allowed to grow, and the remaining are thinned out and utilized as fodder (Fig. 10.1c). However, planted offsets may not always produce many sprouts.
(b) *Intercropping:* During initial 3–4 years in wide spacing (5×5 m), plantation, due to its narrow leaves and light crown, can be intercropped successfully with ginger, pineapple, turmeric, cowpea, peanut, maize and mustard (Fig. 10.1d). However, wide spacing (5×5 m) resulted in more branching, and culms are not erect; rather, they droop at the upper part; so all the culms are tied together with ropes at the mid-position of the clump to keep them erect and straight. After 2 years the young soft culms become strong and erect automatically.

Loosening of Soil, Application of Fertilizer, Mounding and Mulching Soil work and mounding may add FYM (10–25 kg per clump) depending on the fertility status of the land. Finally, 5–7 cm thick mulching has to be done with the litters and bamboo leaves. Often the clumps get uprooted during northwestern storm. Mounding of soil has to be done at the base, and in next season all new culms would grow straight, and thus the clumps are rehabilitated (Fig. 10.1e). The practice of mulching is very important for maintaining healthy and prolonged productive clump condition through conserving soil moisture and maintaining ground temperature around 20–22 °C during draught and winter.

When clumps are not regularly maintained and cleaned, termite attack may be seen on the persistent culm sheath attached on the internodes of current year culms; some clumps were seen partly suppressed by climbers, like *Mikania scandens*, *Eupatorium odoratum*, etc., and the harvested poles lose their attractiveness, and thus market value gets lowered.

10.1.4 Specific Characters for Identification of Thyrsostachys oliveri in the Field

1. The plant grows in dense clumps, with branching near the top, making it attractive (Fig. 10.1a).
2. The culms are straight with persistent culm sheath, and the sheaths wrap the stem from the base to the tip.
3. The culm tip part is thin and narrow (Fig. 10.1f).
4. The culm sheath is fairly large, felted with truncate apex, symmetrical and narrow; auricles are absent; blade is arising from a flat top.
5. In the young shoot, it may have yellow lines on the sheath; blades are linear; lower culm nodes are often bearded.
6. Well-known structural bamboo for furniture industry (Fig. 10.1g).

10.2 *Thyrsostachys siamensis* Gamble

10.2.1 General Information

10.2.1.1 Vernacular and Local Names

Phai ruak (Thailand); Monastery Bamboo (USA)

10.2.1.2 Natural Distribution and General Habitat

Thyrsostachys siamensis is a native bamboo species to Thailand, Myanmar, Yunnan, Laos and Vietnam. The species was introduced in different parts of India, Sri Lanka, Bangladesh and Peninsular Malaysia, which nowadays is being extensively cultivated in these countries, especially in the park, hotels and monastery, and also as patches of commercial plantations.

10.2.1.3 Uses

In Thailand *T. siamensis* yields wood and good edible shoots for the rural people (Ramyarangsi 1987) and also is used as Thai umbrella bamboo and umbrella-handle bamboo. Several kinds of traditional foods such as 'soup nor mai' (or bamboo salad) are as popular as 'som tam' (or papaya salad) nationwide. It is also planted as an ornamental and is one of the well-known bamboo furnitures used as raw material in many factories of Kanchanaburi, Thailand.

10.2.2 Plant Data

10.2.2.1 Vegetative

Plant Habit

Close, compact and erect.

Culm Character

Dull to glaucous green, 6–10 m tall, matured ones yellowish green, glabrous, lower internodes sparsely pubescent (Fig. 10.2a).

Culm Diameter

3–4 cm.

Nodes and Internode

Nodes are slightly greater than internode. Nodal line is single and horizontal, and there is a narrow band of white powdery ring below the nodal line; internode is 12–27 cm long.

Branching

From mid-culm position to top.

Fig. 10.2 (**a**) *Thyrsostachys siamensis* bamboo has a medium-sized clump with dull to glaucous green culms. (**b**) The shape of seeds of *T. siamensis* is cylindrical, surmounted by a yellowish and glabrous tops ending in a long beak formed by the persistent base of the style. (**c**) The clumps of both the species look very similar when grown side by side. However, on freshly detached small branches, wilting of leaves (rolling and drying) may be seen immediately (within 2–3 min) in *T. siamensis*, while the leaves remain fresh and turgid in *T. oliveri*

Leaves

Narrow, linear, 7–15 cm long × 0.8–1.2 cm broad, very short petiole and tip slightly twisted.

Culm Sheath

Generally persistent in imbricate manner towards the base of the culm, covering more than half of the internode to full internode. *Sheath* is properly symmetrical, 15–30 cm long and 8–12 cm wide at base, with silvery hairs along the margin;

auricle consists none; *imperfect blade* is erect, narrowly triangular, 6–10 cm long, 2–3 cm wide at base much less than *T. oliveri*, abaxial surface glabrous and slightly pubescent on the adaxial surface; the edges are recurved; *ligule* is narrow, 0.25 cm broad, glabrous and finely ciliate.

Vegetative Growth

Moist summer to rainy season (May–September) is the growth period.

Physical and Mechanical Properties of Culm

Thyrsostachys siamensis is almost solid at the bottom and middle portion, but hollow at the top portion. In this species the moisture content (%) at cross section along the culm length in green condition is 110.5, 98.8 and 95.3 at the bottom, middle and top portion, respectively. In *Thyrsostachys siamensis* density varies from bottom to top in the range of 700–400 kg/m^3 along with node and internode.

The values of some mechanical properties of air-dried culm (12 % moisture content) of *Thyrsostachys oliveri* and *T. siamensis* are also presented below (Table 10.4).

Narasimhamurthy et al. (2013) studied different mechanical properties of the culm wood collected from the clumps of *T. siamensis* and *Dendrocalamus membranaceus* grown at Tumkur district and Karnataka, India, and found that *T. siamensis* bears the MOR and compressive strength about 15 % more than that of *D. membranaceus*, and MOE of *T. siamensis* was found to be about 5 % more than that of *D. membranaceus* (Table 10.5).

10.2.2.2 Reproductive

Flowering

Five offsets of *Thyrsostachys siamensis* are collected from Kanchanaburi, Thailand, during 1979 and planted in the line no. 49, at BFRI Bambusetum; all the clumps raised in the Bambusetum flowered during 1990 and also flowered in Thailand

Table 10.4 Mechanical properties of air-dried culm of *Thyrsostachys oliveri* and *T. siamensis*

	Static banding (N/mm^2)		Compression parallel to grain (N/mm^2)		Shear (N/mm^2)	
	MS	MS	MS	MOE	MS(N)	MS(WN)
Thyrsostachys oliveri	64.4	22,396	59.1	2322	17.9	16.4
Thyrsostachys siamensis	119.6	30,033	64.0	2349	21.7	19.0

Source: Sint et al. (2008)

Note: *MS* maximum stress, *MOE* modulus of elasticity, *N* specimen with node, *WN* specimen without node

Table 10.5 A comparative data on mechanical properties of *Thyrsostachys siamensis* and *D. membranaceus*

	Compressive strength (MPa)		Modulus of rupture (MOR) (MPa)		Modulus of elasticity (MOE) (MPa)	
Culm position	*T. siamensis*	*D. membranaceus*	*T. siamensis*	*D. membranaceus*	*T. siamensis*	*D. membranaceus*
Top	66.7	64.7	124.3	102.3	8464	6609
Middle	82.2	69.7	119.0	97.3	6402	7458
Bottom	83.6	66.7	129.2	127.7	6407	6414
Mean	77.5	67.0	124.2	109.1	7091	6827

during 2009–2010; a few seeds collected through INBAR and raised two plants and planted in JICA supported Bambusetum at Anandanagar, Tripura. The species usually flower *completely* within 1 year, while a few clumps also showed *part-flowering* nature for 2½ years and then died (Banik 2000).

Inflorescence

It is a large, graceful with many thin branchlets bearing bracteates, clusters or few fertile *spikelets*. *Stamens* are exserted, filaments are free, *anther* is pale yellow and many honeybees were seen to visit florets during blooming.

Seed Character

About 10–82 seeds weigh 1 gm; it varied from locality to locality in Thailand (Anantachote 1987). The species flowered in BFRI Chittagong in 1990, and seeds were described (Banik 2000), 88 seeds in 1 gm; shape was cylindrical, surmounted by a yellowish and glabrous tops ending in a long beak formed by the persistent base of the style (Fig. 10.2b).

Seed Germination

In *Thyrsostachys siamensis* seed germination ranged from 7 to 86 % when collected from different localities of Thailand (Anantachote 1987). Big seeds showed better germination than the smaller ones. Seeds collected from Chaum district of Phetchaburi Province showed 86 % germination, while those from Ngao District of Lampang showed only 7 % germination. The seed sample stored at room temperature lost the viability within 21 months, which can be extended by reducing the initial moisture content before storing. The seeds stored for 27 months under low temperature (2–4 °C), at 5.9–10.2 % moisture content, could maintain a high percentage of viability and were able to germinate (89.2–92.5 %) within 3–4 days (Ramyarangsi 1988).

Seedling

Looks somewhat like seedlings of other bamboo species.

Vegetative Propagation

As rhizomes are comparatively smaller and light, they are commonly used for plantation. Both culm (2-node segment) and branch cuttings treated with rooting hormones, IBA (250 ppm) or NAA (100 ppm), give better rooting inside the sand propagation bed under misting. The guidelines for producing culm and branch cutting, collection and bagging of wild bamboo seedling and multiplication through macroproliferation and their nursery management are provided in *Appendix I*; for details consult the INBAR Technical Report No. 6 (Banik 1995).

Tissue Culture

T. siamensis (explant)	Medium and environments	Results	References
(Node with bud)	(a) Liquid or gelled MS or B_5+BAP 1.0 mg/l by repeated transfer of culture at 3–4 weeks	(a) Shoot proliferation, 20–75 shoots/culture	Banik et.al (1993)
	(b) In same media within 6–7 weeks after the proliferation of shoot	(b) Healthy roots produced	

Cytogenetics

Not seen any documentation.

Diversities

Detail exploration and data collection documentation of both morphological and interseeding periods of different population have to be carried out in the natural belt of the species. It appears from the seed germination data (Anantachote 1987) of *T. siamensis* that some population are fertile, while others are very poor in germination which is similar to the observation of Banik (1987a) on *B. tulda*. Both physiological and genetical studies of such populations need detail investigation. Documentation and sharing of the information among the researchers of different countries of the region are necessary.

10.2.3 Cultivation and Management

Regarding site selection *Thyrsostachys siamensis* is a lowland to medium-altitude species. It is less vigorous in establishment and clump development but shows a steady culm increment once established. The species can be cultivated in a wide range of South Asian climate. Due to the erect culm with light crown and compact clump nature *Thyrsostachys oliveri* and *T. siamensis* are ideal species for road side plantation at closer spacing in 2–3 rows. *T. siamensis* is also seen to grow even in indoor condition of hotel, airport, etc., and can tolerate partial light condition and semidrier sites up to some extent, while *T. oliveri* does not like shady and dry habitat but rather prefers to grow in moist and warm condition.

Both plantation raising and grove managements in *Thyrsostachys siamensis* are more or less similar to that of *T. oliveri.*

In both the species selection felling method is followed—(a) cut the weak and broken, and retain the strong culms; (b) dense clumps are thinned out selectively from the central part of clump by cutting the overmature culms (more than 3 years old and above).

10.2.4 Specific Spot Characters of Thyrsostachys siamensis in the Field for Identification

1. The species is characterized by its compact clumps.
2. Persistent culm sheaths with attenuated rounded top.
3. Branching from upper mid-culm
4. Leaves are small.

Note: The clumps of both the species look very similar. However, I found the quick (within 2–3 min) drying and rolling of leaves on freshly detached small branches of *T. siamensis*, while the leaves on the branches of *T. oliveri* remain fresh and turgid (Fig. 10.2c).

Chapter 11
Ringal Bamboos of the Himalayas

In most parts of the central Himalayas, there are a number of shrubby bamboos, locally known as *ringal*, found in the temperate zones and higher altitudes (snow line).

At the higher altitude of Himachal Pradesh from 1800 to 3600 m (districts Shimla, Solan, Sirmour, Kullu, Chamba and Mandi) and Uttarakhand (Garhwal and Kumaon hills), Nepal and also in western part of Bhutan, people have been cultivating different types of temperate shrubby bamboos on the hill slopes along with natural belt.

On the basis of socio-economic and ecologic benefit, commonly four major types of ringal bamboo are used by the local indigenous people in different areas of the Himalayas. These species are:

Drepanostachyum falcatum (Nees) Keng f. (local name: Gol ringal)
Himalayacalamus falconeri (Munro) Keng f. (local name: Deo ringal)
Thamnocalamus spathiflorus (Trin.) Munro (local name: *Tham* ringal)
Sinarundinaria anceps (Mitford) C.S. Chao & Renvoize (local name: Jamura ringal)

Himalayacalamus falconeri is recorded at highest elevations among all the above bamboo species. *Drepanostachyum falcatum* and *Sinarundinaria anceps* are other two species used in the local system for different articles. The communities that make articles from bamboo and *ringal* are called *Baruree* and *Rudia*, respectively.

11.1 *Drepanostachyum* Keng f.

Stapleton (1994a, b) while studying bamboos in Nepal and Bhutan described some of the species of *Arundinaria* under the genus *Drepanostachyum* Keng f.

Clump-forming thornless bamboos up to 5 m tall, with many branches, are found from 1000 m up to 2200 m in drier subtropical forests and also cultivated. Branch

R.L. Banik, *Silviculture of South Asian Priority Bamboos*, Tropical Forestry,
DOI 10.1007/978-981-10-0569-5_11

buds at the nodes of the culms are shorter than their height and always open. The buds have many initials visible, which will produce many branches at each node; about half of them grow in the first year. The branches are quite uniform in size and spread around the culm. When growing strongly, the upper half of the culm sheaths are very narrow, and culm-sheath ligules are long and ragged. The sheaths are always rough inside at the top, and this distinguishes them from *Himalayacalamus* species. Rhizomes are short and solid, less than 30 cm in length and similar to those of *Thamnocalamus* (Stapleton 1994a, b).

11.1.1 Drepanostachyum falcatum (Munro) Keng f.

(Syn. *Arundinaria falcata* Nees in Linnaea 9:478. 1834; Munro, Trans, Linn. Soc. Lond. 26: 26. 1868; Gamble in Ann. Roy. Bot. Gard. Cal. 7:12–13.1896 and Hook. F., Fl. Brit. India 7:381. 1897.)

Vern. Name: *Gol ringal*, *nirgal*, *nigal*, *nagre* (Uttarakhand, Himachal of India), *gola ringal*, *gol ringal*, *gadh ringal* and *godra ringal* (Garhwal, Uattarakhanda); uombang (Meghalaya); *tite nigalo* and *diu nigalo* (Nepal).

11.1.1.1 Distribution

The species is usually found in drier subtropical forests and forms dense thickets or moderately dense undergrowth in evergreen oak, deodar, chir and rhododendron forests. The species grows naturally on the hills from 1000 to 2000 m from Himachal to Western Nepal. It is also cultivated around farmland near Dehradun and lower hills. It is planted in gullies, along the paths and on wasteland. It is also planted as effective soil stabilizer on terrace risers (Fig. 11.1a).

11.1.1.2 Habit

Drepanostachyum falcatum is a shrubby and clump-forming bamboo (Fig. 11.1a(1)). The culm is about 2.5–6.7 m tall and around 2.0 cm in diameter when cultivated, but according to Stapleton (1994b), they are usually much smaller in forest because of browsing. The culm internodes are 9–15 cm long, smooth surface, cylindrical and green with white scurf when young (Fig. 11.1a(2)). Gol ringal has slightly elevated nodes in the culms, whereas these are smooth in the case of Deo ringal.

Fig. 11.1 (**a**(**1**,**2**)) Gol ringal (*Drepanostachyum falcatum*) is a shrubby and clump-forming bamboo. The culm internodes are 9–15 cm long, smooth surface, cylindrical and green with white scurf when young. (**b**), and *Tham* ringal (*Thamnocalamus spathiflorus*) is a densely tufted shrubby bamboo, has erect culm and is 6 m tall and 2 cm in diameter, and branches are chiefly from the upper nodes. (**c**) *Tham* ringal has the capability to stop moving earths and rolling small stones on slopes. (**d**) A flowering clump of Gol ringal (*D. falcatum*) bamboo; all leafy twigs produced flowers; clump died after flowering. (**e**) Seedlings of Gol ringal (*D. falcatum*) raised and maintained in the nursery. (**f**) Flowering clump of *Tham* ringal bamboo (*Thamnocalamus spathiflorus*). (**g**) *Tham* ringal seeds (caryopsis) are small, grain-like. (**h**) A few plants of Deo ringal (*Himalayacalamus falconeri* syn. *Thamnocalamus falconeri*) was found to produce flowers in the nursery bed. (**i**) A ringal bamboo nursery with many nursery beds under open sky on the flat site of hills of Barkot, Uttarakhand at 1400 m altitude. Both Gol ringal and Deo ringal/Dev ringal plants are seen in nursery beds. (**j**) Within 6–8 months of age, a Dev ringal seedling may attain 45–60 cm height and develop 3–6 shoots (culms) with well-developed rhizome system below. (**k**) Many novelty items are made of Deo ringal/Dev ringal (*Thamnocalamus spathiflorus*). It is the most harvested species for making baskets, mats, flowerpots, etc., and other commercial purposes

Fig. 11.1 (continued)

Fig. 11.1 (continued)

11.1.1.3 Uses

The culms are not very straight, and they have rather swollen nodes, which make them not very suitable as weaving materials. They are also used in construction. The cane is less flexible, durable and comparatively weak, so mainly used for making hats, baskets, fishing rods, etc. *Roof covering* is used for the covering of grassy roof of 'Goth' or single-floor houses (thatch). *Thangra* is mainly used as stakes for vegetable vines—like beans and other leguminous vegetables—and as gourds. The foliage is often fed to animals in winter or browsed in the forest. The new shoots are bitter. The plant is very effectively utilized as soil stabilizer in Nepal.

11.2 *Sinarundinaria*

11.2.1 Sinarundinaria anceps (Mitford) C.S. Chao & Renvoize

[*Sinarundinaria anceps* (Mitf.) Chao & Renv. Kew Bull. 44: 359. 1989. (Fig. 90). *Arundinaria anceps* Freeman Mitford, Bamboo Garden 181.]

(Syn. *Chimonobambusa jaunsarensis* (Gamble) Bahadur & Naithani in Indian J For 1: 40 1978.)

Vern. Name: Jamura ringal (Uttarakhand, Himachal of India) and nigalo (Nepal)

11.2.1.1 Distribution

Mostly found within the altitude of 2100–2700 m and grows naturally in clusters inside the forest. This bamboo is predominantly available under oak and juniper forests of Kumaon hills, distributed mostly in cool temperate zone and sometimes in subalpine zone, formed undergrowth over more than 130 km^2 in Garhwal (Osmaston 1922), but now rare (Tewari 1992).

11.2.1.2 Habit

The species is reed-like, with single culm at intervals, and non-clump forming, and rhizome is long and creeping and 0.7 cm in diameter; joints are about 1–1.5 cm long; scales are straw-coloured and shining. The species is economically very important. This bamboo is more flexible and durable.

11.2.1.3 Uses

The products of bamboos, viz. baskets, mats, etc., are used for domestic purposes in rural areas and are being manufactured by men during their spare time when agricultural activities are at low intensity. The details of utilization are (Arya 2014):

Ghat ki Solti—used for carrying grains and other domestic utility things

Moli or Gobar (*dung*) *Solti*—used for carrying cow dung, fodder, fuel, etc.
Sutari Solti—used for carrying dry leaves from forest
Koluyn—used for storage of all kinds of grains
Jungada (*Palaki*)—used for caring small babies
Bisawa—used for drying and cleaning of grains and pulses
Kauna—used for collection of grains during the harvesting of crops
Doli—used for worship of different goddesses during the Navratri and 'Nanda Devi Yatra'

The species flowered in 1978 from Chamoli district, Garhwal (Tewari 1992).

11.3 *Thamnocalamus* Munro

This is a genus of temperate climate bamboo of Asia. Two species are described from the Himalayas. Previously the species of this genus were included in the genus *Arundinaria*. The genus differs from *Arundinaria* in having many branches of the same level and size in each node.

11.3.1 Himalayacalamus falconeri (Munro) Keng f.

(Syn. *Thamnocalamus falconeri* Munro)

Vern. Name: Deo ringal, *Deo ringal*, *Dev ringal*, *bodputra* (Uttarakhand, Himachal of India): *thudi nigalo and singhane* (Nepal).

11.3.1.1 Distribution

The species grows naturally on the hills and/or under oak forests between 1900 and 2750 m in the hills. In Nepal it is common at the summit of Phulchowki in the Kathmandu valley, but the variety in western Bhutan has slightly ridged culms and light hairs at the base of the culm sheath (Stapleton 1994a).

11.3.1.2 Habit

This bamboo is smooth, flexible, shining and soft in nature; the species is distinguished from other *Himalayacalamus* species by the absence of spines, hairs or auricles on the bullet-shaped culm sheaths and the smooth fairly short culm internodes. The culm sheaths have a short and broad ligule and are often striped with yellow and purple lines (Stapleton 1994a, b). The top of the sheath is narrow, as in species of *Drepanostachyum*, and it has a densely hairy tall, fringed ligule. The young shoots have a thick glutinous covering, which leads to the local name in

Eastern Nepal, *singhane nigalo*. This covering may help to reduce attack by insects such as shoot bores (Stapleton 1994b).

The Deo ringal occurs naturally from Western Himalayas to Central Nepal on the hills from altitudes 1900 to 2750 m near water sources and high-humidity sites. It generally grows to a height of 3.3–6.5 m and is around 1.5–2.3 cm in diameter.

11.3.1.3 Uses

Dev ringal is most preferred for craft purposes since it splits evenly and cleanly. One culm can be split into 8–10 strips to make one basket. This *Dev* (meaning God) *ringal* has been used to make temple-related basketry articles such as fans, umbrellas, crowns, etc., for idols of gods and baskets and trays for offerings. The details of these items are:

Chhapri/Tokari (*basket*)—used for keeping breads (rotis), vegetables and fruits.
Supa—used for cleaning rice, wheat and pulses.
Fancy basket—used as flower pots and other ornamental purposes.
Mat (*Motha/Matula*)—used for drying ripe grains, pulses and mustard during harvesting period.
Goddess umbrella—used for preparation of goddess umbrella which is locally called 'Chhantoli' which is used during the famous 'Nanda Devi Raj Jat' Yatara in Uttarakhand (Arya 2014). It is believed that an uncoated ringal basket is pure and holy. The household items are also made from this bamboo.

11.3.2 Thamnocalamus spathiflorus (Trin.) Munro

(Syn. *Thamnocalamus aristatus* (Gamble) E.G. Camus; *Arundinaria spathiflora*)

Vern. Name: *Tham* ringal (Uttarakhand, Himachal of India) and nigalo (Nepal). The species was introduced to Europe as an ornamental plant.

11.3.2.1 Distribution

Found to grow between 2800 and 3500 m altitude or above in temperate forest, in the Himalayan zone of northwest India, Uttarakhand (Kumaon and Garhwal Hills); Himachal Pradesh (Simla); common bamboo of Central and Eastern to Western Nepal. It is a clump-forming frost-hardy bamboo. It prefers deeply sloping sites.

As a reference it is also to be noted that the *Tham* ringal (*Thamnocalamus spathiflorus* syn. *Thamnocalamus aristatus*) naturally occurs in moist or wet mountain slopes and ravines in association with *Abies*, *Acer*, *Magnolia*, *Betula* and *Rhododendron* between 2400 and 4000 m.

11.3.2.2 Habit

A densely tufted shrubby bamboo; *culm* is erect, 6 m tall and 2 cm in diameter (Fig. 11.1b). Glaucous green when young, shining and yellow with age; internode is short 8–15 cm long; node is much raised, dark coloured and marked by a ring formed by the scars of the fallen sheath; branches are chiefly from the upper nodes(Fig. 11.1b). Culm sheath measures 13 × 5 cm, loose, striate, coriaceous, glabrous and ciliate on the edge. This is a very strong bamboo and the central hole is much smaller in comparison to other ringal species. And thus *Tham* (stop) ringal has the capability to stop moving earths and rolling small stones on slopes (Fig. 11.1c).

11.3.2.3 Uses

The small brittle culm with short internodes and swollen nodes makes *Thamnocalamus spathiflorus* unsuitable for weaving works. The bamboo is long, durable and hard in nature. The species is commonly used in covering and weaving roof of grassy houses (goth or chhani). Mature stick is used for making hookah pipes and making walking sticks.

The *Tham* ringal (*Thamnocalamus spathiflorus*) is usually planted in rows with close spacing on the upper slopes of their huts to stop the rolling of small stones, thus protecting their habitation (Banik et al. 2008).

The panicles and flowers of *Thamnocalamus* remain hidden by sheaths, while the sheaths fall quickly from the panicles of *Drepanostachyum* (Stapleton 1994a).

Chromosome number of *Thamnocalamus spathiflorus*: 48 (Guang-zhu 1987).

11.3.3 Flowering, Seeding Phenology and Natural Regeneration

This ringal group of plant usually flowers frequently, irregularly at times, and gregariously over large areas, while few culms may be found in flowers almost every year. Campbell (1988) stated that 'recorded gregarious dates 1858, 1878/86, 1916, 1950, 1972 suggesting a rough periodicity of about 30 years'. Troup (1921) and later Varmah and Bahadur (1980) reported that the species flower gregariously after 28–35 years interval. However, other sporadic flowering events have also been occurring.

In recent times, *Gol ringal* bamboo, *Drepanostachyum falcatum*, flowered gregariously during 1998 in a stretch of 20 km from Bhatwari to Dabrani in Uttarkashi district at an altitude between 1500 and 2000 m all along the riverbank of Bhagirathi and in mountains (Naithani 1998). It is reported that the flowering cycle in some cohort (populations) of Gol ringal varies from 12 to 18 years (Brandis 1899, Gamble

1896). Therefore, availability of seeds in Gol ringal is not as few and far between as what has happened in other bamboo species. Similarly, while discussing the field foresters, they also informed me that the ringal bamboo, in general, flowers after 12–14 years interval, and then all the seeded clumps die (Fig. 11.1d). Seedlings were raised from the seeds in the seedbed, then pricked up into polythene bags and maintained in the nursery (Fig. 11.1e).

The past incidents of mast flowering in *Tham* ringal (*T. spathiflorus*) were recorded during 1818–1821 from northwestern Himalayas, during 1881–1882 from Jaunsar and Tehri Garhwal (Broun 1886) and during 1942 from Chakrata. Dr. Naithani told me that he observed in June 2001 that few clumps of *T. spathiflorus* in flowering state were found near Deoban, Chakrata. After 1 year, i.e. June 2002, all the clumps burst into gregarious flowering in Kedarnath Forest Division and between Ghangaria and Pushpawati river on the way to National Park, Chamoli. So in recent years, gregarious flowering of *Thamnocalamus spathiflorus* and *T. falconeri* bamboos occurred in the hills of Uttarakhand during 2001–2003 (Naithani et al. 2003), and thus the estimated flowering period in *T. spathiflorus* is about 60 years. Gregarious flowering in *Thamnocalamus spathiflorus* subsp. *spathiflorus* covering an area of 3.5 km^2 at an altitude of 3000 m in Sikkim, India, was recorded during August 2006. Populations of this bamboo are known to completely collapse consequent to gregarious flowering (Fig. 11.1f), and this withering away of dry bamboo culms causes gaps at patches previously occupied by the bamboo clumps.

As flowering cycle, in some population, seems to be short, one can keep record of plantation raised from seeds and its subsequent next flowering date (from seed to seed). This will provide an exact time interval and locality for future availability of seeds. Thus, it appears that the *Tham* ringal bamboo species might have two seeding intervals (short and long) and fertile (seed) and less fertile (few seeds produced) behaviour exhibited by different flowering population (cohorts). The caryopsis is grain-like, small and elongated covered with glumes (Fig. 11.1g).

The Deo or Dev ringal (*Himalayacalamus falconeri* syn. *Thamnocalamus falconeri*) also flowers both sporadically and gregariously. However, sporadically flowering clumps are common, but seed has not been produced. It was reported that about 60 years back this species lastly flowered gregariously in the northwest Himalayan region (Blatter 1929, 1929–1930). A few years back while working in the hills of Pipalkoti on September 12, 2002, Mr. G.S. Rana, DFO, Badrinath Forest Division, informed me that Deo *ringal* has started flowering that year in the hills of Nanda Devi and also in previous year near Badrinath temple. Further on June 05, 2005, I also collected 150 seedlings of Deo ringal from Ranichauri (Hill Agri. Centre), to use in the multilocation trial at Pantnagar Govind Ballabh Pant University of Agriculture and Technology (GBPUNIV) campus. The seedlings were raised in 2004 from the seeds collected from the population growing naturally on the nearby hills. Both *Gol ringal* and *Deo/Dev ringal* bamboo seedlings were observed on March 18, 2007, at Soil Conservation Division Nursery, Barkot, Uttarakhand. A few plants of *Deo* ringal were found to start producing flowers in the nursery bed (Fig. 11.1h), and through discussion I came to know that these flowered plants were raised by planting rhizome collected from the nearby hills. So logically it can be assumed that this species has also simultaneously flowered in the nature from where the rhizomes were collected. Stapleton (1994b) reported that sporadic flowering is common in this species at Nepal, but seed has not been found. Thus, it appears from the above facts that Gol, Deo and *Tham* ringal bamboo have been flowering in recent years in different hills of Uttarakhand.

During flowering time the flower buds become visible in March to April, the flowers opened early in July and the seeds usually ripen in the month of October to November. Under natural conditions the seed lies on the ground (during cold winter) until the following season, June to July, when it germinates (Troup 1921). Thus, it seems the seeds of ringal bamboo may need cold shock (vernalization treatment) for germination.

(i) *Maintenance of natural regeneration through protection of seeding mother and wild seedling*: Flowering of hill bamboos and their subsequent drying creates panic among the local indigenous people, who depend mostly on this resource. Firstly, it creates acute shortage of bamboo sticks to works and fodder for animals. Secondly, there is a fear of forest fire. No harvesting of flowering mothers should be allowed till the seeds are mature and fallen on the ground. Additionally, bamboo inflorescence are also favoured food of sheep, goats and other wild animals. Flowering culms of *T. spathiflorus* were found to be more prone to such animal damage during January to when other sources of fodder were scarce (Goraya et al. 2008). These flowering culms should be protected for obtaining the mature seeds and success of natural regeneration.

During flowering in a ringal plant, seeds fall nearer and below the crown and start germination producing seedlings in profuse number. Seedlings could also be seen in lower slopes and flatlands far below the clumps where the washed away seeds accumulated. These seedlings among themselves start competing for food, light and water. Ultimately a few can survive and develop into adult clumps, but most of them die due to competition. Besides competition, some surrounding factors like *weeds*, *fire* and *shade* also influence the survival and establishment of the regenerating seedlings (Banik 1988b). The condition of the bamboo seedlings is found to be better with elongated culm nodes under partial shade and lower weed conditions than under full weeds and complete shade. In complete shade, almost all seedlings gradually degenerate. A certain amount of overhead cover was found necessary for survival and development of bamboo seedlings during the earliest stage of natural regeneration. The harvesting of dead mother clumps may be delayed for a few months so that seedlings regenerating below the crown can get partial shade. In general bamboo seedlings need to be protected from scorching sunlight. If the area is properly managed, gradually in 3 or 4 years time clusters begin to form, and eventually in 5 years or more the area carries a homogeneous crop of more or less equally spaced young clumps.

11.3.4 Ringal Nursery and Planting Material

(i) *Collection of rhizomes*: This is of similar procedure to other commercial bamboo species—*Bambusa*, *Dendrocalamus* sp., etc. The rhizome planting becomes risky if it is collected from the area where flowering is going on. The plants developed from such rhizome may start flowering and die simultaneously with the mother. A few plants of Deo ringal (*Himalayacalamus falconeri* syn. *Thamnocalamus falconeri*) produced flowers in the nursery bed, and through discussion I came to know that these flowered plants were raised by planting rhizome collected from the hills. So logically it can be assumed that this species has also simultaneously flowered in the nature from where the rhizomes were collected. Thus, it appears that rhizome planting should be discouraged in case of cohort having short flowering cycle. The plants developed from rhizomes will never get full duration of vegetative life like those developed from seeds. Such rhizome-developed plants will flower within short period of time before culms will attain the utilizable/merchantable size.

(ii) *Seed sowing in poly bags*: A temporary nursery is raised on the flat site of hills of Barkot Uttarakhand at 1400 m altitude, with many nursery beds under open sky nearer to the area of flowering so that the collected seeds can get similar habitat condition (especially temperature condition) for germination (Fig. 11.1h, i). The collected ripe seeds are to be sown immediately in poly bags with regular watering. The bags with sown seeds are to be left under the open sky for cold treatment during ensuing winter, and then start germination leading to production of seedlings in the next spring (Fig. 11.1j). Within 6–8 months of age, a seedling of Deo/Dev ringal bamboo may attain 45–60 cm height and produce 3–6 brownish culms (shoots) with well-developed rhizome system below (Fig. 11.1j). After 1 year of age, the seedling may be transplanted in the field for raising plantation. Sun-dried seeds of *Thamnocalamus spathiflorus* were stored at room temperature (20–25 °C) for 4 years and viability was 90–93 % for the entire period. 25 °C was found to be the best temperature in which after 3 days of sowing, seed germination started, and 93.3 % germination was recorded 12 days after sowing (Bag et al. 2013).

The seed bed may be prepared in a greenhouse having temperature of 20 °C. Regular watering is needed. The germination may take place after 4–6 months. Seedlings are to be maintained in partially shaded place inside the greenhouse until large enough to plant out. After germination maintenance of partial sunlight condition at the nursery stage at least for 9 months and then gradually to the full sunlight could be an important practice to produce healthy bamboo seedlings.

(iii) *Thinning, collection and utilization of wild seedlings of ringal*: Such huge number of seedlings would die due to competition. So wild seedlings (wildings) should be thinned out (pricked up) carefully (with minimum damage in roots) from such densely populated natural regeneration areas to ease the competition. Plants of Gol ringal in the nursery were collected as pricks (wildings) from the areas where recent flowering took place. These thinned-out seedlings (pricks) can be brought into the nursery and be transplanted in to poly bags filled in a mixture of 1 (soil): 2(sand): 1 compost. As a good nursery practice, a cemented platform should be constructed on which beds (size: 1.25 m wide, 6.0 m long and 18 cm deep) are made and demarcated by placing three layers of brick on the edge. Each bed is to be filled in with above-mentioned mixture of soil, sand and compost. The depth (18 cm) of the growing mixture inside the bed is enough to hold the plants, it is porous and well drained and the rhizomes do not get damaged, while well-developed seedlings are dug out. The rhizome and roots of the ringal seedlings cannot penetrate the cemented platform below the polythene bags. Though initially this type of expenditure is high, it is one time only and finally brings better success in the nursery and in the field. It is better to avoid the planting of collected seedlings (pricks) into the bed prepared directly on the ground to cut down the risk of rhizome/root damage of seedlings during transplantation to the field.

(iv) *Application of macroproliferation method for multiplication of ringal seedling*: As it appears from the records and conversation with the local hill people that ringal bamboo has *long* as well as *short* flowering populations (cohorts) growing naturally in the hills, *seedlings developed from short flowering cycle population (cohort) should not be multiplied by macroproliferation* method. The seedlings of such population have short (say 15 years) period of vegetative life. Let it be explained; out of 15 years time (from seed to seed), about 1–1½ years time is spent in the nursery and then about 3–5 years in the field for attaining the culms' utilizable and commercial size. In total about 6–7 years is spent out of 15 years, and so the resource can be harvested for only 7–9 years. Now, if such cohort of ringal seedlings are multiplied again by rhizome separation (macroproliferation), another 1½–2 years are lost from the 7 to 9 years commercial life of the bamboo before they die due to flower again. Therefore, it is advisable to collect as much as possible seeds and seedlings of short flowering cycle cohort of ringal bamboo, so that there is no need of further splitting of rhizomes for increasing the seedling number by macroproliferation method. However, the method may be utilized only in cohort/population having long flowering cycle.

(v) *Macro- and micropropagation*: Planting of part clump (large rhizome section with 2–4 culms should be used) is a very successful method of planting the species. Smaller plant may not survive at all. In tissue culture method, the nodal explants with single axillary bud of *D. falcatum* showed good shoot multiplication on Murashige and Skoog (MS) medium supplemented with 3.0 mg per litre of 6-benzylaminopurine (BAP). A regular subculture in every 3–4 weeks increased the multiplication rate. These in vitro shoots rooted well when transferred to MS medium supplemented with auxin indole-3-butyric acid (IBA), NAA or IAA (Arya 2008).

An efficient protocol for in vitro propagation of *Thamnocalamus spathiflorus* through multiple shoot formation from zygotic embryos excised from germinating seeds, as well as from nodal explants taken from a 2 year-old plant, has been developed (Bag et al. 2000). Multiple shoot formation was standardized in both the cases on MS medium supplemented with 5.0 μM BAP and 1.0 μM IBA. Clumps of three to four microshoots when transferred to plant growth regulator-free medium following 14 days' initial culture on medium supplemented with 150.0 μM IBA resulted in 100 % rooting with 5–12 roots per rooted plant. Then, 18 months old *in vitro* propagated plants transferred into earthen pots containing soil and growth under outdoor conditions and compared with seedlings of the same age. The results show that *in vitro* propagated and hardened plants of *T. spathiflorus* are morphologically as well as functionally comparable to seed-raised plants of the same age.

However, a number of multilocation plantation trials of TC plants may be carried out to prescribe this method for raising large-scale plantation in the harsh environment of the Himalayas.

11.3.4.1 Utilization and Socio-economy

In Garhwal Himalayas during winter, the *ringal* bamboos remain green while all other fodder plants become dry and shed leaves. Foliage of hilly bamboos, therefore, forms a very important source of fodder in the time of scarcity. The clumps of hilly bamboo, such as *D. falcatum* and *Thamnocalamus spathiflorus*, are usually lopped at heights of 1–1.5 m giving the clumps bushy appearance (Goraya et al. 2008). However, the fodder production in the clumps of *D. falcatum* is more preferred (34 %) than in *Thamnocalamus spathiflorus* (21 %).

Ecologically *Thamnocalamus spathiflorus* is extremely important ringal bamboo species for wildlife, providing food for animals such as red panda (*Ailurus fulgens*) and bears and shelter for birds such as pheasants (*Lophophorus impejanus*). It is also browsed by livestock in winter. It does not hinder regeneration of trees as seedlings can grow in the gaps between the clumps (Stapleton 1994b).

The *ringal* is one of the oldest weaving materials used by mankind, and in the district Rudraprayag *ringal* weavers are known as *Rudhiya*. It has been a main source of livelihood for local inhabitants due to its multifaceted use. *Drepanostachyum falcatum* and *Thamnocalamus spathiflorus* species of *ringal* are the most harvested species for making baskets, mats, flowerpots, etc., and other commercial purposes (Fig. 11.1k). Due to easy availability of *Drepanostachyum falcatum* (Gol ringal) in lower altitudes (1000–2000 m), it is used maximally (87 %), and on high-altitudinal regions (above 2000 m) *Thamnocalamus spathiflorus* (Dev ringal) is used maximally (83 %); however, Sararu ringal and Bhattputra (locally identified) are also used (70 % and 61 %). Due to less availability of *Tham* ringal, it is used minimally (35 %) in high altitudes (Kumar 2009a). However, *Drepanostachyum falcatum* is not a preferred bamboo by the weavers due to its less-durable nature and roughness of bark (fibre).

It is found that r*ingal* is the primary livelihood activity particularly within the socially backward caste. These families have retained ancient as well as traditional knowledge of ringal weaving work since time immemorial. Ringal bamboo thus plays an important role in enhancing and strengthening their economic as well as socio-economic status as these families are directly or indirectly engaged in ringal weaving work. It serves to fulfil the various day-to-day needs of these communities like housing, utensils, food and agricultural products and thus plays important socio-economic role (Kumar 2009b).

They are totally dependent on this natural resource and have no other source of income. About 78 % annual income of the stakeholders generated from the *ringal* activity and nearly about 14 % and 8 % income come from labour and agricultural activity. Maximum part of the earned money from *ringal* is used for providing food (53 %) and clothes (27 %), and the remaining about 20 % money is used in medicines and marriage and recreations (Kumar 2009b). In all the hilly areas, *ringal* sector contributes the highest to generate income as compared to other sources. *Ringal* bamboo is low-input cost and high-income-generating activity for its weavers. Although *ringal* has played a vital role in the day-to-day activities of the inhabitants, it has never achieved a status of commercially used resource due to lack of knowledge and awareness among the local people and foresters.

11.3.4.2 Diversities and Conservation

There exist lots of morphological variations within a species of ringal bamboo population in the wide range of natural distribution at various heights of altitude. In Kaski district of Nepal, *Drepanostachyum falcatum* has small ciliate leaf sheath auricles, while in Palpa district it has long cilia on one leaf sheath edge and hairs under the leaves. Another species *T. spathiflorus* extends along the Himalayas and has several subspecies and varieties (Stapleton 1994b). The leaf sheaths are quite variable, and some sheaths have no auricles or bristles at all. Other sheaths have small auricles when they are young. In Chautara, Nepal, the leaf sheaths have larger, more persistent auricles with spreading bristles. Techniques of morphological and molecular characterization of *Thamnocalamus spathiflorus* subsp. *spathiflorus* at population level have been developed (Bhattacharya et al. 2009) and can be utilized for studying the genetic variation among the population.

Ringal conservation can be improved through livelihood improvement of the weavers and not by direct ban on *ringal* harvesting. To reduce dependency on forests, weavers should be provided the plantation of extensively used *ringal* species in their wastelands. If the *ringal*-based livelihood activity would be started properly, the migration of some unemployed youth of the area to plains can be reduced, because they can get the job opportunity in *ringal* sector. To increase *ringal* production, scientific techniques about nursery development through seeds and cuttings and information about the time of flowering and seed maturing should be given to the weavers through trainings and meetings. Training and awareness programmes should be conducted for *ringal* weavers for making fancy and modern products of *ringal* like flowerpots, small baskets, pen stands, file covers, fancy bags, etc., to compete the modern market demand.

Appendix I

Dos and Don'ts in Producing Bamboo Cuttings, Collection and Bagging of Wild Bamboo Seedling and Multiplication Through Macroproliferation

A. **Basic environmental conditions required for vegetative propagation in the propagation bed**

1. *High air humidity surrounding the cutting is an essential condition for rooting.*

    - Through experience it has been observed that the use of sand as propagation medium in the propagation bed is efficient as it is cheap, easily available, and has all physical properties to induce better and early rooting (see under B 6).
    - The bamboo cuttings must not be allowed to show wilting for any length of time.
    - Wilting of the cuttings due to drop of humidity may injure the cuttings to such extent that rooting may not occur, even though high-humidity conditions are subsequently resumed.
    - Death due to desiccation before rooting is the major cause of lack of success in propagating bamboos.
    - So the need for adequate watering is discussed as follows:

2. *Watering/misting on cuttings to maintain humidity.*

    - Always use iron-free and normal fresh water during watering and misting operation in propagation nursery including overall nursery management activities. Water containing iron used in misting system would plug the nozzles and delivery pipelines. Such water forms a reddish thin layer on the plant body and surface of the propagation medium and interferes

R.L. Banik, *Silviculture of South Asian Priority Bamboos*, Tropical Forestry,
DOI 10.1007/978-981-10-0569-5

oxygen passage inside media and hinders root initiations and development in the cuttings and other planting materials.

- In misting system, proper nozzles should be used that breaks the water into a fine spray or provide fogging on the cuttings.
- The drained out water from the propagation bed can be collected in a tank and after cleaning may be reused.
- Intermittent misting provides showering/bathing effect that cools the body temperature of the plant piece (cuttings) and thus downs the rate of respiration, so the stored food level remains higher in the plant piece means better chances/amount of rooting.
- Continuous misting usually brings about rotting in the cuttings. So intermittent misting has been found to be better for rooting in bamboos.
- The common weakness/drawbacks are that in most of the propagation nursery, watering/misting is not maintained properly.

3. *Air and rooting media temperature.*

- The optimum temperature for both rooting medium and air should be within 25–35 °C.
- The air temperature about 35 °C accelerates the bud break and sprouting, and thus stored food in the cuttings is utilized rapidly for shooting, leaving far less for rooting.
- The higher temperature (above 38–40 °C) even for short time is likely to result in the death of the cuttings.

4. *Light condition during propagation.*

- Shading (50–60 % sunlight reaches the bed) is beneficial for bamboo species such as *Bambusa cacharensis*, *B. polymorpha*, *B. pallida*, *B. nutans*, *B. tulda*, *T. oliveri*, etc. which take longer time to root. (Fig. A.1a, b)
- Overhead partial shade with side ventilation cuts down the air temperature near the propagation beds. This condition controls the height of new shoots developed from culm/branch cuttings.

B. **Propagation bed and rooting media**

5. *Propagation bed size.*

- Propagation beds are prepared on the cemented levelled platform; bed size are generally 1.2 m in width, 6 m or more in length and 21–28 cm in depth with removable three-layered bricks on top of each other (not fixed with cement at the sides and edges) and filled with medium to fine size of particle of sand (Fig. A.2). In rural village at the absence of cemented platform, beds are made on the properly levelled ground covered with black polythene sheet filled with coarse sand at the bottom layer of 7 cm

Fig. A.1 (**a**) Laths made of bamboo in sizes of 5 cm (*width*) × 1.3 m (*long*) are held by wire or nylon ropes at 5 cm gap from each other, and a 5–7-m-long fence is made. The lath rolls are laid on frames above the propagation bed to allow 50 % light below. Such overhead shade is very suitable for temporary bamboo nursery at rural areas. (**b**) A nursery having propagation beds under the green net shed allows only 60 % light below and cuts down the intense heat that prevents drying, and sprouts produced by cuttings can continue photosynthesis

depth and above it a fine sand layer of 15–20 cm deep (Fig. A.3); the bed sides and edges may also be fixed with bamboo flats (Fig. A.4)

6. *Sand has been found as a suitable type of rooting medium for root production* (Banik 1983b).

- It is chemically inert, maintains uniform media temperature and does not harbour any insect/harmful microorganism.
- It maintains full moist but well-aerated bed condition in the propagation beds.
- Rooted cuttings are dug up easily from sand bed with less root damage.
- It is comparatively a cheap rooting medium and is readily available.

Fig. A.2 Propagation beds are made on a cemented levelled platform to prevent penetration of cuttings roots, and thus rooted cuttings are easy to lift and transplant from beds to polythene bags; each bed is 1.2 m in width, 6 m or more in length and 21–28 cm in depth with removable three-layered bricks at the bed sides and edges and contains medium to fine size of particle of sand as rooting medium which should be fully moist by watering. As the bricks are not jointed below and sidewise with each other by mortar (cement and sand), the excess water from the bed drains out easily

Fig. A.3 In rural village at the absence of cemented platform, beds are made on the properly levelled ground covered with black polythene sheet to prevent root penetration and to facilitate drainage; each bed is filled with coarse sand at the bottom layer of 7 cm depth and above it a fine sand layer of 15–20 cm deep

7. *Drainage condition in propagation media.*

   - Sand media in the propagation bed should be fully moist by watering (Fig. A.2).
   - A propagation bed having misting facilities with sprinkler is efficient (Fig. A.5) in rooting cuttings than a bed without it. Misting may also be done from above through hanging sprinklers (Fig. 3.1h of *B. balcooa*).
   - Waterlogging hinders root initiations and rooting by limiting the oxygen supply to the portion of vegetative pieces buried in the bed.

Fig. A.4 In a temporary nursery, the bed sides and edges may be fixed with bamboo flats

Fig. A.5 A propagation bed having misting facilities with sprinkler is efficient in rooting cuttings than a bed without misting. Water needs to be distributed evenly over the bed with sufficient pressure (at least 30 lb) for creating a fine mist condition

- Waterlogging in rooting medium limits the oxygen supply to the portion of vegetative pieces buried in the bed and thus hinders biological activities and results less or no roots.
- Waterlogging also promotes moulds and rotting of cuttings. So, a well-drained condition in the propagation bed should be maintained.
- Adequate drainage should be ensured.

8. *Rooting media temperature.*

- The optimum temperature for rooting media should be within 25–35 °C.

C. **Types and conditions of cuttings for producing planting materials**

9. *Major types of cuttings.*

- *Culm cuttings:* cuttings produced from the culm segments
- *Branch cuttings:* cuttings produced from the branches

10. *Optimum age of cutting materials.*

	Dos	Do not use
Culm cuttings	Take culm segments from 1½–2-year-old culm root well	Less than 1- and more than 2-year-old culm
Branch cuttings	Take from 1½–2-year-old culm	Branches from too young and old culm

11. *Position and right type of cutting materials on the standing bamboo clump.*

	Species	Collect and use	Do not use
Culm cuttings	*B. vulgaris, B. balcooa, D. hamiltonii* and *D longispathus*	Culm segments from the lower to mid portion of culm (1½–2 years old)	Upper narrow culm portion
	B. polymorpha and *B. cacharensis*	culm segments from the mid to upper portion of culm (1½–2 years old)	Upper narrow culm portion
	B. bambos, B. tulda, Thyrsostachys oliveri and *T. siamensis*	Culm segments mid to lower culm nodal positions	Upper narrow culm portion
Branch cuttings	Species mentioned under D. Preparation of cuttings *B. balcooa, B. polymorpha, B. vulgaris, D. hamiltonii, D. asper,* etc.	Fresh creamy coloured root initials developed at swollen rhizomatous branch base (Fig. A.6) In some species profuse aerial roots and rhizomes are produced at branch base, called as pre-rooted and pre-rhizomed branch (Fig. A.7)	Blackish or old root at swollen rhizomatous branch base

12. *Collection time of plant parts for producing rooted cuttings.*

	Suitable time	Not suitable time
Culm cutting	Later part of February to October (only four batches)	Cool and dry months
Branch cuttings	Later part of April (after a few showers of rain) to October (only three batches)	Cool and dry months

Note: There is *a 1-week gap in between the batches* of cutting production for cleaning the sand, bricks and mist nozzles and preparation in the nursery

D. Criteria for Producing Quality Planting Material (QPM)

13. *Always identify and select quality mother clumps of desired bamboo species as sources of collecting plant parts (culm-segments, nodes, branches, offsets, rhizomes, etc.) for propagation.*

- Selected clumps should be healthy (not diseased) and adult (usually minimum 8 - 10 year old), so that all characters are expressed.
- Clumps should be high-yielding ---- those produce tall, large diameter more culms per year. Select more number of such mother clumps from a

population for each of the desired bamboo species, and this would ensure wider genetic base in raised bamboo planting materials (BPM). All the BPMs raised from a mother clump are genetically similar and belong to one clone only. BPMs from more number of clones of a species provide wider genetic base in the plantation of that species.

- Mark the selected clumps as for example as Species Name - Bambusa tulda/Clump serial Number/Locality source, in short BT/CL001/Asm. The raised BPMs of the same clone are to be labeled in the nursery beds as similar to the code number of source mother clump.
- Maintain the BPMs in the nursery under healthy conditions and proper care till the field planting.

E. **Preparation of cuttings**

14. *Sizing and condition of cutting.*
The pieces and segments of culms and collected branches are brought into nursery by packing and covering with jute/cloth sacs soaked in water to prevent from getting dried and injured during transportation (Fig. A.8). The following points are important:

	Species	Dos and use	Do not use
Culm cutting	*B. vulgaris*, *B. balcooa, B. bambos, B. nutans, B. tulda*, *D. asper*, *D. giganteus*, *D. hamiltonii, D. membranaceus*, *D. strictus*, *D. stocksii*, *Gigantochloa andamanica*, *T. oliveri* and *T. siamensis*	Two-node culm segments (from 1½- to 2-year-old culm) with viable healthy buds at nodes Sharp tools, e.g. saw; keep 5–10 cm on either side of the node, and dip the culm segments of all species in water with 0.1 % Bavistin solution for 3–5 h (Fig. A.9)	Culm node having dead (blackish) or injured buds Splitted ends in a culm segment
	B. polymorpha, *B. cacharensis*, *D. longispathus* and *Schizostachyum dullooa*	Mostly one-node culm segments but may use two-node segment if culm internodes are not too elongated	Same as above
Branch cuttings	*B. balcooa*, *B. bambos*, *B. cacharensis*, *B. polymorpha*, *B. nutans*, *B. tulda*, *B. vulgaris*, *D asper*, *D. giganteus*, *D. hamiltonii*, *D longispathus*, *D. membranaceus*, *D. strictus*, *D. stocksii*, *G. andamanica*, *T. oliveri*, and *T. siamensis*	Use sharply toothed saw to excise the pre-rooted branches from a culm node so that branch bases, roots and rhizomes are not damaged (Fig. A.10a, b) Trim leaves' side branches and main branch tips, keeping 3–4 nodes by secateurs Dip the branch base in water with 0.1 % Bavistin (a fungicide) solution for 3–5 h (Fig. A.10) Expose the soaked base of branches by removing thin papery sheath	Black or old root, dead buds at swollen rhizomatous branch base Splitted or damaged rhizomatous branch base

Fig. A.6 Most of the bamboo species having stout and thick branches develop root initials and roots only near the buds at swollen rhizomatous base

15. *The use of rooting hormones.*
Rooting hormones have been found very useful in rooting of cuttings for the bamboo species such as *Bambusa bambos*, *B. cacharensis*, *B. polymorpha*, *B. tulda*, *B. nutans*, *D. asper*, *D. giganteus*, *D. hamiltonii*, *D. longispathus*, *D. membranaceus*, *D. strictus*, *D. stocksii*, *G. andamanica*, *T. oliveri and T. siamensis* and thin-walled species like *S. dullooa* which is difficult to root.

16. *Treatment of cuttings with rooting hormones.*
    - The most common rooting hormones are IBA (indole butyric acid) and NAA (naphthalene acetic acid).
    - Right dose of hormone— 200–250 ppm *(parts per million) IBA* is a good concentration of the hormone for dipping the cuttings. A combination of the two hormones, *i.e. NAA (100 ppm)* and *IBA (200 ppm)*, can also be used.
    - How to make 200 ppm solution—dissolve 200 mg of the IBA or NAA in 10 ml of ethyl alcohol in a 1-litre volumetric flask. Add distilled water to make the volume of 1 litre.
    - How to use rooting hormone—a quick dip of branch base (for branch cutting) in 0.1 % Bavistin solution for a few seconds following an IBA

Fig. A.7 In some species like *B. vulgaris*, *D. hamiltonii*, etc., profuse roots and rhizomes are produced at branch base, called as pre-rooted and pre- rhizomed branch; these can be successfully utilized for producing planting materials

Fig. A.8 The collected pieces and segments of culms and branches are brought into nursery by packing and covering with jute/cloth sacs soaked in water to prevent from getting dried and injured during transportation

Fig. A.9 The culm segment and branches are sized and pruned with a sharp saw, and all are dipped in water mixed with 0.1 % Bavistin (a fungicide) solution for 3–5 h for healthy rooting of cutting

Fig. A.10 (**a, b**) A beak-shaped sharply toothed saw is used to excise the pre-rooted and pre-rhizomed branches from the culm node; similarly a young branch with roots and root initials at the rhizomatous base is also collected by this tool to minimize the damage of roots and branch base

treatment (rooting hormone powder/hormone solution) should be done. Similarly the buds on the node of culm segment (for culm cutting) have to be treated either by powder or tissue paper soaked with solution.

- Otherwise, the IBA in talc (root-promoting powder) may be mixed with a little amount of Bavistin powder. However, most of the bamboo species root satisfactorily without any hormonal treatment.

Fig. A.11 Air inside the hollow portion at both the ends of the culm segment has been replaced by filling with sand rooting media for uninterrupted absorption of moisture through lumen to have better rooting at propagation bed

Fig. A.12 The culm segments are placed horizontally in a row inside the propagation bed, 1–2 cm below the surface of sand medium with 10–15 cm gap side by side

17. *Placing of cuttings in propagation beds.*

- Fill the hollow portion at both the ends of the culm segment with sand rooting media for better rooting in cuttings at propagation bed (Fig. A.11).
- The culm cutting segments are to be placed horizontally in a row inside the propagation bed, 1–2 cm below the surface of sand medium with 5–10 cm gap (Fig. A.12).
- The branch cuttings need to be placed erect putting only the rhizomatous base, 4–5 cm deep below the surface of sand medium with 2.0 cm gap (Fig. A.13), and medium has to be pressed at cutting base to keep it erect at place (Fig. A.14).
- Misting of water has to be started immediately after placing the culm segments and branches in the propagation bed.

Fig. A.13 The branch cuttings need to be placed erect putting only the rhizomatous base, 4–5 cm deep below the surface of sand medium with 2.0 cm gap

F. **Rooting and management of cuttings in the nursery**

18. *Shoot and root development in cuttings.*
    The cutting starts producing leaves and branches within 7–10 days and produces profuse active roots in the propagation bed, depending on the season and species within 4–8 weeks (for cuttings taken during April to August) or 8–10 weeks (for cuttings taken during October to November) (Fig. 3.1h of *B. balcooa*; Fig. 3.3h of *B. cacharensis*; Fig. 3.5f of *B. polymorpha*; Fig. 3.7j of *B. vulgaris*).

19. *Management of cuttings while in propagation beds.*
    Leaves that drop on the rooting medium and as well as any dead cuttings should be removed promptly from the propagation beds. These materials harbour pathogens and can destroy other cuttings. Proper sanitation measures should be taken.

Harmful pathogens/pests are best eliminated by dealing with the following measures:

- Jungle/weeds near the propagation beds and nursery should be cleared to minimize the insect/pest population. Integrated pest management has to be followed to control disease problems in the cuttings.
- Occasionally the propagation area has to be washed with a 2 % formaldehyde solution or sodium hypochlorite solution (1 part in 9 parts water).

Fig. A.14 The sand medium has to be pressed at cutting base to keep the branch cutting erect at place

- The misting water, as far as possible, must be free from pathogens.
- The sand rooting media should be washed and cleaned by hot water in each time of use.
- There should not be any stagnation of water on the pucca platform, drain and nearby the propagation beds.
- Infected leaves or twigs of a cutting have to be collected and burnt away from the nursery.

G. **Shifting of rooted cuttings from sand bed to polythene bags containing soil mix**

20. *Handling and removal of rooted cuttings from the propagation bed.*

Dos	Don'ts
Remove the cutting from the propagation bed when only adequately (sufficiently) rooted Can feel the amount of root by inserting fingers inside the rooting medium Remove the rooted cuttings batch by batch; almost every day, as all the cuttings in the propagation bed do not root at a time Afternoon or rainy/cloudy days are the best time of removal of rooted cuttings from the propagation bed	*Do not* keep the rooted cuttings for longer period in the sand medium as it is not a growth medium

21. *How are the rooted cuttings from sand propagation beds shifted?*

- While removing, the above-ground portion of cutting is to be held (not pulled) by the left hand and simultaneously the finger tips of the right hand be pushed slowly inside the rooting medium (sand); the fingers and palm are to be placed below the roots of the cuttings, and lift up with ease so that minimum damage will take place to the roots.
- After adequate rooting in cuttings, remove the bricks from the edges of the bed, and then slowly dig out and lift (not pulled out) the cuttings from the sand rooting medium (Fig. 3.3h of *B. cacharensis*).
- Immediately after taking out from the sand bed, the rooted cuttings should be placed/kept in a container (bucket/bowl) having water to prevent desiccation.
- The freshly removed rooted cuttings (soaked/dipped in water) are to be carried in a container with water nearer to the poly bags (half filled with growth mixture) already kept ready for transplanting.

22. *Preparation of polythene bags.*

- The poly bag's size for branch cuttings is 15 × 25 cm; and for culm cuttings, the size is 21 × 30 cm.
- There should be drainage holes at the lower portion and sides of the poly bags.
- The bags should be filled in half with growth mixture. It is a mixture of sandy loam and FYM at the ratio 3:1 or hill soil, sand and FYM (2:2:1) (Fig. A.15).
- Bags should be kept ready (half filled with growth mixture) before removing the rooted cuttings from propagation bed, so that transplanting is not delayed.

Fig. A.15 The rooted cuttings collected from propagation beds, carried in water bucket to the nursery shed and transplanted in already filled bags in half with growth mixture(GM) of sandy loam and FYM at the ratio 3:1 or hill soil, sand and FYM (2:2:1). The rooted cutting has to be placed straight at the centre of the bag so that the roots are placed properly, then GM is poured from all the sides of the root inside the bag, and thus the upper half of the bag is filled up completely

23. *Transplanting of cutting in polythene bags.*

- The freshly removed rooted cutting has to be placed straight at the centre of the bag so that the roots are placed properly above; the growth mixture is poured from all the sides of the root inside the bag, and thus the upper half of the bag is filled up completely (Fig. A.15).
- Then the bag should be shaken slowly so that the growth mixture tightly fills the bags and no air pockets exist in side.
- Some amount of soil mixture may be added in the top of the bag to fill the space that formed due to shaking.
- After completing the transplantation of all the cuttings of one batch in polythene bags, start the next transplantation.

H. **Hardening of cuttings in polythene bags and gradual shifting to the nursery beds**

- The bags (containing freshly planted rooted cuttings) need careful handling and placement under shade-net roof (for providing 50–60 % partial shade).
- It is essential to water in the form of fogging the cutting's leaves and stems with a spray nozzle (use the hand sprayer by pressing the fingers if there is no electricity) at frequent intervals; excess watering will start the rotting of cuttings.
- When a new growth in the cuttings is visible, likely within a week or so, the bags should be shifted in the nursery beds under open sky.

In the absence of automatic overhead fogging equipment, moisten the leaves with a spray nozzle (may use hand sprayer by pressing the fingers) at frequent intervals, especially during hot weather.

I. **Placing and maintenance of polythene bags in open nursery**

24. *Maintenance of cuttings in polythene bags in the nursery beds.*

- When cuttings in polythene bags are left for 6–8 months in an open nursery, they produce strong rhizome system and roots which penetrate the ground below and neighbouring polythene bags.
- During lifting/transportation of polythene bags, cuttings with rhizomes and roots are damaged.
- Shifting of bags from one bed to another at 2–3-month interval is a good practice to control the root and rhizome growth.
- Providing cemented platform is a better option, where rhizome and roots do not penetrate; and as a result, there is no or less damage of cuttings (Fig. A.16).
- Maintain regular watering to the cuttings two times a day (not in rainy days) till the field planting.
- Clean the weeds/grass from the poly bags.
- Occasionally do hoeing the soil, and add powdered FYM in poly bags.
- There should not be any stagnation of water on the cemented platform, drain and nearby the propagation beds.
- Infected leaves or twigs of a cutting have to be collected and burnt away from the nursery.

J. **Yearly time schedule of production of bamboo planting materials by cutting in a nursery**

Dates	Cutting type
Feb4wk–Apr2wk, Apr4wk–Jun2wk, Jun4wk–Aug2wk, Aug3wk–Oct4wk (so *four batches*)	Culm cutting
Apr4wk–Jun2wk, Jun4wk–Aug2wk, Aug3wk–Oct 4wk (so *three batches*)	Branch cutting

Note: wk week

Wild Bamboo Seedling Collection and Bagging in Nursery

Wild seedlings of bamboo look like rice or wheat seedlings and are often seen as a thick mat on the ground just below the flowering mother clumps (Figs. 3.6i and 5.1g). These densely populated seedlings compete strongly in the wilderness for survival and should be thinned out to minimize competition and thus can be utilized in raising plantation.

- Wild seedlings of different bamboos can be collected and transplanted in the polythene bags and used as planting stocks.
- Two- to four-leaved stage of wild seedlings (1–3 months old) of *Bambusa cacharensis*, *B. tulda*, *B. bambos*, *D. hamiltonii*, *D. strictus*, *D. longispathus*, *Gigantochloa andamanica* and *Schizostachyum dullooa* is the best for collection (Fig. 9.1g(1), g(2)), while in *M. baccifera* germinating seeds are best. However, seedlings up to 18 months old can be collected.
- If the seedlings are very leafy and more than 30 cm tall, chop off the top just after transplanting in polythene bags; this reduces transpiring surface, and thus drying of seedlings gets delayed.
- Always select healthy, tall and vigorously growing seedlings; avoid dwarf, bushy and lanky ones.

Fig. A.16 Cuttings in polythene bags are maintained on cemented platform for 6–8 months in an open nursery till the next planting season and produce roots and strong rhizome system which cannot penetrate the ground below; and as a result, there is no or less damage of cuttings during transplantation

Dos	Don'ts
A. Best time for wild seedling collection	
From June to September	November to February
B. Steps of operation	
1. (a) Collect wild seedlings at 4–6-leaves stage carefully without damaging rhizome and roots (Fig 9.1 (g1, 2)); just prick up one by one	1. (a) Dragging or pulling of seedlings from the ground
1. (b) However, 6–9-month-old wild seedlings can also be collected, but collection and transplanting in the nursery have to be done during May to June after rains. Rain makes the soil in forest floor soft and moist, and as a result, the shock due to injury in the rhizomes and roots during collection becomes less. Immediately after collection, these should be transported to the nursery for bagging and hardening. Care should be taken during *transportation* of these bare-rooted planting stocks. Seedlings can be damaged by overheating, too little moisture and physical damage during transportation. Protect seedlings from sun, wind and excessive drying. The bare root and rhizome portion of the collected seedlings may be bundled (one bundle contains 10–20 seedlings depending on size) by wrapping with water-soaked paper towel/jute sheet. Side branches and few leaves may be trimmed off to cut down the transpiration surface of the transporting seedlings. All bundles then are staked side by side on the floor in trucks and railway wagon. During long distance transportation through truck, time to time sprinkle cool water on them (excess watering rot the seedlings), and put insulated cover to protect against the sun with well-ventilated condition. Prefer to move in the night, early morning or in rainy days	1. (b) Transportation of bare-rooted seedlings under direct sunlight will kill most of them
2. Bring quickly to the nursery and immediately transplant to polythene bags containing soil and sand mixed with FYM/cow dung (2:1:1) (Fig. 3.3e of *B. cacharensis*)	2. Delay in transplanting will increase the mortality
3. Water the plant body through intermittent fogging (may use hand spray for atomizing the water into finer drops) and keep under shade net (50 % shade/light) for 1 week for hardening 4. Transplanted seedlings may get dry as if dead, but eventually produce new shoots from the base and survive 5. After 1 week, gradually increase the overhead light (70 % light) by replacing overhead shade net for healthy growth and maintain fogging 6. Maintain in this way for 3–4 weeks with regular weeding 7. Gradually shift them under open sky	3. Watering through pipe or by Jhajri or bucket will shake the seedlings and delay the establishment and new root production
C. Caution	
Avoid operational activities in drought and under scorching sunlight	

- Frequent shifting of seedlings from one bed to another helps in minimizing the root rhizomes intermingling.
- Nine- to 12-month-old seedlings/cuttings survive at higher rate (about 80–90 %) in the field after transplantation. The rainy season is the best time for planting of seedlings/cuttings.

Use Macroproliferation Method

- The method is applicable in multiplication of seedlings and also in cuttings having more than three shoots.
- Multiply a 6–9-month-old seedling/cutting having more than three to four shoots by macroproliferation method. Depending on the availability of shoots in the mother cutting, we can get two or three or four number of cuttings from one mother cutting.
- Multiplied individual will be always in less number shoots than those of the mother seedlings/cuttings. Say a seedling/cutting has four shoots, the number of produced individual will be less than four, may be two or three.
- Suppose a nursery has 10,000 seedlings, you can multiply it into 30,000 numbers. Plant 20,000 and multiply the rest by 10,000 after 6 months. Continue to do *macroproliferation* of your stock not more than 3 years. Steps of operation are as follows:

Dos	Don'ts
D. Best time for doing macroproliferation	
From February 2nd week to October	November to mid February
E. Steps of operation	
1. Tear off the poly bags longitudinally (Fig. A.17), carefully clean the soil around the rhizome part and roots of the seedling or cuttings by jetting water from a water tap (Fig. A.18) 2. Separate the shoots (culms) by cutting the rhizome with secateurs so that *each unit includes old and young shoots with or without rhizome buds and some amount of roots* (Figs. 3.3g(1) and 3.3g(2) of *B. cacharensis*); also Figs. A.19 and A.20 3. Each of the separated individual is transplanted into a polythene bag and hardened for 2–3 weeks under shade net (50 % shade) and intermittent fogging condition 4. Then maintain the polythene bags in open nursery under full sun and normal misting (no need of fogging) till the plantation in the field	*Avoid operational activities in drought and under scorching sunlight*
F. Caution	
In continuous application of *macroproliferation* technique, the rhizome vitality gets reduced over time, and the individuals produced from last multiplication may start flowering due to physiological maturity and thus have comparatively short productive life	

Fig. A.17 Tearing of a polythene bag containing soil and a 5-month old seedling of *B.bambos*

Fig. A.18 Exposing of root & Rhizome of the seedling by washing and clearing soil in water (Source: Banik 1995)

Fig. A.19 Shoots of the seedling are separated by cutting the exposed rhizome system by secateurs (Source: Banik 1995)

Fig. A.20 Each separated unit should have old and young rhizome, rhizome bud and roots (Source: Banik 1995)

References

Adewuyi AP, Otukoya AA, Olaniyi OA, Olafusi OS (2015) Comparative studies of steel, bamboo and rattan as reinforcing bars in concrete: tensile and flexural characteristics. Open J Civ Eng 5:228–238, http://dx.doi.org/10.4236/ojce.2015.52023

Adhikari AK (1928) Flowering of *Bambusa arundinacea*. Indian Forester 54(7):424

Adhikari R, Shrestha K (2008) Infraspecific variation of *Bambusa nutans* subspecies *nutans* from six different sites of central Nepal. Sci World 6(6):81–84

Alam MK (1995) Systematic studies in *Bambusoides* of Bangladesh. PhD thesis, Department of Botany, University of Dhaka, Dhaka

Alam MK (2001) Bamboos of Bangladesh: a field identification manual. Bangladesh Forest Research Institute (BFRI), Chittagong, pp 1–35

Ali AH, Nirmala C, Badal T, Sharma ML (2009) In vitro organogenesis and simultaneous formation of shoots and roots from callus in *Dendrocalamus asper*. VIII world bamboo congress, Bangkok 16–19 Sept, 2009 proceedings, vol 6, Massachusetts, World Bamboo Organization, pp 32–41

Anantachote A (1987) Flowering and seed characteristics of bamboo in Thailand. In: Rao AN, Dhanaranjan G, Sastry CB (eds) Recent research on bamboos. Proceedings of the international bamboo workshop, IDRC Canada, Hangzhou, 6–14 Oct 1985, pp 136–145

Anantachote A (1988) Flowering characteristics of some bamboos in Thailand. In: Ramanuja Rao IV, Gnanaharan R, Sastry CB (eds) Bamboos – current research. Proceedings of the international bamboo workshop, KFRI, Cochin, pp 66–75

Anon (1881) The flowering of the bamboo. Indian Forester 7(2):162

Anon (1916) Ann Rep For Admin Burma 1915–1916: para 50

Anon (1922) Flowering of bamboos. Indian Forester 48(7):404

Anon (1946) Flowering of bamboos. Indian Forester 72(10):280

Anon (1983) Bamboo cultivation. Editorial. The Bangladesh Times (The daily news paper). 24th Nov, Dhaka

Anon (1994) Partner country – overview: Sri Lanka: Bamboo and Rattan. INBAR News Letter No 5, Dec, New Delhi

Anon (1995a) Bhutan: non-wood forest products. INBAR News Letter 3(2):9–10. New Delhi

Anon (1995b) Nepal: fodder value of bamboo species. INBAR News Letter 3(2):14. New Delhi

Anon (1996a) Nepal: INBAR partner country. INBAR News Letter 3(4). New Delhi

Anon (1996b) India-flowering of *Dendrocalamus hamiltonii*. INBAR News Letter 3(4):9–10. New Delhi

Anon (2007) Bamboo flowering in Mizoram – a paper presented in the State Bamboo workshop on "Bamboo, the ultimate resource for prosperity", 4–5 Oct 2007. State Bamboo Steering Committee, Art & Culture Auditorium, Berawtlang

R.L. Banik, *Silviculture of South Asian Priority Bamboos*, Tropical Forestry,
DOI 10.1007/978-981-10-0569-5

Anon (2012) Bamboo flowers across Karnataka after 40 years. The New Indian Express, 16th May 2012

Anon (2014) Bamboo clumps withering away in Kodagu forests. The Hindu (The English News Paper). http://www.thehindu.com/todays-paper/tp-national/tp-karnataka/Oct11

Arya S (2008) Micropropagation of Himalayan bamboos (*Drepanostachyum falcatum* and *Bambusa balcooa*). In: International conference on improvement of bamboo productivity and marketing for sustainable livelihood (Abstracts), NBM, Department of Agriculture and Cooperation, Ministry of Agriculture, Govt of India, New Delhi, p 8 (Abstract No: 01.07)

Arya D (2014) Dwarf bamboo (ringal): a traditional livelihood option for scheduled caste families of Garhwal Himalaya. Int J Adv Res Eng Appl Sci 3(7):67–73

Arya ID, Rana PK, Satsangi R, Muzaffar FS, Sharma S, Arya S (2002a) Rapid and mass multiplication of bamboos through tissue culture techniques. In: Nandi SK, Palni LMS, Kumar A (eds) Role of plant tissue culture in biodiversity conservation and economic development. Gyanodaya Prakashan, Nainital, pp 29–39

Arya S, Satsangi R, Arya ID (2002b) Rapid mass multiplication of edible bamboo *Dendrocalamus asper*. J Sustain For 14:103–109

Arya S, Rana PK, Sharma R, Arya ID (2006) Tissue culture technology for rapid multiplication of Dendrocalamus giganteus Munro. India n Forester 132:345–357

Arya S, Kaur B, Arya ID (2009) Micropropagation of economically important bamboo *Dendrocalamus hamiltonii* through axillary bud and seed culture. VIII World Bamboo Congress Proceedings, vol 6. pp 122–130

Arya ID, Kaur B, Arya S (2012) Rapid and mass propagation of economically important Bamboo *Dendrocalamus hamiltonii*. Indian J Energy 1(1):11–16

Ashaduzzaman M, Rana R, Khan MNH, Shams MI (2011) Suitability of Muli Bamboo (*Melocanna baccifera*) for making bamboo mat plywood. Bangladesh J Sci Ind Res 46(4):543–548

B., TF (1887) Seeding of bamboos. Indian Forester 13:579 [mentioned in Janzen DH (1976) Why bamboos wait so long to flower. Annu Rev Ecol Syst 7:347–391]

Bag N, Suman C, Man L, Palni S, Nandi SK (2000) Micropropagation of dev-ringal [*Thamnocalamus spathiflorus*(Trin.) Munro] – a temperate bamboo, and comparison between in vitro propagated plants and seedlings. Plant Sci 156(2):125–135

Bag N, Man L, Palni S (2013) Seed germination studies of dev-ringal *Thamnocalamus spathiflorus* (TRIN.) Munro – a temperate bamboo. Indian Forester 139(7):610–614

Bahadur KN (1979) Taxonomy of bamboos. Indian J For 2(3):222–241

Bahadur KN (1980) A note on the flowering of *Bambusa nutans*. Indian Forester 106(4):314–316

Bahadur KN, Naithani HB (1983) On the identity, nomenclature, flowering and utility of the climbing bamboo *Melocalamus compactiflorus*. Indian Forester 109(8):566–568

Banerjee M, Gantait S, Pramanik BR (2011) A two step method for accelerated mass propagation of *Dendrocalamus asper* and their evaluation in field. Physiol Mol Biol Plants 17(4):387–393. doi:10.1007/s12298-011-0088-0

Banik RL (1979) Flowering in biajja bansh (*Bambusa vulgaris*). Letter to the editor, Bano Biggyan Patrika 8(1&2):90-91

Banik RL (1980) Propagation of bamboos by clonal methods and by seeds. In: Lesserd G, Chouinard A (eds) Bamboo research in Asia. Proceedings of a workshop Singapore, IDRC, Ottawa, IUFRO, 28–30 May, pp 139–150

Banik RL (1983a) Emerging culm mortality at early developing stage in bamboos. Bano Biggyan Patrika 12(1&2):47–52

Banik RL (1983b) Macroporpagation of bambusoid grass (*Phragmites communis* (cav.) Trin and Stend.) and micro-propagation of bamboo (*Bambusa glaucescens* Siebold). MS thesis, Department of Horticulture, University of Saskatchewan, Saskatoon, pp 1–155

Banik RL (1984) Macropropagation of bamboos by prerooted and prerhizomed branch cuttings. Bano Biggyan Patrika 13(1&2):67–73

Banik RL (1986) Observations on special features of flowering in some bamboo species of Bangladesh. In: Higuchi T (ed) Bamboo production and utilization. Proceedings of the project Group PS.04., 18th IUFRO World Congress, Ljubljana, 7–21 Sept, pp 56–60

Banik RL (1987a) Seed germination of some bamboo species. Indian Forester 113(8):578–586

Banik RL (1987b) Techniques of bamboo propagation with special reference to prerooted and prerhizomed branch cuttings and tissue culture. In: Rao AN, Dhanaranjan G, Sastry C B (eds) Recent research on Bamboos Proceedings of the international Bamboo Workshop, 6–14 Oct 1985, The Chinese Academy of Forestry, Hangzhou/IDRC, Ottawa, pp 160–169

Banik RL (1988a) Investigation on the culm production and clump expansion behaviour of five bamboo species of Bangladesh. Indian Forester 114(9):576–583

Banik RL (1988b) Management of wild bamboo seedlings for natural regeneration and reforestation. In: Ramanuja Rao IV, Gnanaharan R. Sastry CB (eds) Bamboos - current research. Proceedings of the international bamboo workshop, 14–18 Oct 1988, KFRI, Cochin, pp 92–95

Banik RL (1989) Recent flowering of muli bamboo (*Melocanna baccifera*) in Bangladesh: an alarming situation for bamboo resource. Bano Biggyan Patrika 18(1&2):65–68

Banik RL (1991a) Studies on seed germination, seedling growth and nursery management of *Melocanna baccifera* (Roxb.) Kurz. In: Bamboo in Asia Pacific. Proceedings of 4th international bamboo workshop, Chiangmai, IDRC, Ottawa/Forestry Research Support Programme for Asia and the Pacific(FORSPA), Bangkok, pp 113–119

Banik RL (1991b) Trial on the tissue culture of *Melocanna baccifera* (Roxb.) Kurz. In: Bamboo in Asia Pacific. Proceedings of 4th international bamboo workshop, Chiangmai, IDRC, Ottawa/ FORSPA, Bangkok. (Abstract), p 135

Banik RL (1992) Bamboo: forestry master plan of Bangladesh. Asian Dev Bank (TA. No. 1355-BAN). UNDP/FAO BGD 88/025. (8 Appendix) Dhaka/Manila, pp 1–62

Banik RL (1993a) Periodicity of culm emergence in different bamboo species of Bangladesh. Ann For 1(1):13–17

Banik RL (1993b) Morphological characters for culm age determination of different bamboo species of Bangladesh. Bangladesh J For Sci 22(1&2):18–22

Banik RL (1993c) Selection and multiplication of bamboos for rural and industrial planting programmes. In: Vivekanandan K et al (eds). Proceedings of a workshop on the production of genetically improved planting materials for afforestation programmes. ICFRE/FAO/UNDP-FORTIP, June 1993, Coimbatore, pp. 76–97

Banik RL (1994a) Distribution and ecological status of bamboo forests of Bangladesh. Bangladesh J For Sci 23(2):12–19

Banik RL (1994b) Diversities, reproductive biology and strategies for germplsm conservation of bamboos. In: Ramanatha Rao V, Rao AN (eds). Bamboo and rattan genetic resources and use. Proceedings of the first INBAR biodiversity, genetic resources and conservation working group, 7–9 Nov 1994, Singapore, IPGRI/INBAR, pp 1–22

Banik RL (1994c) Review of conventional propagation in bamboos and future strategy. In: Constraints to production of bamboos and rattan. Report of a consultation held in Bangalore, 9–13 May 1994. INBAR Technical Report No 5. New Delhi, pp 115–142

Banik RL (1995) A manual of vegetative propagation of bamboos. INBAR Tech Report No. 6. International Network for Bamboo and Rattan, New Delhi, pp 1–66

Banik RL (1997a) The edibility of shoots of Bangladesh bamboos and their continuous harvesting effect on productivity. Bangladesh J For Sci 26(1):1–10

Banik RL (1997b) Domestication and improvement of Bamboos. INBAR/UNDP/FORTIP, New Delhi/Guangzhou/Eindhovan, pp 1–53

Banik RL (1997c) *Melocanna baccifera* (Roxb.) Kurz – a priority bamboo resource for denuded hills of high rainfall zones in South Asia. In: Karki M, Rao AN, Rao VR, Williams JT (eds). The role of bamboo, rattan and medicinal plants in mountain development Proceedings of a workshop held at the Institute of Forestry, Pokhara. INBAR Technical Report No.15, INBAR/ IPGRI/ ICIMOD/IDRC, pp 79–86

Banik RL (1998a) Ethnobotany of bamboo and rattan and their indigenous conservation practices in Chittagong Hill Tracts. In: Banik RL, Alam MK, Pei SJ, Rastogi A (eds) Applied ethnobotany. Proceedings of the subregional training workshop on applied ethnobotany. BFRI-UNESCO-ICIMOD Nepal, 17–22 Dec 1997, Published by BFRI, Chittagong, pp 93–102

Banik RL (1998b) Reproductive biology and flowering populations with diversities in muli bamboo *Melocanna baccifera* (Roxb.) Kurz. Bangladesh J For Sci 27(1):1–15

Banik RL (1999a) Flowering in *Dendrocalamus hamiltonii* Nees & Arn. ex Munro and *Schizostachyum dullooa* (Gamble & Majumdar) in Chittagong, Bangladesh. Bangladesh J For Sci 28(2):69–74

Banik RL (1999b) Annual growth periodicity of culm and rhizome in adult clumps of *Melocanna baccifera* (Bambusoidae:Gramineae). Bangladesh J For Sci 28(1):7–12

Banik RL (2000) Silviculture and field-guide to priority bamboos of Bangladesh and South Asia. BFRI, Chittagong, pp 1–187

Banik RL (2004a) Fatal flowers, World bamboo Congress, & CIBART/INBAR Communiqué (News Letter), vol 1(1), Feb 2004, New Delhi, pp 6–7

Banik RL (2004b) Bamboos of Tripura. Indian Forester 130(9):1081–1083

Banik RL (2008) Issues on production of bamboo planting materials—Lessons and Strategies. Indian Forester 134(3):291–304

Banik RL (2010) Biology and silviculture of muli (*Melocanna baccifera*) bamboo. Published by NMBA (National Mission on Bamboo Applications), TIFAC, Department of Science & Technology, Govt of India, New Delhi, pp 1–237

Banik RL (2014) An introductory note on Kanak Kaich bamboo of Tripura. NTFP Centre of Excellence, Tripura. Manjari 3(3), pp 5–9

Banik RL (2015a) Morphology and growth. In: Liese W, Kohl M (eds) Bamboo – the plant and its uses, Tropical forestry series. Springer, Hamburg, pp 43–90. doi:10.1007/978-3-319-14133-6

Banik RL (2015b) Bamboo silviculture. In: Liese W, Kohl M (eds) Bamboo – the plant and its uses, Tropical forestry series. Springer, Hamburg, pp 113–174. doi:10.1007/978-3-319-14133-6

Banik RL (2015c) Harvesting technique. In: Liese W, Kohl M (eds) Bamboo – the plant and its uses, Tropical forestry series. Springer, Hamburg, pp 193–226. doi:10.1007/978-3-319-14133-6

Banik RL (2015d) Flowering nature of *Dendrocalamus longispathus* (Kurz) Kurz with notes on seedling and other planting materials for raising plantation. Indian Forester 141(9):920–929

Banik RL (2015e) Growth, behavior and silviculture of Bamboos. In: Kaushik S, Singh YP, Kumar D, Thapliyal M, Barthwal S (eds) Bamboos in India. ENVIS Centre on Forestry, Govt of India, ICFRE Dehradun, pp 27–88

Banik RL, Alam MK (1987) A note on the flowering of *Bambusa balcooa* Roxb. Bano Biggyan Patrika 16(1&2):25–29

Banik RL, Das S (1996) Studies on culm diameter, wall thickness and the internode length in relation to habitat condition of some bamboo species of Bangladesh. J Non-Timber For Prod 3:126–131

Banik RL, Islam SAMN (2005) Leaf dynamics and above ground biomass growth in *Dendrocalamus longispathus* Kurz. J Bamboo Rattan 4(2):143–150

Banik RL, Sharma DK (2009) Management of gregariously flowered bamboo areas in Tripura NTFP centre of excellence. Tripura-JICA Project Forest Development, Agartala/Tripura, pp 1–80

Banik RL, Islam SAMN, Hadiuzzaman S (1993) In vitro regeneration of multiple shoots in three bamboo species. Plant Tissue Cult 3(2):101–106

Banik RL, Tiwari S, Kaushal R (2008) Bamboo in homestead and agroforestry system of India. In: Kundu SS, Dagar JC, Chaturvedi OP (eds) Environment, agroforestry & livestock management. International Book Distributing Company, Lucknow, pp 351–365

Bansal AK, Nath SK (2002) Sustainable utilization of bamboo resource (Technology development). In: Proceedings of expart consultation on strategies for sustainable utilization of bamboo resource subsequent to gregarious flowering in the northeast, Rain Forest Research Institute (RFRI), Jorhat, CBTC, Guwahati, UNIDO 24–25 Apr 2002, pp 47–56

Baruah D (1930) Bamboo seeds. Indian Forester 56(8):361

Basha CS (1991) *Ochlandra* (Bamboo Reed) a vanishing asset of forests in Kerala – South India. In: Bamboo in Asia Pacific. Proceedings of 4th international bamboo workshop, Chiangmai. International Development Research Centre (IDRC), Canada/Forestry Research Support Programme for Asia and the Pacific (FORSPA), Bangkok, pp 18–25

Beena VB (2011) Reproductive biology and biochemical changes associated with flowering of *Dendrocalamus stocksii* and *Ochlandra travancorica*. PhD thesis, Cochin University of Science and Technology, KFRI, Kerala, pp 1–147

Bejoy M, Anish NP, Radhika BJ, Nair GM (2012) In vitro propagation of *Ochlandra wightii* (Munro) Fisch.: an endemic reed of southern western ghats India. Biotechnol 11(2):67–73

Bennet SSR (1989) The climbing bamboos. *Dinochloa* and *Melocalamus* in India. Van Vigyan 27:119–123

Bennet SSR, Gaur RC (1990) Thirty seven bamboos growing in India. Controller of Publications, Government of India, New Delhi, pp 17–18

Bhandari MS, Kaushal R, Banik RL, Tewari SK (2015) Genetic evaluation of nutritional and fodder quality of different bamboo species. Indian Forester 141(3):265–274

Bhattacharjee NJ, Chakravarthy KGKD (2008) Indian bamboo industry market overview & outlook. International conference on improvement of bamboo productivity and marketing for sustainable livelihood, 15–16 Apr 2008, New Delhi, pp 97–104

Bhattacharya S, Das M, Bar R, Pal A (2006) Morphological and molecular characterization of *Bambusa tulda* with a note on flowering. Ann Bot 98(3):529–535

Bhattacharya S, Ghosh JS, Das M, Pal A (2009) Morphological and molecular characterization of *Thamnocalamus spathiflorus* subsp. *spathiflorus* at population level. Plant Syst Evol 30. doi:10.1007/s00606-008-0092-1

Bhodthipuks P (1981) Bamboo plantation in Thailand. In: Higuchi T (ed) Bamboo production and utilization. Proceedings of the congress group. 5.3A, 17th IUFRO world congress, Kyoto, 6–17 Sept 1981, pp 165–168

Blackwell GFR (1902) An interesting bamboo. Indian Forester 28(12):432–433

Blatter E (1929) Indian bamboos brought upto date. Indian Forester 55:541–562, 586–612.

Blatter E (1929–1930) Flowering of bamboos, I-III parts. J Bombay Nat Hist Soc 33:899–921, 34:135–141, 447–467

Blatter E (1931) Some notes on the flowering of bamboo. J Bombay Nat Hist Soc 34:1097–1099

Boa ER, Rahaman MA (1984) Bamboo blight in Bangladesh. Overseas Development Administration (ODA), London, pp 1–24

Bodekar FWT (1930) A few observations on the flowering of Kyathaungwa (*Bambusa polymorpha* Munro). Indian Forester 56(9):404–405

Bor NL (1941) *Thyrsostachys oliveri* Gamble. Indian Forest Rec (New series) Botany 2(Pt.2):221–225

Bose RB, Pandey HS, Banerjee AK (1987) Bamboos of the Indian botanic garden. Bull Bot Surv Ind 29:29–42

BPG [Bamboo Phylogeny Group] (2012) An updated tribal and subtribal classification of the bamboos (Posceae:Bambusoideae). In: Gielis J, Potters G (eds) Proceedings of the 9th world bamboo congress, Antwerp, 10–12 Apr 2012, pp 3–27

Bradley JW (1914) Flowering of Kya-thaung bamboo (*Bambusa polymorpha*) in the Prome Division, Burma. Indian Forester 40:526–529

Brandis D (1899) Biological notes on Indian bamboos. Indian Forester 25(1):1–25

Brandis D (1906) Indian trees. Periodical experts. Book Agency, Delhi, pp 1–767

Brar J, Anand M, Sood A (2012a) In vitro seed germination of economically important edible bamboo *Dendrocalamus membranaceus* Munro. Indian J Exp Biol 51:88–96

Brar J, Shafi A, Sood P, Sood A, Anand M (2012b) Micropropagation of *Dendrocalamus membranaceus* Munro. through axillary shoot proliferation and confirmation of clonal fidelity of in vitro raised plants. J Bamboo Rattan 11(1–4):13–29

Brink M (2008) *Dendrocalamus giganteus* Munro. In: Louppe D, Oteng-Amoako AA, Brink M (eds) Prota 7(1): Timbers/Bois d'œuvre 1. [CD-Rom]. PROTA, Wageningen. http://database.prota.org/PROTAhtml/Dendrocalamus%20giganteus_En.htm

Broun AF (1886) Seeding of bamboos. Indian Forester 12:413–414

Campbell JJN (1988) Notes on the Sino-Himalayan bamboo species. School of Biological Sciences, Lexington. Unpublished manuscript, pp 1–105. [Reported in: Naithani HB, Chandra S (1998) Gregarious flowering of a bamboo (*Drepanostachyum falcatum*), Research Note. Indian Forester 124(8):663–666]

Camus A (1935) Classification des Bambusees. Archives Museum National Historie Naturelle. Paris. Series 6(12):601–605

Cavendish FH (1905) A flowering of *Dendrocalamus hamiltonii* in Assam. Indian Forester 31(8):479

Chatterjee D (1960) Bamboo fruits. J Bombay Nat Hist Soc 57(2):451–453

Chaturvedi AN (1986) Bamboos for farming. UP Forest Bulletin No 52, Lucknow, pp 1–36

Chaturvedi HC, Sharma M (1988) Micropropagation of *Dendrocalamus strictus* through *in vitro* culture of single-node segments. In: Proceedings 75th Indian science congress: part III: abstracts of section VI, Botany, p 297

Chaubey OP, Sharma A, Prakash R (2013) Eco-silvicultural interventions for rehabilitation of gregariously flowered bamboo forests with special reference to *Dendrocalamus Strictus* (Roxb.) Nees. Glob J Sci Front Res Agric Vet 13(13):31–38, Version 1.0 [Online ISSN: 2249–4626 & Print ISSN: 0975–5896]

Chaubey OP, Sharma A, Krishnamurthy G (2015) Plant diversity, edaphic status and population structure in different forest types of Madhya Pradesh and Chhattisgarh states in India. Int J Bio-Sci Bio-Tech 7(2):115–124

Clements ID (1956) Flowering of *Dendrocalamus strictus* at Atkins Garden Soledad, Ceinfueges, Cuba. Science 124:1291

Coffey TM (1931) Seeding of mitenga bamboos (*Bambusa tulda*). Indian Forester 67(9):418

Conservator of Forests, West Bengal (1976) Gregarious flowering of bamboos – *Dendrocalamus hamiltonii*. Indian Forester 102(11):83

Das T (1969) *Bambusa arundinacea* in Nowrangpur Division, Koraput District, Orissa. Indian Forester 95:279

Das T (1976) Thorny bamboo bloomed-gregarious flowering of Kanta Bans death knell for *Bambusa arundinacea* clumps occurrence in South Orissa. Indian Forester 102:473

Das AN (1988) Bamboo research in Nepal. In: Ramanuja Rao IV, Gnanaharan R, Sastry CB (eds) Bamboo – current research. Proceedings of the international bamboo workshop, Cochin, pp 1-5

Das R (2014) Studies on edible bamboo shoot production as influenced by various cultural practices and clump management of some selected species of Jharkhand. PhD thesis, Forest Research Institute University Dehra Dun, Uttarakhand, pp 1–304 (Appendices I–VI)

Das M, Pal A (2005) Clonal propagation and production of genetically uniform regenerants from axillary meristems of adult bamboo. J Plant Biochem Biotechnol 13:185–188

Das M, Bhattacharya S, Basak J, Pal A (2007) Phylogenetic relationships among the bamboo species as revealed by morphological characters and polymorphism analyses. Biol Plant 51(4):667–672

Das MC, Singnar P, Nath AJ, Das AK (2014) Gregarious flowering in a climbing bamboo *Melocalamus compactiflorus* in Assam. (Research Notes). Indian Forester 140(9):935–936

De RN (1932) Flowering of *Bambusa nutans* in Garo hills. Indian Forester 58:630

Deekers AJ, Rao AN (1989) Tissue culture of four bamboo genera. In: Rao AN, Yusoff AM (eds) Proceedings of the seminar on tissue culture of forest species FRIM, Malaysia, IDRC, pp 83–90

Deogun PN (1937) The Silviculture and management of the bamboo *Dendrocalamus strictus* Nees. Indian For Rec 2(n. s., silviculture):75–173

Desai AP, Subramanian K (1980) A note on bamboos in Maharashtra. 3rd conference of the Silviculturist, Dharwad, Mar 1980, pp 66. quoted in [Bennet SSR, Gaur RC (1990) Thirty seven bamboos growing in India. FRI, Dehra Dun]

Dransfield S (1980) Bamboo taxonomy in the Indo-Malesian Region. In: Lesserd G, Chouinard A (eds) Bamboo research in Asia. Proceedings of a workshop, Singapore, 28-30 May, IDRC, Ottawa, IUFRO, pp 121–130

Dransfield S, Widjaja EA (1995) Plant resources of South-East Asia, (No. 7): Bamboos; Backhuys Publishers, Leiden, pp 85–87

Du F, Xue JR, Yang YM, Hui CM, Wang J (2000) Study on flowering phenomenon and its type of bamboo in Yunnan in past fifteen years. Sci Silvae Sin 3:57–68

Dutra J (1938) Bambusees de Rio Grande du sud Revista. Sudamerica de Botanica 5:145–152

Dwivedi AP (1988) Gregarious flowering in *Dendrocalamus strictus* in Shahdol (Madhya Pradesh) – some management considerations. Indian Forester 114(9):532–538

Elizabeth A, Widjaja, Risyad Z (1987) Anatomical properties of some bamboos utilized in Indonesia. In: Rao AN, Dhanaranjan G, Sastry CB (eds) Recent research on Bamboos proceed-

ings of the international bamboo workshop, 6–14 Oct 1985, The Chinese Academy of Forestry, Hangzhou/IDRC, Canada, pp 244–246
Farah NR, Shamsi FQ, Ramanuja Rao IV, Usha Rao I (1991) Clonal propagation of mature *Bambusa balcooa* in-vitro. Proceedings of IV international bamboo workshop, IDRC, Chiangmai, p 71 (Abstract)
Gadgil M (1980) News and notes. Indian Forester 106:316
Gadgil M, Prasad SN (1984) Ecological determinants of life history evolution of two Indian bamboo species. Biotropica 16:161–172
Gamble JS (1894) A handsome new Burmese bamboo. Indian Forester 20:1
Gamble JS (1896) The Bambuseae of British India, vol 7, Annals of the Royal Botanic Garden, Calcutta. Bengal Secretariate Press, Calcutta/London, pp 1–133
Gaur RC (1987) Bamboo research in India. In: Rao AN, Dhanaranjan G, Sastry CB (eds) Recent research on Bamboos proceedings of the international bamboo workshop, 6–14 oct 1985, The Chinese Academy of Forestry, Hangzhou/IDRC, Ottawa, pp 26–32
Gopakumar B, Motwani B (2013) Factors restraining the natural regeneration of reed bamboo *Ochlandra travancorica* and *O. wightii* in Western Ghats India. J Trop For Sci 25(2):250–258
Gopal R (1989) Natural regeneration of bamboo (*Dendrocalamus strictus*) after gregarious flowering and its effect on the forage and browse availability in Bandhavgarh National Park Madhya Pradesh. J Trop For 5(4):330–342
Goraya GS, Pal M, Kapoor KS, Jishtu V (2008) Hill bamboos – socio-economic significance and conservation imperatives: a case study from Himachal Pradesh. Indian Forester 134(3):351–362
Guang-zhu Z (1987) Studies on the chromosome number of some bamboo species with clump rhizomes. In: Rao AN, Dhanaranjan G, Sastry CB (eds) Recent research on bamboos. Proceedings of the international bamboo workshop. IDRC, Hangzhou, pp 175–178
Gupta MLS (1952) Gregarious flowering of *Dendrocalamus*. Indian Forester 78:547–550
Gupta KK (1972) Flowering of different species of bamboos in Cachar dist of Assam in recent times. Indian Forester 98:83–85
Gupta KK (1982) Note on bamboo flowering in North East India. Indian Forester 108:596
Gupta KK (1988) Flowering and fruiting of bamboo *Melocanna baccifera* in north Cachar Hills, Assam (a letter to the Editor). Indian Forester 114(9):602
Hadfield W (1958) The raising of bamboo from seeds. Two A Bud 5(2):9–10
Han-Qi Y, An M-Y, Zhi-Jia G, Bo T (2012) Genetic diversity and differentiation of *Dendrocalamus membranaceus* (Poaceae: Bambusoideae), a declining bamboo species in Yunnan, China, as based on Inter-Simple Sequence Repeat (ISSR) analysis. Int J Mol Sci 13:4446–4457. doi:10.3390/ijms13044446
Haque AMAJ (2010) Studies on development of regeneration protocol of five bamboo species through *in vitro* culture and evaluation of their growth performance in the field. PhD thesis, Department of Botany, Jahangirnagar University, Savar, pp 1–206
Hasan SM (1973) Seeding behaviour of Bangladesh bamboos. Bano Biggyan Patrika 5(2):21–36
Hasan SM (1979) Observation on culms and culm – sheaths of Ora bamboo raised from seeds. Bano Biggyan Patrika 8(1&2):13–26
Hasan SM (1980) Lessons from the past studies on the propagation of bamboo. In: Lessard G, Chouinard A (eds) Bamboo Research in Asia. Proceedings of a workshop, Singapore. 28, 30 May. IDRC, Canada, pp 131–138
Holttum RE (1956) The classification of bamboos. Phytomorphology 6:73–90
Holttum RE (1958) The bamboos of the Malay Peninsula. The Garden Bulletin 16, Singapore, pp 1–135
Honnareddy N (2014) Bamboo flowering at Kalajari hill range. Government of Tripura Office of the District Forest Officer, Dhalai District, Ambassa, an official letter (No. F.4-41/Dist. For. Off./DLI.2013/4124-26; dt.22.02.2014) addressed to The Chief Conservator of Forests, (Territorial) O/o the PCCF, Tripura, pp 1–1
Hossain KMI (1962) Bamboos of East Pakistan with particular reference to muli bamboo and its flowering. Pak J For 19:194–201

Htun N (1999) Bamboos of Myanmar. In: Rao AN, Rao VR (eds) Bamboo conservation, diversity, ecogeography, germplasm resource utilization and taxonomy. Proceedings of a training course cum workshop, 10–17 May 1998. Kunming and Xishuangbanna, Yunnan. IPGRI-APO, Serdang, pp 201–214

Huang LC, Huang BL (1995) Loss of the species distinguishing trait among regenerated *Bambusa ventricosa* McClure plants. Plant Cell Tiss Org Cult 42:109–111

Islam SAMN, Rahman MM (2005) Micro-cloning in commercially important six bamboo species for mass propagation and at a large scale. Plant Tiss Cult Biotech 15:103–111

Jain VK, Kambo AS (1991) A new approach to seasoning of round bamboo. J Indian Acad Wood Sci 22(1):29–34

Janzen DH (1976) Why bamboos wait so long to flower. Annu Rev Ecol Syst 7:347–391

Jha LK (2010) Bamboo based agroforestry systems to reclaim degraded hilly tracts (jhum) land in North Eastern India: study on uses, species diversity, distribution and growth performance of *Melocanna baccifera, Dendrocalamus hamiltonii,* D. *longispathus* and *Bambusa tulda* in natural stands and in stands managed on a sustainable basis. Bamboo Sci Cult 23(1):1–28

Jindal UC (1984) Development and testing of bamboo-fibers reinforced plastic composites. J Compos Mater 20:19–29

Kadambi K (1949) On the ecology and silviculture of *Dendrocalamus strictus* in the bamboo forests of Bhadravati Division, Mysore state, and comparative notes on the species *Bambusa arundinacea, Ochlandra travancorica*, *Oxytenanthera monostigmata* and *O. stocksii*. Indian Forester 75:289–299, 334–349, 398–426

Kamthai S (2003) Alkaline sulfite pulping and ECF – bleaching of sweet bamboo (*Dendrocalamus asper* Backer, MS thesis). Kasetsart University, Bangkok

Kant A, Arya S, Arya ID (2009) Micropropagation protocol for *Melocanna baccifera* using nodal explants from mature Clump (edited). VIII World Bamboo Congress Proceedings, vol 6, Horticulture, Thailand. [WORLD BAMBOO.NET; WORLD BAMBOO CONGRESS.ORG; World Bamboo Organization]

Kariyawasam D (1999) Bamboo resources and utilization in Sri Lanka. In: Rao AN, Rao VR (eds) Bamboo conservation, diversity, ecogeography, germplasm resource utilization and taxonomy. Proceedings of a training course cum workshop, 10–17 May 1998. Kunming and Xishuangbanna Yunnan China IPGRI-APO Serdang, Malaysia, pp 235–247

Kasim J, Jalil AA (1994) Fibre and chemical properties of *Bambusa vulgaris* Schrad. In: Bamboo in Asia and the Pacific. Proceedings of 4th international bamboo workshop IDRC/FAO/UNDP, Chiangmai, 27–30 Nov 1991, pp 218–221

Kennard WC, Freyre RH (1957) The edibility of some bamboos growing at Puerto Rico. Econ Bot 11(3):235

Khedkar MH (2005) Research note – Gregarious flowering of bamboos, letter to the Editor. Indian Forester 131(11):1512

Kirpekar CS (1956) Problem of clump formation in bamboo (*Dendrocalamus strictus*) forest of Chanda, Madhyapradesh. 9th Silvicultural Conference. DehraDun, India. Pt.II:125–127

Kiruba S, Jeeva S (2010) Flowering of bamboos in two biodiversity hotspots of India. Indian Forester 136:137–140

Kondas S (1981) Bamboo biology, culm potential and problem of cultivation. In: Higuchi T (ed) Bamboo production and utilization. Proceedings of the congress group. 5.3A, 17th IUFRO world congress, Kyoto, 6–17 Sept, pp 184–190

Koshy KC, Harikumar D (2001) Reproductive biology of *Ochlandra scriptoria*, an endemic reed bamboo of Western Ghat, India. Bamboo Sci Cult 15(1):1–7

Koshy KC, Dintu KP, Gopakumar B (2010) The enigma of leaf size and plant size in bamboos. Curr Sci 99:1025–1027

Kumar M (1988) Reed Bamboos (*Ochlandra*) in Kerala: distribution and management. In: Rao Ramanuja IV, Gnanaharan R, Sastry CB (eds) Bamboos – current research. Proceedings of the international bamboo workshop, Cochin, KFRI, IDRC, pp 39–43

Kumar B (2009a) Ringal (a dwarf bamboo): Resource Use Pattern. Deptt of Botany, Kumaun University, Nainital. Report and Opinion1(4):1–5

Kumar B (2009b) Conservation of Ringal (a dwarf bamboo) through economic development in Rudraprayag district Garhwal (Uttarakhand), India. J Am Sci 5(6):23–26
Kumar M (2011) Grasses and bamboos. (Part- II), Bamboos of Peninsular India, All India coordinated Project on Taxonomy (AICOPTAX). KFRI Research Report No 399, pp 1–140
Kumar S, Nath S (2011) Flowering in *Bambusa tulda* Roxb. at Ranchi Jharkhand. Indian Forester 137(6):802–803
Kumari N, Singh P, Srivastava RJ, Mishra CM, Dubey P, Burfal BS (2001) Using Genetic diversity of *Dendrocalamus strictus* Nees for increasing productivity of bamboo plantations in Vindhyan Region of Uttar Pradesh. Indian forester 127(12):1343–1347
Kurz S (1875) Preliminary report on the forest and other vegetation of Pegu. Appendix B. Baptist Mission Press, Calcutta; bamboos, pp 91–95 and index
Kurz S (1876) Bamboo and its use. Indian forester, 1(3):219–269 and 1(4): 355–362
Kurz S (1877) Forest flora of British Burma, vol 2. Royal Botanical Gardens, Calcutta, pp 547–571, [Reprinted in 1974 by M/s Bishen Singh Mahendra Pal Singh Dehradun-1, and M/s Periodical Experts New Delhi-32.]
Lahiry AK (1974) Sporadic flowering of *Dendrocalamus giganteus*, Munro in Bamonpokhri, Kurseong Forest Division. Indian Forester 105(8):532
Lalnuntluanga F, Mawia L, Jha LK (2003) Bamboo flowering and famine in Mizoram (abstract). In: Kamesh S (ed) Cane and Bamboo News, BAMTECH, 11–13 Mar 2003, Bamboo Technology Seminar, Guwahati, p 26
Latif A-M, Liese W (1995) Utilization of bamboo. In : Razak A-O, Latif A-M, Liese W, Norini H (eds) Planting and utilization of bamboo in Peninsula Malaysia. FRIM research pamph let no.118. Forest Research Institute Malaysia Kuala Lampur, Malaysia, pp 50–102
Li DZ, Hsueh CJ (1988) A study on the genus *Dendrocalamus* Nees from China. J Bamboo Res 7:12–13
Liese W (1980) Anatomy of bamboo. In: Lessard G, Chouinard A (eds) Bamboo research in Asia. Proceedings of a workshop, 28–30 May, Singapore, IDRC, pp 161–164
Liese W (1985) Bamboos-biology, silvics, properties, utilization. Deutsche Gesellschaft für Technische Zusammenarbeit (GTZ) GmbH, Eschborn, pp 1–132
Liese W, Ding Y (1994) Strucutre and functions of the nodes in bamboos. In: Proceedings 4th international bamboo workshop on bamboo in Asia and the Pacific, Chiangmai, 27–30 Nov 1991, pp 213–217
Limaye VD (1952) Strength of bamboo (*Dendrocalamus strictus*). Indian For Rec 78:558–575
Lohani DN (1973) *Dendrocalamus membranaceus* Munro in flower. Indian Forester 99(2)
MacMillan HF (1907) Flowering of *Dendrocalamus giganteus* the 'Giant bamboo'. Ann Roy Bot Gard Peradeniya 4:123–129
Mahzuz HMA, Ahmed M, Dutta J, Rose RH (2015) Use of bamboo (*Bambusa balcooa*) as reinforcement in concrete beam. Int J Struct Eng 6(2):89–108. doi:10.1504/IJSTRUCTE.2015.069684
Majumder RB (1985) Three new taxa of Indian bamboos. Bull Bot Surv India 25:235–238
Malick KC (1974) *Melocalamus compactiflorus* (Kurz) Benth.& Hook. f. a new record in India. Bull Bot Surv India 16:166–167
Manoharan TM, Trivedi BNV (2008) Forest policy and laws governing cultivation, harvesting, transport and trade of bamboo in Kerala. In: Choudhary ML, Salam K (eds) Proceedings on international conference on improvement of bamboo productivity and marketing for sustainable livelihood, 15–17 Apr 2008, New Delhi, pp 182–192
Mathauda GS (1952) Flowering habits of bamboo--*Dendrocalamus strictus*. Indian Forester 78:86–88
McClure FA (1936) The generic type, and a new species of bamboo genus *Schizostachyum* from Java. Blumea 2(2):86–94
McClure FA (1937) "Tree grass", a tribute to bamboo published in program announcing the Lingnan University Research Lecture for 1936–1937, entitled "Bamboo as a field for research." The lines reproduced herein were composed as an "invitation" to the lecture
McClure FA (1966) The bamboos: a fresh perspective. Harvard University Press, Cambridge MA, pp 1–347

Medhi P, Borthakur SK (2013) Sacred groves and sacred plants of the Dimasas of North Cachar Hills of Northeast India. African J Plant Sci 7(2):67–77. http://www.academicjournals.org/AJPS. doi: 10.5897/AJPS12.038

Mehta U, Rao I V, Mohan Ram HT (1982) Somatic embryogenesis in bamboo. In: Fujiwara A (ed) Plant tissue culture 1982. Proceedings 5th international congress of plant tissue and cell culture, July11-16, Japan. Abe Photo Printing Co Ltd, Tokyo, pp 109–110

Menon KG (1918) Flowering and after of *Bambusa arundinacea*. Indian Forester 44:519–520

Mishra PK, Krishna Rao B (2012) Hydrologic and economic evaluation of bamboo plantations in gullied lands under major ravine systems of India. Central Soil & Water Conservation Research & Training Institute, Dehradun, NBM/Department of Agric & Cooperation, Ministry of Agriculture, Government of India Krishi Bhawan, New Delhi, pp 1–46

Mishra Y, Patel PK, Yadav S, Shirin F, Ansari SA (2008) A micropropagation system for cloning of *Bambusa tulda* Roxb. Sci Hortic 115:315–318

Moebius M (1898) Ueber ein eigentümliches Biühen von *Bambusa vulgaris* Wendl., SenckenSenckengische Naturforsch. Ges. (Frankfurt a. M.), pp 81–89

Mohan D (1992) Gregarious flowering of *Bambusa arundinacea* Wild introduced in Shiwalik Forest Division, Uttar Pradesh. Indian Forester 118(4):310

Mohapatra S (1969). Observation on bamboo flowering. Indian Forester 95:215

Munro W (1868) A Monograph of the *Bambuseae* including description of all the species. Trans Linn Soc Lond 26:1–157 [Reprinted in 1966 S.R. Publishers Ltd., England]

Nadgauda RS, John CK, Mascarenhas AF (1993) Floral biology and breeding behavior in the bamboo *Dendrocalamus strictus* Nees. Tree Physiol 13(4):401–408

Nadgir AL, Phadke CH, Gupta PK, Parsharami VA, Nair S, Mascarenhas AF (1984) Rapid multiplication of bamboo by tissue culture. Silvae Genet 33(6):219–223

Naithani HB (1990) Nomenclature of Indian species of *Oxytenanthera*. J Bombay Nat Hist Soc 87:439–440

Naithani HB (1998) Gregarious flowering of a bamboo (*Drepanostachyum falcatum*). Indian Forester 124(8):663–666

Naithani HB (1999) Nomenclature and Identity of some Bamboo Species. Indian Forester 125(11):1130

Naithani HB (2007) Survey report on the distribution of bamboo species in Meghalaya, India. Department of Forest and Environment, Government of Meghalaya, Shillong, pp 1–55

Naithani HB, Biswas S (1992) Gregarious flowering of *Dendrocalamus membranaceus*. Indian Forester 118(4):300

Naithani HB, Pal M, Lepcha STS (2003) Gregarious flowering of Thamnocalamus spathiflorus and T. falconeri bamboos from Uttaranchal, India. Indian Forester 129:517–526

Narasimhamurthy Maya C, Nadanwar A, Pandey CN (2013) A study on physico-mechanical properties of *Thyrsostachys siamensis* (Kurz) Gamble and *Dendrocalmus membranaceus* (Munro) in Tumkur district, Karnataka. India Int J Curr Microbiol App Sci 2(2):62–66, ISSN: 2319–7692

Nath JC (1930) Sale of *Bambusa tulda* and *Melocanna bambusoides* seeds (a letter to the editor). Indian Forester 56(9):414

Nath GM (1959) A letter to the editor. Indian Forester 85(1):72

Nath GM (1960) A letter to the editor. Indian Forester 87(3):184

Nath GM (1962) Flowering of daloo bamboos in Cachar District (a letter to the editor). Indian Forester 88:523

Nath GM (1968) Flowering in muli bamboo (*Melocanna bambusoides*), a letter to the editor. Indian Forester 94(4):346

Nath AJ, Das AK (2008) Bamboo resources in the homegardens of Assam: a case study from Barak. J Trop Agric 46(1–2):58–61

Nath AJ, Das AK (2010) A correspondence – gregarious flowering of a long-lived tropical semelparous bamboo *Schizostachyum dullooa* in Assam. Curr Sci 99(2):154–155

Nath AJ, Lall R, Das AK (2015) Ethnopedology and soil properties in bamboo (*Bambusa* sp.) based agroforestry system in North East India. Catena Elsevier 135:92–99

Nayak SGR, Das P (2003) Evaluation of the genetic variability in bamboo using RAPD markers. Plant Soil Environ 49(1):24–28

Nees CG (1829) Agrostologia brasiliensis in C. F. P. Martius, Flora brasiliensis (Engelmann, Leipzig), vol 1, part 1; bamboos, pp 520–538

Negi AS (1986) Note on flowering of bamboo (*Dendrocalamus strictus*) in Rajpipla (E)Division, Gujarat. Indian Forester 112(11):1034–1036

Nicholls J (1895) The flowering of thorny bamboo. Indian Forester 21:90–95

Nicholson JW (1945) Flowering of *Bambusa arundinacea* in Orissa---a letter to the editor. Indian Forester 71(12):435–436

Nirula R, Rajbhandary SB (1987) *In-vitro* propagation of *Dendrocalamus strictus* and its establishment in soil. Proceedings of regional workshop on tissue culture of tropical crop plants, Dhaka, pp 17–20

NMBA (National Mission on Bamboo Applications) (2009) Flowering records of bamboo 2009. TIFAC, Govt of India, Delhi

Noushad KAM (2008) Industrial utilization and value addition of reed bamboo in Kerala. In: Proceedings of international conference on improvement of bamboo productivity and marketing for sustainable livelihood, 15–17 Apr 2008, New Delhi, pp 390–396

Ohrnberger D (1999) The bamboos of the world. Elsevier Science BV, Amsterdam. Wiesenstr. 5, D-86462 Langweid am L., Germany, pp 1–596 [e Book ISBN 9780080542386]

Olive JE (2005) The Encyclopedia of world climatology. From Wikipedia, the free Encyclopedia, Springer, pp 115–117. ISBN 9781402032646

Osmaston BB (1918) Rate of growth of bamboos. Indian Forester 44(2):52–57

Osmaston BB (1922) Working plan for the north Garhwal Forest Division, UP, India, p 20

Oye R, Mizuno T (1967) Studies on rayon grade bamboo pulp: removal of hemicellulose of bamboo and bamboo pulp. J Soc Fibre Sci Technol 23(5):213–221

Pande VC, Kurothe RS, Rao BK, Kumar G, Parandiyal AK, Singh AK, Kumar A (2012) Economic analysis of bamboo plantation in three major ravine systems of India. Agric Econ Res Rev 25(1):49–59

Pandey BN, Singh NB (2012) Micropropagation of *Dendrocalamus strictus* Nees from mature nodal explants. J Appl Nat Sci 4(1):5–9

Parry NE (1931) On the flowering of bamboos. J Bombay Nat Hist Soc 34:1099–1101

Pathak KC (2002) Gregarious bamboo flowering in northeastern region and some management considerations. In: Pattanaik S, Sing AN, Kundu M et al (eds) Proceedings of expert consultation on strategies for sustainable utilization of bamboo resources subsequent to gregarious flowering in the north-east, 24–25 Apr 2002. RFRI Jorhat, Book No-12-2002. India, UNIDO, pp 131–135

Patil VC, Patil SV (1988) Performance of bamboo under varying spacing and fertility levels. In: Ramanuja Rao IV, Gnanaharan R, Sastry CB (eds) Bamboos – current research. Proceedings of the international bamboo workshop, 14–18 Nov 1988, Cochin. KFRI India, pp 107–111

Pattanaik S, Hall JB (2014) Patterns of morphometric variability in *Dendrocalamus hamiltonii* Munro populations across East Khasi Hills, Northeast India. Indian Forester 140(9): 868–874

Pattanavibool R (1999) Bamboo research and development in Thailand. In: Rao AN, Rao VR (eds) Bamboo—conservation, diversity, ecogeography, germplasm, resource utilization and taxonomy. Proceedings of a training course cum workshop 10–17 May 1998. Kunming and Xishuangbanna Yunnan China IPGRI-APO, Serdang, pp 248–259

Paudel P, Kafle G (2012) Assessment and prioritization of community soil and water conservation measures for adaptation to climatic stresses in Makawanpur district of Nepal. J Wetl Ecol (6):44–51. http://www.nepjol.info/index.php/jowe

Pillai PT (1901) The flowering of bamboo in Travancore. Indian Forester 7(8):429

Poudyal PP (1991) Utilization of bamboo in the Kathmandu valley of Nepal. Proceedings of 4th international bamboo workshop, IDRC, Chiangmai, pp 258–262

Prasad J (1948) Silviculture of ten species of bamboo suitable for paper manufacture. Indian Forester 74(3):122–130

Prasad R (1986) Bamboo plantation. For Bull No. 22. SFRI, Jabalpur

Prawirohatmodjo S (1998) Comparative strengths of green and air-dry bamboo. In: Ramanuja Rao IV, Gnanaharan R, Sastry CB (eds) Bamboos - current research. Proceedings of the international bamboo workshop, 14–18 Nov 1988, Cochin. KFRI India, pp 218–222

Prutpongse P, Gavinlertvatana P (1992) In vitro micropropagation of 54 species from 15 genera of bamboo. Hortscience 27:453–454

Pukittayacamee P (1999) Bamboo seed technology for viability testing and germplasm conservation. In: Rao AN, Rao VR (eds) Bamboo and rattan genetic resources and Uuse and research reports. Proceedings of the third INBAR-IPGRI biodiversity, genetic resources and conservation working group, 24–27 Aug 1997, IPGRI-APO, Serdang, pp 169–194

Qureshi IM (1956) Development of bamboo forests after gregarious flowering. Proceeding 9th all India silver conference, Dehra Dun. Part 1, pp 127–129

Raitt W (1929) The Burma bamboo pulp survey. Indian For Rec 24(1):1–48

Ram HYM, Gopal BH (1981) Some observations on the flowering of bamboos in Mizoram. Curr Sci 50(16):708–710

Ramanayake SMSD (2006) Flowering in bamboo: an enigma! (review). Cey J Sci (Bio Sci) 35(2):95–105

Ramanayake SMSD, Meemaduma VN and Weerawardene TE (2006), In vitro shoot proliferation and enhancement of rooting for the large-scale propagation of yellow bamboo (*Bambusa vulgaris* 'Striata'). Sci. Hortic. 110, 109–113

Ramanayake SMSD, Weerawardene TE (2003) Flowering in a bamboo, *Meloccana baccifera* (Roxburgh) Kurz ex Skeels (Bambusoideae: Poaceae). Bot J Linn Soc 143(3):287–291

Ramanayake SMSD, Yakandawala K (1997) Micropropagation of the giant bamboo (*Dendrocalamus giganteus* Munro) from nodal explants of field grown culms. Plant Sci 129:213–223

Ramanayake SMSD, Yakandawala K (1998) Incidence of flowering, death and the phenology of the giant bamboo (*Dendrocalamus giganteus* Wall. Ex Munro). Ann Bot 82(6):779–785

Ramyarangsi S (1987) Bamboo research in Thailand. In: Rao AN, Dhanarajan G, Sastry CB (eds) Recent research on bamboos. Proceedings of the international bamboo workshop, IDRC, Hangzhou, pp 67–69

Ramyarangsi S (1988) Techniques for seed storage of *Thyrsostachys siamensis.* In: Bamboos *current research.* Proceedings of the international bamboo workshop, Cochin, 14–18 Nov 1988, pp 133–135

Rao MS (1988) Gregarious flowering of *Bambusa arundinacea* and *Dendrocalamus strictus.* Indian Forester 114(9):601

Rao KS, Ramakrishnan RS (1987) Comparative analysis of the population dynamics of two bamboo species, *Dendrocalamus hamiltonii* and *Neohouzeana dullooa*, in a successional environment. For Ecol Manage 21(3–4):177–189

Rao KS, Ramakrishnan PS (1988) Architechtural plasticity of two bamboo species (*Nehouzeua dullooa* A Camus and *Dendrocalamus hamiltonii* Nees and Arn.) in successional environments in north-east India. Proc Indian Acad Sci (Plant Sci) 98(2):121–133

Rao IU, Rao R IV, Narang V (1985) Somatic embryogenesis and regeneration of plants in the bamboo *Dendrocalamus strictus*. Plant Cell Rep 4:191–194

Rao AN, Rao VR, Williams JT (eds) (1998) Priority species of bamboo and rattan. INBAR/IPGRI-APO, Serdang, pp 1–78

Rao VR, Gairola SC, Shashikala S, Sethy AK (2008) Bamboo utilization in Southern India. Indian Forester 134(3):379–386

Rao BK, Kurothe RS, Pande VC, Gopal K (2012) Throughfall and stemflow measurement in bamboo (*Dendrocalmus strictus*) plantation. Indian J Soil Conserv 40(1):60–64

Rhind D (1945) The grass of Burma. Published by Calcutta, Baptist Mission Press, Calcutta, pp 1–99

Rogers CG (1900) Flowering of bamboos in the Darjeeling District. Indian Forester 26(7):331–332

Roxburgh W (1814) Bamboo. Hortus Bengalensis 25, Serampore [Mentioned in Gamble JS 1896]

Roxburgh W (1819) Plants of the coast of Coromandel, Published under the direction of Sir Joseph Banks, London, George Nicol.; bamboos, vol 3. pp 37–38

Ryan G (1901) Flowering and seeding of Manwell bamboo (*Dendrocalamus strictus*) in Central Thana Division Bombay Presidency. Indian Forester 27:428–429

For S. (1882) Flowering of the ringal bamboo in Jaunsar. Indian Forester 7:258

Sagereiya KP (1941) *Dendrocalamus strictus* seed. Indian Forester 67:649–650

Sarma H, Sarma AM, Sarma A, Borah S (2010) A case of gregarious flowering in bamboo dominated lowland forest of Assam India: phenology, regeneration, impact on rural economy and conservation. J For Res 21(4):409–414. doi:10.1007/s11676-010-0090-3

Sattar MA, Kabir MF, Bhattacharjee DK (1991) Effect of age and height position of muli (*Melocanna baccifera*) and barak (*Bambusa balcooa*) bamboo on their physical and mechanical properties. In: Bamboo in Asia Pacific. Proceedings of 4th international bamboo workshop, Chiangmai, Thailand, 27–30 Nov 1991, IDRC, Ottawa. Forestry Research Support Programme for Asia and the Pacific (FORSPA), Bangkok, pp 183–187

Sattar MA, Kabir MF, Bhattacharjee DK (1992) Physical and mechanical properties of six important bamboo species of Bangladesh. In: Zhu S, Li W, Zhang X, Wang Z (eds) Bamboo and its use. international symposium on industrial use of bamboo. 7–11 Dec 1992, Beijing, Chinese Academy of Forestry, China, pp 112–117

Saxena S (1990) In vitro propagation of the bamboo (*Bambusa tulda* Roxb.) through shoot proliferation. Plant Cell Rep 9:431–434

Saxena S, Bhojwani SS (1991) Regeneration and mass production of bamboo species through tissue culture. Proceedings of IV international bamboo workshop IDRC, Chiangmai (Abstract), p 64

Seethalakshmi KK (2006) Flowering of bamboos and management of flowered bamboo. In: International training workshop on bamboo propagation, management and harvesting; Methods, Policy issues & strategies, 27 Feb–05 Mar 2006. KFRI/INBAR/CIBART, Peechi, pp 26–30

Seethalakshmi KK, Gnanaharan R (1998) Two decades with the giant grass. Evergreen 41:1–6

Seethalakshmi KK, Kumar M (1998) Bamboos of India: a compendium. KFRI, Peechi; International Network for Bamboo and Rattan (INBAR), Beijing

Seethalakshmi KK, Surendran T, Somen CK (1988) Vegetative propagation of *Ochlandra travancorica* and *O. scriptoria* by culm cuttings. In: Ramanuja Rao IV, Gnanaharan R, Sastry CB (eds) Bamboos – current research. Proceedings of the international bamboo workshop, 14–18 Nov. KFRI, Cochin, pp 136–143

Shah NC (1968) Flowering of the bamboo *Dendrocalamus hookeri* and *Dendrocalamus strictus* in Assam and Bihar States. Indian Forester 94(9):717

Shaji KA (2011) Bamboo in bloom triggers famine fears. The Times of India (Daily news paper), Chennai, 27 Jan 2011, 05.33AM IST

Sharma YML (1982) Some aspects of bamboos in Asia and the Pacific. FAO Regular programme. No. Rapa 57, Bangkok, pp 1–56

Sharma YML (1987) Inventory and resource of bamboos. In: Rao AN, Dhanarajan G, Sastry CB (eds) Recent research on bamboos. Proceedings of the international bamboo workshop, IDRC, Hangzhou, pp 1–17

Sharma A (1992) Muli bamboo flowers in FRI. Indian Forester 118:862–864

Sharma TK (2004) Tuting (Upper Siang, Arunachal Pradesh, India) – sporadic flowering, Records of the Government of Arunachal Pradesh (February, SRF BSI)

Sharma TK (2005) Roing, Dibang Valley, Arunachal Pradesh – gregarious flowering, Records of the Government of Arunachal Pradesh (April, SRF BSI), India

Sharma P, Sarma KP (2011) *In vitro* propagation of *Bambusa balcooa* for a better environment. International conference on advances in biotechnology and pharmaceutical sciences (ICABPS'2011), Bangkok, pp 248–252

Sharma HR, Yadav S, Deka B, Meena RK, Bisht NS (2014) A short communication: sporadic flowering of *Dendrocalamus longispathus* (Kurz) Kurz in Mizoram, India. Trop Plant Res 1(1):26–27

Shibata S, Ikeda K, Lulmuanpuia C, Suyama Y, Saito T, Hasegawa H, Nishiwaki A, Makita A (2008) Mautam-*Melocanna baccifera* flowering—ecological characteristics and influence to the juhm agriculture. In: Proceedings of international conference on improvement of bamboo productivity and marketing for sustainable livelihood, 15–17 Apr 2008, New Delhi, pp 155–163

Shrestha K (2009) The first report of flowering and fruiting phenomenon in *Melocanna baccifera* (Roxb.) Kurz in Nepal. Power point presentation in the 8th world bamboo congress, Bangkok, 2009 (51 slides)

Singh NB (1993) Analysis of genetic diversity in *Bambusa tulda* Roxb. from north-east India. Adv Hortic Forestry 3:187–191

Singh GS, Singh LJ (1994) The study on the biochemical changes during the fermentation of *soibum*. BIC-India Bull 4(1/2):1–5

Singh HB, Kumari B, Singh RS (2003) Bamboo resources of Manipur: an overview for management and conservation. J Bamboo Rattan 2(1):43–55

Singh PK, Devi SP, Devi KK, Ningombam DS, Athokpam P (2010) *Bambusa tulda* Roxb. in Manipur State, India: exploring the local values and commercial implications. Notulae Scientia Biologicae 2(2):35–40

Singha LB, Bhatt BP, Khan ML (2003) Flowering of *Bambusa cacharensis* Mazumder in the southern part of North-East India: a case study. J Bamb Rattan 2(1):57–63. doi:10.1163/156915903321908567

Sinha SDN (1967) Gregarious flowering of bamboo (*Dendrocalamus strictus*). Indian Forester 93(4):210

Sint KM, Hapla F, Myint CC (2008) Investigation on physical and mechanical properties of some Myanmar bamboo species. J Bamb Rattan 7(3&4):183–192

Sitaram RM (1988) Gregarious flowering of *Bambusa arundinacea* and *Dendrocalamus strictus*. Indian Forester 114(9):601

Smythies A (1901) Flowering of bamboo in the CP. Indian Forester 27:126–127

Soderstrom TR, Calderon CE (1979) A commentary on the bamboos (Poacea: Bambusoideae). Biotropica 11(3):161–172

Soderstrom TR, Ellis RP (1987) The position of bamboo genera and allies in a system of grass classification. In: Soderstrom TR, Hilu KW, Campbell CS, Barkworth ME (eds) Grass Systematics and evolution. Smithsonian Institution, Washington, DC, pp 225–238

Soderstrom TR, Ellis RP (1988) The woody bamboo (Poacea: Bambuseae) of Sri Lanka: a morphological- anatomical study. Smithson Contrib Bot 72:66–73

Stapleton CMA (1982) Bamboo in East Nepal: preliminary findings. Forest Research & Information Centre Report, Department of Forests, Kathmandu

Stapleton C (1994a) Bamboos of Bhutan: an illustrated guide. Royal Botanic Garden Kew. ODA, London. Forestry Research Programme, University of Oxford, pp 1–64

Stapleton C (1994b) Bamboos of Nepal: an illustrated guide. Royal Botanic Garden Kew. ODA, London. Forestry Research Programme, University of Oxford, pp 1–66

Stapleton CMA (1994c) The bamboos of Nepal and Bhutan, part I: *Bambusa, Dendrocalamus, Melocanna, Cephalostachyum, Teinostachyum,* and *Pseudostachyum* (Gramineae: Poaceae, Bambusoideae). Edinb J Bot 51(1):1–32

Stapleton CMA, Rao VR (1996) Progress and prospects in genetic diversity studies on bamboo and its conservation. In: Bamboo, people & the environment 2: Proceedings of the IVth international bamboo congress, Bali, 19–22 June 1995. IDRC, Ottawa, pp 23–44

Stapleton CMA, Xia NH (1997) A new combination in *Bambusa* (Gramineae: Bambusoideae). Kew Bull 52(1):235–238

Sujatha MP, Thomas T, Pand Sankar S (2008) Influence of reed bamboo (*Ochlandra travancorica)* on the soils of the Western Ghats in Kerala – a comparative study with adjacent non-reed bamboo areas. Indian Forester 134(3):403–416

Sundriyal M, Sundriyal RC (2011) Prospects in the Central Himalaya: a case study of the traditional artisans of Uttarakhand, India. Ethnobot Res Appl 9:445–454, www.ethnobotanyjournal.org/vol9/i1547-3465-09-445.pdf

Sunita D, Sinha S, Sinha RK (2013) Flowering, vivipary and natural seedling of Bambusa cacharensis, Majumdar – a socio economically important endemic species of North East India. Indian J Plant Sci 2(2):56–60. [ISSN: 2319–3824 (Online) at http://www.cibtech.org/jps.htm]

Sur K, Lahiri AK, Basu RN (1988) Hydration dehydration treatments for improved seed storability of bamboo (*Dendrocalamus strictus*). Indian Forester 114(9):560–563

Suraj K and Nath S (2011). Flowering in *Bambusa tulda* Roxb. at Ranchi, Jharkhand. Indian Forester 137(6):802–803

Surendran T, Seethalakshmi KK (1985) Investigations on the possibility of vegetative propagation of bamboos and reeds by rooting stem cuttings. KFRI Res. Report No. 31, Kerala Forest Research Institute, Peechi, pp 1–47

Tamang B, Tamang JP (2009) Traditional knowledge of biopreservation of perishable vegetable and bamboo shoots in Northeast India as food resources. Indian J Tradit Knowl 8(1):89–95

Tammincha S (1996) Bamboo shoot industry and development. In: Proceedings of the IVth international bamboo congress, 19–22 June 1995, Bali, pp 33–39

Tang C, Jeng IC, Hong YM, Sung WP (2013) The amount estimation of interception in the thorny bamboo plantation. In: Sung, Kao, Chen (eds) Frontier of energy and environmental engineering. Taylor & Francis Group, London, pp 12–16

Tatwawadi HN, Kali BG (1983) Gregarious flowering of bamboo *Dendrocalamus strictus* in Jarida range of East Melghat Division, Amravati Circle - Maharashtra State. Indian Forester 109(2):111–112

Tewari DN (1992) A monograph on bamboo. International book distributors, Dehra Dun, pp 1–498

Thakur ANS (2005) Study report of bamboo flowering in East Kamong dist.—Official correspondence, 17 May 2005. Division of Entomology, ICAR Research Complex for NEH Region, Meghalaya

Tickell C (1881) Flowering of large bamboo. Indian Forester 7:58

Trevor CG (1927) Bamboo seeds (editorial). Indian Forester 53(12):718

Trevor CG (1928) Bamboo seeds (editorial). Indian Forester 54(10):545

Trevor CG (1929) Bamboo seeds (editorial). Indian Forester 55(8):450

Trigg CT (1930) Bamboo seeds (editorial). Indian Forester 56(5):228

Trinius CB (1821) Agrostographische Beyträge. In: Sprengel K (ed) Ncue entdeckungèn Pflanzenkunde, vol 2, Fleischer, Leipzig. pp 33–94 [mentioned in McClure F A(1966)]

Tripathi KC (2002) Gregarious bamboo flowering in northeastern region and some management considerations. In: Proceedings of expart consultation on strategies for sustainable utilization of bamboo resources subsequent to gregarious flowering in the northeast, 24–25 Apr 2002. Rain Forest Research Institute (RFRI), Book No-12-2002. UNIDO, Jorhat, pp 131–135

Troup RS (1921) The Silviculture of Indian trees, vol II. The Clarendon Press, Oxford, pp 785–1195

Tsvelev NN (1976) *Zlaki SSSR.* (Grasses of USSR.). Editio Nauka, Leningrad

Tyson PD (2002) Global-regional linkages in the earth system, Springer-Verlag Berlin Heidelberg New York, Printed in Germany. p 76

Uppin SF (1978) Flowering of *Dendrocalamus strictus*. Indian Forester 104(7):525

Vaid KM (1972) Vivipary in bamboos, *Melocanna bambusoides* Trin. J Bombay Nat Hist Soc 59:696–697

Varmah JC, Bahadur KN (1980) India – a country report. In: Lessard G, Chouinard A (eds) Bamboo research in Asia. Proceedings of a bamboo workshop, Singapore, 28–30 May. IDRC, Ottawa, IUFRO, pp 19–46

Venkatesh CS (1984) Dichogamy and breeding system in a tropical bamboo *Ochlandra travancorica*. Biotropica 16:309–312

Vishwanath S, Arade S (2014) Kanak kaich: a promising multipurpose bamboo species for peninsular India. NTFP Centre of Excellence Tripura India. Manjari 3(3):16–18

Vivekanandan K (1987) Bamboo research in Sri Lanka. In: Rao AN, Dhanarajan G, Sastry CB (eds) Recent research on bamboos. Proceedings of the international bamboo workshop, IDRC, Hangzhou, pp 61–66

Wang TT, Chen MY (1971) Studies on bamboo flowering in Taiwan. Tech Bull Exp For. Taiwan Univ No 87, p 27

Watanabe M, Hamada H (1981).How long is the flowering interval of bamboo. In: Higuchi T (eds) Bamboo production and utilization. Proceedings of the congress group. 5.3A, 17th IUFRO world congress, Kyoto, 6–17 Sept, pp 77–83

Wathon (1903) The flowering of *Bambusa polymorpha*. Indian Forester 29(6):244–245

Wendland JC (1810) Collectio plantarum, vol 2. Hahn, Hannover, p 26, pl 47

Widjaja EA (1995) *Dendrocalamus giganteus* Wallich ex Munro. In: Dransfield S, Widjaja EA (eds) Plant resources of South-East Asia No 7. Bamboos. Backhuys Publishers, Leiden, pp 85–87

Win UN (1951) A note on Kyathaung (Bambusa polymorpha) flowering in Pyinmana Forest Division. Burmese For 1:52–56

Wood ES (1881) Flowering of large bamboo at Dehradun – letter to the editor. Indian Forester 7(11):59

Xie N, Chen L-N, Wong K-M, Cui Y-Z, Yang H-Q (2016) Seed Set and Natural Regeneration of *Dendrocalamus membranaceus* Munro after Mass and Sporadic Flowering in Yunnan, China. PLoS ONE 11(4):e0153845. doi:10.1371/journal.pone.0153845

Xue JR, Yang YM, Hui CM, Li R (1995) Bamboo resources and development research of Yunnan. Yunnan Science and Technology Publishers, Kunming, pp 7–11

Yadava MR (2002) Flowering of bamboo and management issues arising out of it, with special reference to *Melocanna baccifera* (Muli bamboo) in the southern parts of Assam. In: Proceedings of expart consultation on strategies for sustainable utilization of bamboo resources subsequent to gregarious flowering in the northeast, 24–25 Apr 2002. RFRI, Jorhat, pp 28–33. Book No-12-2002, UNIDO

Yang Y, Xue J (1999) Bamboo resources and their utilization in China. In: Rao AN, Rao VR (eds) Bamboo – conservation, diversity, ecogeography, germplasm, resource utilization and taxonomy, Proceedings of a training course cum workshop, 10–17 May 1998, Kunming and Xishuangbanna Yunnan China. IPGRI-APO, Serdang, pp 9–13

Yasodha R, Sumathi R, Malliga P, Gurumurthi K (1997) Genetic enhancement and mass production of quality propagules of *Bambusa nutans* and *Dendrocalamus membranaceus*. Indian Forester 23:303–306

Yasodha R, Kamala S, Ananda KSP, Durai KP, Kalaiarasi K (2008) Effect of glucose on in vitro rooting of mature plants of *Bambusa nutans*. Sci Hortic 116:113–116

Yudodibroto H (1987) Bamboo research in Indonesia. In: Rao AN, Dhanaranjan G, Sastry CB (eds) Recent research on bamboos. Proceedings of the international bamboo workshop, P.R. China, IDRC, Hangzhou, pp 33–44

Ziegler AD, Fox JM, Xu JC (2009) The rubber juggernaut. Science 324:1024–1025

Zoysa Neela de (1994). *Ochlandra stridula* Moon ex Thw-a profile. Proc 4th Intl Bamboo Workshop on Bamboo in Asia and the Asia Pacific, Nov 27–30 1991, Chiangmai Thailand IDRC/FAO (FORSPA)-UNDP, Bangkok, pp 41–47

Zoysa Neela de, Hettige U, Vivekanandan K (1988) Some aspects of bamboo and utilization in Sri Lanka. In: Bamboo – current research. Proceedings of the international bamboo workshop, Cochin. Published by KFRI, Kerala India and IDRC Canada, pp 6–11

Index

R.L. Banik, *Silviculture of South Asian Priority Bamboos*, Tropical Forestry,
DOI 10.1007/978-981-10-0569-5

MIX
Papier aus verantwortungsvollen Quellen
Paper from responsible sources
FSC® C105338

If you have any concerns about our products,
you can contact us on
ProductSafety@springernature.com

In case Publisher is established outside the EU,
the EU authorized representative is:
Springer Nature Customer Service Center GmbH
Europaplatz 3, 69115 Heidelberg, Germany

Printed by Libri Plureos GmbH
in Hamburg, Germany